Developing Smart Sustainability With Artificial Intelligence

Bhupinder Singh
Sharda University, India

Christian Kaunert
University of South Wales, UK & Dublin City University, Ireland

Sahil Lal
Sharda University, India

Manmeet Kaur Arora
Sharda University, India

Anjali Raghav
Sharda University, India

IGI Global
Scientific Publishing
Publishing Tomorrow's Research Today

Vice President of Editorial	Melissa Wagner
Managing Editor of Acquisitions	Mikaela Felty
Managing Editor of Book Development	Jocelynn Hessler
Production Manager	Mike Brehm
Cover Design	Phillip Shickler

Published in the United States of America by
IGI Global Scientific Publishing
701 East Chocolate Avenue
Hershey, PA, 17033, USA
Tel: 717-533-8845
Fax: 717-533-8661
E-mail: cust@igi-global.com
Website: https://www.igi-global.com

Library of Congress Cataloging-in-Publication Data

CIP PENDING

ISBN13: 9798369386842
Isbn13Softcover: 9798369386859
EISBN13: 9798369386866

British Cataloguing in Publication Data
A Cataloguing in Publication record for this book is available from the British Library.

Table of Contents

Detailed Table of Contents

AI is acknowledged as an essential tool in the fight against climate change for developing mitigation and adaptive measures. AI mechanisms remain riddles wrapped in an enigma, but their potential is clear. AI can make sense of vast amounts of data to improve climate modeling, leading to more accurate forecasts for extreme weather or environmental changes. This empowers communities and policymakers to plan effective adaptation strategies, such as timely disseminating warnings about impending natural disasters and judicious resource management. AI applications are fast shaping a new, sustainable global energy landscape, from optimizing renewable sources to shrinking carbon footprints. However, the application of AI in climate action also raises ethical dilemmas like biased data that further widen existing inequalities. To address these challenges, it is crucial that AI systems are built and governed ethically and transparently.

In light of our increasingly rapid urban expansion and the far-reaching impact that this has upon climate change, resource depletion and numerous other issues which affect sustainability within cities: here integrates AI technologies as a key tool for optimizing urban sustainability. Drawing on insights in key sectors such as transportation management, energy efficiency, and environmental monitoring, this paper explores how AI can enable resource optimization and service delivery improvements. Through this study, successful use-cases from various global cities were enunciated where AI solutions have been deployed to cater to urban

An massive, unexplored renewable resource that has the potential to make an important contribution to the world's energy mix is ocean wave energy. In order to maximize the efficacy and efficiency of wave energy conversion, this chapter focuses on the incorporation of artificial intelligence (AI) and machine learning technologies. It looks at these technologies' capability to support clean energy solutions and build a sustainable energy environment particularly in the context of urban living.

This chapter seeks to establish the usefulness of deep learning methods in heatwave prediction with the help of the ERA5 dataset. Heat waves, which are defined as periods of elevated temperature and humidity, are a critical concern because they affect the health of people, farming, water supply, and energy consumption. This study relies on state-of-the-art deep learning models such as Convolutional Neural Networks (CNNs), Recurrent Neural Networks (RNNs), and Long Short-Term Memory (LSTM) to predict heatwave cases. The data set utilized in this work is the ERA5-HEAT dataset; this is a high-resolution atmospheric reanalysis data set. In this paper, the process of data cleaning, temporal alignment, and spatial aggregation in handling large climate data sets has been described. More importantly, the idea of using deep learning models in capturing the temporal and spatial dependencies from meteorological data is well supported through real-life applications. Deep learning methods provide a higher level of accuracy and flexibility than traditional meteorological models.

This paper argues that only artificial intelligence can confront climate change, as it offers unique methods for adaptation and mitigation. Climate change threatens the protection of the environment and the future of society, making the use of AI crucial in supporting sustainable development. This paper seeks to discuss the use of AI, particularly in understanding climate change and the various tasks in this area. In order to promote ecological conservation, AI improves weather forecasting and regulates the environment. Legal systems like the UN Climate Change Agreements incorporate AI to reduce emissions. However, ethical questions and regulatory issues are crucial for ensuring a more sensible use of AI that does not pose a threat.

Chapter 9

There is no universal definition of the term Blue Economy. However, the term is being widely used in the recent times in the context of climate change and environmental degradation. The rapidly increase in the plastic pollution and damage caused to the aquatic lives including coral reefs have now been a prominent cause of concern. Worldwide all the notable economies derive a good amount of revenue from the marine ecosystem. However, the increase in the environmental damage have impacted on the livelihoods of many individuals relying on the marine ecosystem for their survival. All the economies are looking into all possible forms of remedies to save the degradation caused to the blue economy. Artificial Intelligence (AI) is being gradually adopted by all the economies in the world for generation of clean energy, monitoring the data used for examining the marine ecosystem and minimizing the harm done through destructive fishing practices and detecting the damage from plastic pollution. The use of AI will help to minimize such activities by alerting the authorities in a timely manner.

Chapter 10

The growing complexity of urban life has led to the emergence of the notion of "smart cities." These pioneering urban areas are leading the way in terms of efficiency and sustainability, utilizing technology to enhance the overall standard of living. The integration of advanced IoT technology with wireless sensors is reasonable, as it serves as the basis for intelligent parking and transit systems. Real-time data analytics serves as the fundamental basis for intelligent transportation systems. With empowering communities to make decisions based on data, this trend reduces traffic and enhances the overall efficiency of transportation. The incorporation of wireless sensor technology with parking systems has led to a novel approach to urban mobility. This chapter examines the interaction between Highly Intelligent IoT and wireless

Preface

Rapid changes to technology have opened new frontiers of sustainable development and Artificial Intelligence is leading the charge in this transformation. With the resource-conserving measures listed in change, the reality of threats from climate change and resource depletion, population growth and urbanization are some of the unprecedented challenges facing our world today. This book on "Developing Smart Sustainability with Artificial Intelligence" breaks down the deep implications AI can have on feeding intelligent systems delivered, that supports sustainability and makes advances for economic growth operating consonance fully with environmental stewardship. Today, an emerging interconnected world is rapidly spread around web-based technologies that are able to view and probe with precision and fluency within the perplexity of both natural habitats and human condition. In this light, AI is not a tool, but an innovation engine. The book covers a multitude of ways in which AI can aid sustainability from energy optimization to carbon offsetting, as well as waste management and biodiversity conservation. This can deliver smarter cities, more efficient industries and communities that are empowered to make their own data-driven choices about how to live better within the limits of our planet using AI.

This book hints on how AI can be exploited in United Nations Sustainable Development Goals by crossing the boundaries among technology, environmental science and public policy. The book provides a range of case studies, theoretical frameworks and practical applications that can be used as comprehensive guidelines for academics, professionals or policymakers who wishes to engage AI in their sustainability strategies. At a time when intelligent solutions have never been more urgent and we face the ever-increasing pressures of our planet desperately in need of change, Developing Smart Sustainability with Artificial Intelligence hopes to awaken a landscape for innovation that can motivate collective transformation for all layers of society and the Earth.

In this edited volume, we explore the transformative intersection of artificial intelligence (AI) and sustainability, with each chapter contributing unique insights on how AI can foster environmental resilience, improve urban systems, and facilitate sustainable practices across various sectors. The authors present forward-thinking solutions and emphasize the importance of ethical AI governance as we navigate the path to a smarter, more sustainable future.

Chapter 1: AI in Climate Change with Advancing Mitigation Tactics and Adaptive Strategies

Pavapanunkul, Raghav, and Raghav emphasize the critical role of AI in tackling climate change. This chapter investigates AI's capacity to enhance climate modeling, improve weather forecasts, and inform adaptation strategies for communities facing extreme environmental conditions. While AI offers vast potential, the chapter addresses the ethical concerns associated with biased data and calls for the development of transparent and accountable AI systems to ensure fairness in climate action.

Chapter 2: Artificial Intelligence and Urban Sustainability: A Governance Perspective on Smart City Policies

Hind, Lal, and Arora shift the focus to urban sustainability, where AI is increasingly integrated into smart city frameworks. They discuss the ways AI can optimize transportation, energy use, and environmental monitoring in cities, offering real-world examples. The chapter stresses the importance of a governance model that includes ethical considerations and stakeholder inclusion to ensure resilient urban developments capable of addressing sustainability challenges.

Chapter 3: Digital Transformation in School Education: Machine Learning Insights into Marketing Strategies

Makwana, Band, Naidu, and Kaushik explore how digital transformation, underpinned by machine learning (ML), is reshaping educational marketing strategies. By analyzing large datasets, AI helps schools tailor their marketing campaigns, predict student preferences, and optimize engagement. This chapter highlights how AI-driven insights can enhance school promotion, streamline communication, and contribute to more sustainable educational practices.

Chapter 4: Featuring Wave and Tidal Energy Conversion with Artificial Intelligence and Machine Learning

Janjua discusses the role of AI and machine learning in revolutionizing wave and tidal energy conversion. With a focus on the potential of ocean energy as a significant, untapped renewable resource, this chapter examines how AI enhances the efficiency of energy conversion technologies and supports the development of sustainable, clean energy solutions, particularly in urban settings.

Chapter 5: Heatwave Prediction on Era5 Data Using Deep Learning Techniques

Basak and Pal delve into the predictive capabilities of deep learning techniques, specifically Convolutional Neural Networks (CNNs), Recurrent Neural Networks (RNNs), and Long Short-Term Memory (LSTM) networks, to forecast heatwaves. By leveraging the high-resolution ERA5 dataset, the chapter demonstrates how AI models can enhance the accuracy of climate predictions, providing timely alerts for public health and environmental management.

Chapter 6: How Can Artificial Intelligence Contribute to Mitigating Climate Change?

Nayak, Bag, and Maria argue that AI is indispensable in addressing climate change. The chapter explores AI's role in improving weather forecasting, reducing emissions, and supporting ecological conservation efforts. It also highlights the need for regulatory frameworks that guide AI applications in climate action, ensuring that technological advancements align with ethical and legal standards to prevent unintended consequences.

Chapter 7: Integrating Policy, Governance, and Law in Regulating AI for Sustainable Environmental Protection

Sahu, Mallick, and Patnaik focus on the regulatory challenges surrounding AI in environmental protection. They advocate for comprehensive legal, policy, and governance frameworks that ensure AI applications in sustainability are ethically deployed. The chapter critiques current regulatory gaps and calls for stronger, enforceable guidelines to align AI with long-term sustainability goals.

Chapter 8: Machine Learning on Digital Consumer Transforming Choice Recommendation Model

Singh, Raghav, and Jermsittiparsert explore how AI-driven consumer recommendation models are transforming digital markets. By analyzing consumer behavior through machine learning, businesses can optimize their marketing strategies and enhance customer engagement. This chapter examines how these changes not only impact consumer decisions but also contribute to the development of more sustainable business practices in urban contexts.

Chapter 9: Role of Artificial Intelligence in the Study of the Blue Economy and Marine Ecosystem

Mukherjee, Wamane, and Pundkar investigate the role of AI in preserving the marine ecosystem and supporting the blue economy. The chapter highlights AI's potential to monitor marine pollution, track the health of aquatic life, and mitigate destructive fishing practices. By offering real-time data and predictive analytics, AI helps safeguard marine environments, crucial for sustainable economic development.

Chapter 10: Uplifting Smart Sensors for Transforming Urban Transport and Parking

Srivastava, Mange, and Kaushik examine the integration of smart sensors and IoT in urban transport systems. The chapter discusses how these technologies enable real-time data analytics to optimize transportation efficiency, reduce traffic congestion, and enhance parking systems. By fostering smarter urban mobility solutions, AI plays a pivotal role in the development of sustainable and resilient urban infrastructures.

Each chapter in this volume contributes a unique perspective on AI's potential to advance sustainability in various sectors, emphasizing the importance of ethical governance, technological integration, and proactive policy frameworks in realizing a sustainable future. Through these insights, the book offers a comprehensive roadmap for leveraging AI in addressing the world's most pressing environmental challenges.

In conclusion, *Developing Smart Sustainability with Artificial Intelligence* brings together a diverse array of perspectives that highlight the transformative potential of AI in fostering sustainability across various domains. The chapters presented in this volume illustrate how AI can be harnessed not only as a tool for technological innovation but as a powerful engine driving a sustainable future. From addressing the complex challenges of climate change and urban sustainability to enhancing energy systems, education, and marine conservation, the authors collectively demonstrate that AI's capacity to improve resource management and environmental stewardship is unparalleled.

As we navigate an era marked by environmental degradation, rapid urbanization, and a growing demand for efficient, sustainable practices, the integration of AI becomes not just a technological necessity but a moral imperative. The book's emphasis on ethical AI governance, stakeholder inclusion, and the alignment of technological advancements with regulatory frameworks offers a roadmap for ensuring that AI's benefits are distributed fairly and responsibly. In this context, *Developing Smart Sustainability with Artificial Intelligence* serves as both a call to action and a comprehensive guide for policymakers, scholars, and industry leaders alike, as they work to create resilient, equitable, and sustainable systems that can withstand the pressures of the future.

Through the case studies, theoretical frameworks, and practical applications shared in this volume, the book presents AI not merely as an innovation but as a cornerstone of global sustainability strategies. With continued collaboration, forward-thinking policies, and a commitment to ethical practices, AI can indeed unlock a more sustainable, prosperous future for all.

Chapter 1
AI in Climate Change With Advancing Mitigation Tactics and Adaptive Strategies

Shayut Pavapanunkul
Universitas Muhammadiyah Sidenreng Rappang, Indonesia

Anjali Raghav
https://orcid.org/0009-0004-0248-7956
Sharda University, India

Richa Raghav
https://orcid.org/0009-0009-8482-6232
Sharda University, India

ABSTRACT

AI is acknowledged as an essential tool in the fight against climate change for developing mitigation and adaptive measures. AI mechanisms remain riddles wrapped in an enigma, but their potential is clear. AI can make sense of vast amounts of data to improve climate modeling, leading to more accurate forecasts for extreme weather or environmental changes. This empowers communities and policymakers to plan effective adaptation strategies, such as timely disseminating warnings about impending natural disasters and judicious resource management. AI applications are fast shaping a new, sustainable global energy landscape, from optimizing renewable sources to shrinking carbon footprints. However, the application of AI in climate action also raises ethical dilemmas like biased data that further widen existing inequalities. To address these challenges, it is crucial that AI systems are

DOI: 10.4018/979-8-3693-8684-2.ch001

built and governed ethically and transparently.

I. INTRODUCTION

Climate change, a severe and complex global challenge, is one of the most significant barriers to worldwide development. Its effects are felt globally, from rising sea levels to more frequent extreme weather events, changing precipitation patterns, and disrupting ecosystems and biodiversity (SaberiKamarposht et al., 2024). These issues, which threaten the environment and lead to food and water scarcity, mass migrations, and economic vulnerability, are not just environmental problems. They also pose significant humanitarian and financial challenges. The urgency of the climate crisis cannot be overstated. This act swiftly and decisively, as the future of our planet and all its inhabitants depends on it (Kang & Nam, 2024)

A government and industry-wide call to arms for the necessary action to combat climate change has come with unprecedented urgency (Hadeed et al., 2024). Mitigation means efforts to reduce or prevent the emission of greenhouse gases (GHGs) causing climate change and adaptation, also known as preparing society for some of the effects already unavoidable due to past emissions. However, the scope and scale of the climate crisis have made traditional methods insufficient. What is called for is a game-changing method that could leverage cutting-edge technology to propel mitigation and adaptation strategies. This is where Artificial Intelligence (AI) comes in as a game-changer (Kalogiannidis et al., 2024). AI and ML mean artificial intelligence, which refers to an artificially created human being capable of thinking independently, learning from experiences, and making self-decisions. Within the realm of climate change, however, because of AI's ability to sift through mountains of data and uncover patterns that would be invisible to a human eye, generate efficiencies in resource utilization, and provide more projections for outcomes, this is where the value at stake lies in relying upon AI as a weapon against climate emergency (Beitelmal, et al., 2024). AI is transforming how we fight and understand problems, like extreme weather events, to deliver efficiency in energy (Bacelar et al., 2024).

II. SCOPE OF AI IN CLIMATE CHANGE IN MITIGATION AND ADAPTATION

AI will bring about new ways of advancing climate action in different industries and sectors. Insights based on big data: Climate change generates enormous datasets from satellites, weather stations, environmental sensors, etc. (Raihan, 2024). There are too many datasets here for manual analysis. AI can examine data and significant trends with actionable insights that enable policymakers, scientists, or industries to be informed and empowered to make effective decisions (Dwivedi et al., 2024).

Also, ML-enabled Forecast: Collectively, it is a dimension of AI, and it provides a predictive analytics layer with probabilistic range estimates of factors influencing climate action. Today, machine learning algorithms can predict climate change patterns, help in energy consumption forecasts, and even in ancient life recovery and the spread of climate-related diseases (Satpathy et al., 2024). This is a helpful predictive tool that allows us to mitigate against the impacts of climate change before it is too late. Furthermore, by optimizing essential resources, AI can maximize dominant resources such as water, energy, and raw materials due to their decreasing availability and increasing environmental degradation (Garg, 2024). With AI-powered tools, renewable energy sources can be used more efficiently, industrial processes can be streamlined to reduce emissions, and water usage for agriculture could come just in time (Han et al., 2024).

Smart Infrastructure and disaster response: AI can be used to design climate-resilient infrastructure and manage the impact of natural disasters (Akter et al., 2024). By incorporating AI in urban planning, smart cities may be devised to withstand the ravages of climate change and lessen their environmental impact. AI can help in emergency response by predicting disasters like floods, wildfires, and hurricanes, allowing people time to plan and respond specifically (Ali & Thakkar, 2023).

III. GROWING IMPORTANCE OF MITIGATION AND ADAPTION

Also, the Paris Agreement, signed in 2015, established global goals for limiting warming to well below 2°C above pre-industrial levels and striving to limit it even more to 1.5°C; this is mitigation. Unfortunately, sound mitigation also fails to suffice in the real world. Changes have already occurred and will worsen (Zhao et al., 2023). Consequently, adaptation to the ecological, social, and economic shifts in response to climate effects is now equally critical. Whereas mitigation addresses the causes of climate change, adaptation seeks to assist societies in adapting to the effects that are already happening or expected to occur. This is a two-sided coin that is essential for making ourselves resilient against climate change and thereby sustaining this growth (Grigorieva et al., 2023)

IV. INTEGRATION OF TECHNOLOGICAL INNOVATION

Due to the scope and complexity of climate change, traditional approaches are not suited to tackle the problem. Renewable energy technologies, conservation efforts, and international climate policies are part of the answer to the first two challenges. Still, they cannot be expected to do all the work without the backing of more complex solutions (Alsamhi et al., 2024). New technologies, especially artificial intelligence (AI), can potentially speed up developments for both mitigation and adaptation (Cortez, 2023). In the last few years, AI has shown its power to change everything from healthcare to finance and transportation. This technology has also led to a revolution in environmental science. AI has the potential to massively increase efficiency, prediction, and intelligent solutions where needed, making it a possible major player in climate action. It can inform better climate policies by governments, support emissions reductions within industries, and help communities climate-proof themselves (Kalogiannidis et al., 2024).

V. AI-BASED CLIMATE MODELING AND PREDICTION

Also, accurate predictions and taking necessary steps to respond to climate change are armed with algorithms that might fail from the surface (Mehta et al., 2023). However, traditional climate models are critical but disproportionately constrained in spatial resolution, computing capacity, and the representation of the climate system feedback responses (Beitelmal et al., 2024). Emerging methods to incorporate artificial intelligence (AI) into climate modeling show promise in addressing these limitations. Using advances in Artificial Intelligence (AI) can improve the accuracy and rate of climate predictions by utilizing vast datasets constantly being developed, enhanced model parameterizations, and a more comprehensive assessment of climate impacts that may result from these (Haleem et al., 2023).

These AI techniques, which include machine learning and deep learning, are increasingly being applied to improve climate models (Satpathy et al., 2024). These technologies identify trends and correlations in comprehensive datasets from satellite imagery and historical weather patterns, and they are based on oceanographic data, which are not visible to the "naked eye" analysis methods. For example, AI can be useful for downscaling the Global Climate Models (GCM) to regional scales. It is essential to aid effective planning and adaption strategies, as it could lead to more localized climate prediction outputs. For instance, Machine Learning Techniques-Machine Learning (ML) algorithms can process large datasets, identifying patterns used to build climate projections (Ahmed et al., 2023).

These algorithms could be run on historical climate data and then used to predict what would happen in conditions similar to various levels of greenhouse gas emissions (Akter et al., 2024). Thus, capability is precious in studying how climate change may affect multiple regions and ecosystems. Regression Models - these can also be used to predict continuous outcomes (e.g., temperature changes, precipitation levels) based on input variables such as greenhouse gas concentrations or land use changes (Nyangon, 2024). Classification Algorithms- An excellent application would be to classify zones into the risk profile from predicted climate impacts by the government to prioritize regions that will probably need assistance most (Sharma et al., 2024). Neural Networks- Neural networks are structures far more complex than polynomials that can give insights into many systems that are reportedly not amenable to simpler models, including the non-linear interactions of variables within the climate system (Haldorai et al., 2024).

VI. USE CASE AND METHODOLOGY OF AI IN CLIMATE MODELLING

AI is a groundbreaking tool for researchers and organizations to address the environmental issues we face. There have been numerous attempts to use AI in climate modeling, and some of them have become successful examples, including Google Earth Engine (Google, 2017), IBM's Green Horizons Project (Bertil), and Climate (Lawyer et al., 2023). AI. They all use advanced AI to sift through tremendous amounts of data, predict changes in the environment, and develop strategies for the response to combat climate change (Ojo, 2024). In this section, we will further analyze these case studies, including their methods, usages, and consequences of climate action (Darjee et al., 2023).

The Definitive Resource on Google Earth Engine for Monitoring of the Environment

Google Earth Engine (GEE) is the enhanced cloud-based geospatial analysis platform developed by Google that opens up a sizeable public catalog of satellite images and powerful computational capabilities. GEE makes using over 90 petabytes of data possible to understand change across the Earth's entire surface. The platform processes long-term datasets rapidly, enabling monitoring of deforestation, land use changes, and the impacts of climate-full screens on these forests (Maddireddy & Maddireddy, 2023).

Data Catalog: GEE A vast number of satellite imagery and geospatial datasets, one of the largest repositories available except for those only accessible on hard drives, removed from any computerized access. This includes an over 50-year-long archive of historical data from sensors such as Landsat, MODIS, and Sentinel satellites. They can access daily, high-resolution imagery necessary to analyze environmental changes promptly (Ijiga et al., 2024).

Integration with Machine Learning GEE supports many machine learning algorithms for various analytical tasks. For instance, users may want to leverage supervised/unsupervised classification techniques to categorize land cover types or track changes over time. This affords significant leverage in the credibility of environmental evaluations (Eruaga, 2024).

Scalability: Since the infrastructure is cloud-based from both availability and workload standpoints, we can perform large-scale analysis without relying on large local computing resources. This scalability is pivotal in dealing with the enormous datasets created during climate research (Rezvani et al., 2023).

Applications: Monitoring Deforestation GEE is an excellent tool for tracking deforestation activities across the globe. Scientists have used Land Trend to investigate Tropical forests, and help gauge how well conservation efforts and anti-deforestation policies are working (Alordiah, 2023).

Management of Water Resources GEE utilizes satellite data on water bodies and precipitation patterns to enhance water resource management. This is especially critical in water-stressed areas due to climate change (Cao, 2023).

GEE Analysis of Urban Heat Island Effects: The GEE allows researchers to study urban heat islands by analyzing city land surface temperatures. Such information is crucial in urban planning to guide strategies that reduce the heat burden among vulnerable groups (Rao & Korres, 2023).

Monitoring Deforestation in the Amazon Rainforest A well-known example of the use of GEE is its application to monitor deforestation in Amazon rainforests (Shaik et al., 2024). Researchers and GEE used machine learning algorithms to process and analyze satellite imagery for forest canopy loss and gain over time (Raska & Bitzinger, 2023). Historical data was compared with current observations to measure deforestation rates and pinpoint illegal logging hotspots (Pandey et al., 2023). These results yielded essential insights for policy-makers and the various conservation agencies tackling issues related to preserving the Amazon rainforest. With an understanding of where deforestation was happening most quickly, the situation could be targeted for intervention (Rawat et al., 2024).

VII. CHALLENGES AND LIMITATIONS OF AI IN CLIMATE MODELING

While integrating Artificial Intelligence (AI) into climate modeling holds great promise for enhancing our understanding and response to climate change, it is accompanied by various challenges and limitations (Rane, 2023). These obstacles can hinder the effectiveness of AI-driven solutions and must be addressed to maximize their potential benefits (Danish, 2023). Below, we delve deeper into the critical challenges associated with AI in climate modeling, including data quality and availability, the complexity of climate systems, the interpretability of models, and ethical considerations (Lücking, 2023). Data Quality and Availability: High-quality data is the backbone of effective AI models. Machine learning algorithms to produce accurate predictions require large datasets that are representative, comprehensive, and free from significant errors or biases. In climate modeling, this means accessing extensive historical climate data, satellite imagery, ground-based measurements, and other relevant environmental information (Elwahsh et al., 2023).

However, insufficient Data in Developing Regions: One of the most significant challenges in utilizing AI for climate modeling is the lack of sufficient data in many regions, particularly developing countries (Wu et al., 2024).

- Limited Monitoring Infrastructure: Many developing nations lack the necessary infrastructure for collecting reliable climate data. This includes weather stations, remote sensing capabilities, and data management systems that store and analyze large datasets (Khaspuria et al., 2024).
- Inconsistent Data Collection: Even when available, data may be collected inconsistently or not at all. For instance, some regions may have sporadic weather observations due to financial constraints or political instability (Babu, 2024).
- Quality Issues: The quality of available data can vary significantly. Historical datasets may sometimes contain inaccuracies due to outdated measurement techniques or human error during data entry (Sarker, 2024). These issues make it challenging to train effective AI models to predict climate trends or impacts in these regions accurately. AI-driven insights may be misleading or inaccurate without reliable data (Kwiatkowski et al., 2023).

Intricacies of Climate Dynamics: Earth's climate system is extraordinarily complex and influenced by many interconnected factors, from atmospheric conditions to ocean currents and terrestrial ecosystems (Khan et al., 2024). This complexity presents several challenges for AI modeling: Nonlinear Interactions: Climate processes often exhibit nonlinear relationships, meaning that small changes in one variable can lead to disproportionately large effects on others (Aslam & Rasheed, 2023). For example, a slight increase in temperature can significantly impact precipitation patterns or vegetation growth (Adnan et al., 2024). Also, Multiscale Phenomena: Climate phenomena occur at various spatial and temporal scales—from local weather events to global climatic shifts. Capturing these multiscale interactions within a single model is inherently challenging. Thus, Feedback Loops: The presence of feedback loops where an effect amplifies itself (e.g., melting ice reduces albedo, leading to further warming) adds another layer of complexity. AI models may struggle to represent these feedback mechanisms accurately without extensive training on diverse datasets (Balyan et al., 2024). Due to these intricacies, even advanced AI models may fail to capture all relevant interactions accurately. As a result, predictions made by these models could be oversimplified or misrepresentative of actual climatic conditions (Bakker, 2024).

Many AI models, especially those based on deep learning techniques—operate as "black boxes." While they can produce highly accurate predictions based on input data, how they arrive at those conclusions is often unclear (Ali et al., 2024). The lack of transparency in model decision-making poses several issues. Trust Issues stakeholders, including policymakers, scientists, and the general public, may hesitate to rely on AI-generated predictions if they do not understand how they were derived (Cruz et al., 2024). Trust is essential for effective decision-making in climate action. Difficulty in Validation: Without interpretability, validating model outputs becomes challenging. Researchers may struggle to determine whether a model's predictions are based on sound reasoning or reflect noise in the data. Limited Insight into Climate Processes: One of the goals of climate modeling is to enhance our understanding of climate processes. If AI models do not provide insights into how different variables interact within the climate system, they may fall short of contributing valuable knowledge to the field (Birchall & Kehler, 2023).

Hence, AI researchers are increasingly focusing on developing interpretable AI models that explain their predictions to address these concerns (Lee et al., 2024). Techniques like feature importance analysis and model visualization can help demystify AI systems' operations.

- **Ethical Considerations:** The deployment of AI must consider ethical implications related to biases present in training data. If historical datasets reflect existing inequalities such as socioeconomic disparities or geographic imbalances, AI models trained on this data may perpetuate these biases (Ghatrehsamani et al., 2023)
- **Equitable Outcomes**: Biased models can lead to inequitable outcomes where specific populations, particularly marginalized communities, are neglected or

adversely affected by climate policies informed by flawed predictions (Afzal et al., 2023).

- **Vulnerable Populations**: Vulnerable populations are often disproportionately impacted by climate change yet may lack adequate representation in datasets used for training AI models (Waheeb & Wheib, 2024). This oversight can result in policies that fail to address their specific needs or risks (Maiti & Kayal, 2024).

Accountability and Governance: As AI technologies become more integrated into climate action strategies, questions surrounding accountability arise:

- **Decision-Making Responsibility**: Determining who is responsible for decisions based on AI-generated insights can be complex. Identifying accountability becomes challenging if an AI model produces inaccurate predictions, leading to adverse outcomes (Ahmed et al., 2024).
- **Regulatory Frameworks**: Effective governance structures must be established to ensure AI technologies are deployed responsibly and ethically in climate action initiatives (Santosh et al., 2024). This includes guidelines for data collection practices, model validation protocols, and stakeholder engagement processes (Kurniawan et al., 2024).

VIII. CONCLUSION

Integrating AI into climate modeling offers tremendous potential for enhancing our understanding of complex environmental systems and improving decision-making processes related to climate action. However, significant challenges must be addressed to realize this potential fully. Ensuring high-quality data availability is critical for training effective AI models; therefore, investments in monitoring infrastructure and consistent data collection practices are necessary—especially in developing regions where these resources are often lacking. Additionally, researchers must continue exploring methods for improving model interpretability while addressing biases inherent in training datasets. Ultimately, navigating these challenges will require collaboration among scientists, policymakers, technologists, and communities affected by climate change. By fostering an inclusive approach that prioritizes transparency and equity in deploying AI technologies for climate modeling, we can harness their power responsibly while striving toward a more sustainable future for all.

REFERENCES

Adnan, M., Xiao, B., Ali, M. U., Bibi, S., Yu, H., Xiao, P., Zhao, P., Wang, H., & An, X. (2024). Human inventions and its environmental challenges, especially artificial intelligence: New challenges require new thinking. *Environmental Challenges*, 16, 100976. DOI: 10.1016/j.envc.2024.100976

Afzal, S., Mubeen, M., Hussain, S., Ali, M., Javeed, H. M. R., Al-Ashkar, I., . . . Jatoi, W. N. (2023). Modern breeding approaches for climate change. In *climate change impacts on agriculture: Concepts, Issues and Policies for Developing Countries* (pp. 299-313). Cham: Springer International Publishing. DOI: 10.1007/978-3-031-26692-8_17

Ahmed, Z., Gui, D., Murtaza, G., Yunfei, L., & Ali, S. (2023). An overview of smart irrigation management for improving water productivity under climate change in drylands. *Agronomy (Basel)*, 13(8), 2113. DOI: 10.3390/agronomy13082113

Akter, S., Babu, M. M., Hani, U., Sultana, S., Bandara, R., & Grant, D. (2024). Unleashing the power of artificial intelligence for climate action in industrial markets. *Industrial Marketing Management*, 117, 92–113. DOI: 10.1016/j.indmarman.2023.12.011

Ali, A. H., & Thakkar, R. (2023). Climate changes through data science: Understanding and mitigating environmental crisis. *Mesopotamian Journal of Big Data*, 2023, 125–137. DOI: 10.58496/MJBD/2023/017

Ali, G., Mijwil, M. M., Buruga, B. A., Abotaleb, M., & Adamopoulos, I. (2024). A Survey on Artificial Intelligence in Cybersecurity for Smart Agriculture: State-of-the-Art, Cyber Threats, Artificial Intelligence Applications, and Ethical Concerns. *Mesopotamian Journal of Computer Science*, 2024, 71–121. DOI: 10.58496/MJCSC/2024/007

Alordiah, C. O. (2023). Proliferation of Artificial Intelligence Tools: Adaptation Strategies in the Higher Education Sector. *Propellers Journal of Education*, 2(1), 53–65.

Alsamhi, S. H., Rowan, N. J., Hawbani, A., Nashwan, F., & O'Brolchain, N. (2024, August). AI-Empowered Peatland Management for Climate-Resilient and Sustainable Future: Challenges and Opportunities. In *2024 4th International Conference on Emerging Smart Technologies and Applications (eSmarTA)* (pp. 1-8). IEEE.

Aslam, R., & Rasheed, H. (2023). *Integrating AI.* Cloud Computing, and IoT for Robust Cybersecurity Ecosystems and Adaptive Threat Mitigation.

Babu, C. S. (2024). Adaptive AI for Dynamic Cybersecurity Systems: Enhancing Protection in a Rapidly Evolving Digital Landscap. In *Principles and Applications of Adaptive Artificial Intelligence* (pp. 52-72). IGI Global.

Bacelar, E., Pinto, T., Anjos, R., Morais, M. C., Oliveira, I., Vilela, A., & Cosme, F. (2024). Impacts of climate change and mitigation strategies for some abiotic and biotic constraints influencing fruit growth and quality. *Plants*, 13(14), 1942. DOI: 10.3390/plants13141942 PMID: 39065469

Bakker, K. (2024). *Gaia's Web: How Digital Environmentalism Can Combat Climate Change, Restore Biodiversity, Cultivate Empathy, and Regenerate the Earth.* MIT Press. DOI: 10.7551/mitpress/14912.001.0001

Balyan, S., Jangir, H., Tripathi, S. N., Tripathi, A., Jhang, T., & Pandey, P. (2024). Seeding a Sustainable Future: Navigating the Digital Horizon of Smart Agriculture. *Sustainability (Basel)*, 16(2), 475. DOI: 10.3390/su16020475

Bayomi, N., & Fernandez, J. E. (2023). Eyes in the sky: Drones applications in the built environment under climate change challenges. *Drones (Basel)*, 7(10), 637. DOI: 10.3390/drones7100637

Beitelmal, W. H., Nwokolo, S. C., Meyer, E. L., & Ahia, C. C. (2024). Exploring Adaptation Strategies to Mitigate Climate Threats to Transportation Infrastructure in Nigeria: Lagos City, as a Case Study. *Climate (Basel)*, 12(8), 117. DOI: 10.3390/cli12080117

Birchall, S. J., & Kehler, S. (2023). Denial and discretion as a governance process: How actor perceptions of risk and responsibility hinder adaptation to climate change. *Environmental Science & Policy*, 147, 1–10. DOI: 10.1016/j.envsci.2023.05.017

Cao, L. (2023). AI and data science for smart emergency, crisis and disaster resilience. *International Journal of Data Science and Analytics*, 15(3), 231–246. DOI: 10.1007/s41060-023-00393-w PMID: 37035277

Cortez, F. (2023). Artificial Intelligence, Climate Change and Innovative Democratic Governance. *European Journal of Risk Regulation*, 14(3), 484–503. DOI: 10.1017/err.2023.60

Cruz, L., Gutierrez, X. F., & Martínez-Fernández, S. (2024). Innovating for Tomorrow: The Convergence of SE and Green AI. *arXiv preprint arXiv:2406.18142*.

Danish, M. S. S. (2023). AI and expert insights for sustainable energy future. *Energies*, 16(8), 3309. DOI: 10.3390/en16083309

Darjee, K. B., Neupane, P. R., & Köhl, M. (2023). Proactive Adaptation Responses by Vulnerable Communities to Climate Change Impacts. *Sustainability (Basel)*, 15(14), 10952. DOI: 10.3390/su151410952

Dwivedi, D. N., Mahanty, G., & nath Dwivedi, V. (2024). Intelligent Conservation: A Comprehensive Study on AI-Enhanced Environmental Monitoring and Preservation. In *The Convergence of Self-Sustaining Systems With AI and IoT* (pp. 215-226). IGI Global.

Elwahsh, H., Allakany, A., Alsabaan, M., Ibrahem, M. I., & El-Shafeiy, E. (2023). A deep learning technique to improve road maintenance systems based on climate change. *Applied Sciences (Basel, Switzerland)*, 13(15), 8899. DOI: 10.3390/app13158899

. Eruaga, M. A. (2024). Policy strategies for managing food safety risks associated with climate change and agriculture.

Garg, A. (2024). AI for a Better World: Sustainability and Technology. *Shodh Sagar Journal of Artificial Intelligence and Machine Learning*, 1(1), 33–38.

Ghatrehsamani, S., Jha, G., Dutta, W., Molaei, F., Nazrul, F., Fortin, M., Bansal, S., Debangshi, U., & Neupane, J. (2023). Artificial intelligence tools and techniques to combat herbicide resistant weeds—A review. *Sustainability (Basel)*, 15(3), 1843. DOI: 10.3390/su15031843

Grigorieva, E., Livenets, A., & Stelmakh, E. (2023). Adaptation of agriculture to climate change: A scoping review. *Climate (Basel)*, 11(10), 202. DOI: 10.3390/cli11100202

Hadeed, M. Z., Ali, A., Malik, A., Raza, A., & Shoaib, M. (2024). Harnessing AI and GIS Technologies for Climate-Resilient Agriculture and Environmental Sustainability. In *Maintaining a Sustainable World in the Nexus of Environmental Science and AI* (pp. 333–364). IGI Global. DOI: 10.4018/979-8-3693-6336-2.ch013

Haldorai, A., Babitha Lincy, R., Suriya, M., & Balakrishnan, M. (2024). Climate Resilience Via Smart Technologies Over Natural Disaster. In *Artificial Intelligence for Sustainable Development* (pp. 383–406). Springer Nature Switzerland. DOI: 10.1007/978-3-031-53972-5_20

Haleem, A., Javaid, M., Khan, I. H., & Mohan, S. (2023). Significant Applications of Artificial Intelligence Towards Attaining Sustainability. *Journal of Industrial Integration & Management*, 8(4), 489–520. DOI: 10.1142/S2424862223500331

Han, H., Liu, Z., Li, J., & Zeng, Z. (2024). Challenges in remote sensing based climate and crop monitoring: Navigating the complexities using AI. *Journal of Cloud Computing (Heidelberg, Germany)*, 13(1), 34. DOI: 10.1186/s13677-023-00583-8

Ijiga, A. C., Peace, A. E., Idoko, I. P., Ezebuka, C. I., Harry, K. D., Ukatu, I. E., & Agbo, D. O.Amina Catherine IjigaAbutu Ehi PeaceIdoko Peter IdokoChijioke Charles EzebukaKimberly D. HarryIfeanyi Emmanuel UkatuDaniel Obekpa Agbo. (2024). Technological innovations in mitigating winter health challenges in New York City, USA. *International Journal of Science and Research Archive*, 11(1), 535–551. DOI: 10.30574/ijsra.2024.11.1.0078

Kalogiannidis, S., Kalfas, D., Patitsa, C., & Chalaris, M. (2024). Relationship between climate change and business risk: strategies for adaptation and mitigation: evidence from a Mediterranean country. *WSEAS Transactions on Environment and Development*, 20, 276–294. DOI: 10.37394/232015.2024.20.28

Kang, B. G., & Nam, Y. (2024). Responsible Artificial Intelligence for Climate Action: A Theoretical Framework for Sustainable Development. *Sustainable Machine Intelligence Journal*, 8, 1–1. DOI: 10.61356/SMIJ.2024.88101

Khan, M. R., Haider, Z. M., Malik, F. H., Almasoudi, F. M., Alatawi, K. S. S., & Bhutta, M. S. (2024). A comprehensive review of microgrid energy management strategies considering electric vehicles, energy storage systems, and AI techniques. *Processes (Basel, Switzerland)*, 12(2), 270. DOI: 10.3390/pr12020270

Khaspuria, G., Ranjan, A., Soni, P., & Dadhich, K. (2024). Natural Disaster Mitigation Strategies: A Comprehensive Review. *Journal of Scientific Research and Reports*, 30(8), 20–34. DOI: 10.9734/jsrr/2024/v30i82221

Kwiatkowski, C. A., Pawłowska, M., Harasim, E., & Pawłowski, L. (2023). Strategies of Climate Change Mitigation in Agriculture Plant Production—A Critical Review. *Energies*, 16(10), 4225. DOI: 10.3390/en16104225

Lawyer, C., An, L., & Goharian, E. (2023). A Review of Climate Adaptation Impacts and Strategies in Coastal Communities: From Agent-Based Modeling towards a System of Systems Approach. *Water (Basel)*, 15(14), 2635. DOI: 10.3390/w15142635

Lee, S., Park, Y., & Park, G. (2024). Using AI chatbots in climate change mitigation: A moderated serial mediation model. *Behaviour & Information Technology*, 1–17. DOI: 10.1080/0144929X.2024.2402532

Lücking, H. (2023). *The use of AI technology for sustainability: The envisioned meaning and role of sustainable AI in climate political discourses in Denmark and Germany* (Bachelor's thesis, University of Twente).

Maddireddy, B. R., & Maddireddy, B. R. (2023). Adaptive Cyber Defense: Using Machine Learning to Counter Advanced Persistent Threats. *International Journal of Advanced Engineering Technologies and Innovations*, 1(03), 305–324.

Mehta, P., Dhaliwal, L. K., Baweja, P. K., Jangra, M. S., & Bhardwaj, S. K. (2023). Concept of climate smart villages using artificial intelligence/machine learning. In *Visualization Techniques for Climate Change with Machine Learning and Artificial Intelligence* (pp. 359–377). Elsevier. DOI: 10.1016/B978-0-323-99714-0.00010-8

Nyangon, J. (2024). Climate-proofing critical energy infrastructure: Smart grids, artificial intelligence, and machine learning for power system resilience against extreme weather events. *Journal of Infrastructure Systems*, 30(1), 03124001. DOI: 10.1061/JITSE4.ISENG-2375

Ojo, B. (2024). Strategies for the Optimization of Critical Infrastructure Projects to Enhance Urban Resilience to Climate Change. *The Journal of Scientific and Engineering Research*, 11, 107–123.

Pandey, D. K., Hunjra, A. I., Bhaskar, R., & Al-Faryan, M. A. S. (2023). Artificial intelligence, machine learning and big data in natural resources management: A comprehensive bibliometric review of literature spanning 1975–2022. *Resources Policy*, 86, 104250. DOI: 10.1016/j.resourpol.2023.104250

Raihan, A. (2024). A review of recent advances, challenges, and potential future directions of climate-smart agriculture. In *Proceedings of The International Conference on Climate-Smart Agriculture*.

Rane, N. (2023). Roles and challenges of ChatGPT and similar generative artificial intelligence for achieving the sustainable development goals (SDGs). *Available atSSRN* 4603244. DOI: 10.2139/ssrn.4603244

Rao, A. N., & Korres, N. E. (2023). Climate Change and Ecologically Based Weed Management. *Ecologically-Based Weed Management: Concepts, Challenges, and Limitations*, 23-48.

Raska, M., & Bitzinger, R. A. (Eds.). (2023). *The AI Wave in Defence Innovation: Assessing Military Artificial Intelligence Strategies, Capabilities, and Trajectories*. Taylor & Francis. DOI: 10.4324/9781003218326

Rawat, A., Kumar, D., & Khati, B. S. (2024). A review on climate change impacts, models, and its consequences on different sectors: A systematic approach. *Journal of Water and Climate Change*, 15(1), 104–126. DOI: 10.2166/wcc.2023.536

Rezvani, S. M., de Almeida, N. M., & Falcão, M. J. (2023). Climate Adaptation Measures for Enhancing Urban Resilience. *Buildings*, 13(9), 2163. DOI: 10.3390/buildings13092163

SaberiKamarposhti, , MWhy, , N. KYadollahi, , MKamyab, , HCheng, , JKhorami, , M. (2024). Cultivating a sustainable future in the artificial intelligence era: A comprehensive assessment of greenhouse gas emissions and removals in agriculture. *Environmental Research*, 118528. PMID: 38403150

Sarker, I. H. (2024). *AI-driven cybersecurity and threat intelligence: cyber automation, intelligent decision-making and explainability*. Springer Nature. DOI: 10.1007/978-3-031-54497-2

Sarma, H. H., Paul, A., Kakoti, M., Talukdar, N., & Hazarika, P. (2024). Climate resilient agricultural strategies for enhanced sustainability and food security: A review. *Plant Archives*, 24(1), 787–792. DOI: 10.51470/PLANTARCHIVES.2024.v24.no.1.108

Satpathy, I., Nayak, A., Jain, V., & Padmadas, S. S. (2024). Applying Data Into Action: AI-Powered Solutions for Mitigating Climate Change and Fostering Sustainable Future. In *AI Applications for Clean Energy and Sustainability* (pp. 50-74). IGI Global.

Shaik, A. S., Alshibani, S. M., Jain, G., Gupta, B., & Mehrotra, A. (2024). Artificial intelligence (AI)-driven strategic business model innovations in small-and medium-sized enterprises. Insights on technological and strategic enablers for carbon neutral businesses. *Business Strategy and the Environment*, 33(4), 2731–2751. DOI: 10.1002/bse.3617

Wu, C., Zhang, R., Kotagiri, R., & Bouvry, P. (2023). Strategic decisions: Survey, taxonomy, and future directions from artificial intelligence perspective. *ACM Computing Surveys*, 55(12), 1–30. DOI: 10.1145/3571807

Zhao, J., Liu, D., & Huang, R. (2023). A review of climate-smart agriculture: Recent advancements, challenges, and future directions. *Sustainability (Basel)*, 15(4), 3404. DOI: 10.3390/su15043404

Chapter 2
Artificial Intelligence and Urban Sustainability:
A Governance Perspective on Smart City Policies

Hammouch Hind
https://orcid.org/0000-0002-5897-1649
University Sidi Mohamed Ben Abdellah, Morocco

Sahil Lal
https://orcid.org/0000-0001-9827-3717
School of Law, Sharda University, India

Manmeet Kaur Arora
https://orcid.org/0009-0002-5071-117X
Sharda University, India

ABSTRACT

In light of our increasingly rapid urban expansion and the far-reaching impact that this has upon climate change, resource depletion and numerous other issues which affect sustainability within cities: here integrates AI technologies as a key tool for optimizing urban sustainability. Drawing on insights in key sectors such as transportation management, energy efficiency, and environmental monitoring, this paper explores how AI can enable resource optimization and service delivery improvements. Through this study, successful use-cases from various global cities were enunciated where AI solutions have been deployed to cater to urban challenges. The study provides urban policy with a thorough framework for how to implement AI technologies in addressing challenges at the intersection of AI and urban sustainability through collaborative governance efforts. This paper calls for a pre-emptive

DOI: 10.4018/979-8-3693-8684-2.ch002

policy making attitude with an emphasis on ethics and stakeholder inclusion for resilience-derived sustainability in urban developments.

1. INTRODUCTION

In this context, the writing of Roberto Polli comments the introduction of Artificial Intelligence (AI) in urban development and gives an allowance for a process that is essentially beginning to see smart cities that think about sustainability and efficiency (Ji, 2021). But now, with as many as 68 percent of the world's people expected to live in cities by 2050, population growth is the least of our urban concerns (Huang et al., 2021). The enormous expansion of our cities necessitates creative solutions that boost efficiency in resource use, lessen environmental degradation and provide better living conditions for urban dwellers (Singh & Kaunert, 2024). Here AI has a crucial role to play, in shaping sustainable urban development through streamlining city operations, aiding decision making and enabling data driven governance (Tang et al., 2021).

Fundamentally, a smart city uses different types of IOT sensor with artificial intelligence to gather and examine huge data from every urban system and generates knowledge that automatically supports the planning or operation of individual products (Dai et al., 2018). These systems involve transportation networks, power grids, waste disposal services and environmental monitoring mechanisms. This data that is being created by the connected citizen can be processed in real-time through AI algorithms and cities can identify patterns, predict outcomes, put proactive measures which helps address some of the urban challenges. For instance, AI can help to alleviate congestion on city streets by analyzing the data from sensors or cameras in real time and then it offers an alternative route that also minimizes pollutant emissions (Halilaj et al., 2022). Likewise, in energy management, we use AI for resource allocation by smarter grids which get adapted over reading load changes. Urban sustainability is intrinsically a part of smart city development ideas. Sustainability in urban form is [by definition] to allow for transferred responsibility from environmental degradation, not only for today but also for future generations. It includes an extensive range of elements associated with economic viability, environmental stewardship, and social justice. Sustainable cities and communities are also one of the United Nations Sustainable Development Goal 11 on inclusive urbanization and public space, which we all need to contribute our quota. AI is pivotal in attaining these objectives by allowing cities to integrate of sustainable procedures into different fields (Singh, 2023). Transportation:AI plays a big role in urban sustainability specially when it comes to transportation management. AI-driven intelligent transport systems can make transportation more accessible by providing instant traffic updates and information on various modes of public transportation. Not only do these systems make for a better commuter experience, they also help to cut down on carbon emissions by directing vehicles more efficiently and minimizing idle times. Additionally, AI-powered predictive analytics can predict transportation trends from prior data and inform the city planning for infrastructure expenses and public transport upgrades (Halilaj et al., 2022).

With the help of AI, how to manage an enormous amount of energy besides transportation. Together with AI features, small and compact smart grids help cities to control energy usage more efficiently and adapt renewable generators in their supply chains (Singh, 2023). This shift to intelligent energy systems supports sustainability by moving away from fossil fuels and cutting greenhouse gas emissions. Working with smart meters that give real-time feedback on consumption, AI can even drive buildings to be more energy-efficient; into the bargain (Luettin et al., 2022). AI has become crucial in the arena of environmental quality monitoring, as well, helping to advance urban sustainability initiatives. There is air pollution, waste management, and a strong need for the reduction of the impact of climate change on us – all issues in which our cities play a big role. With AI, technology can analyze data from things like IoT sensors to watch over air quality levels and identify different areas where pollution is worst. Local governments can use these data to take tailored action to enhance environmental health practices. Plus, AI has the potential to reduce costs through the operational efficiencies related to real-time waste generation data being used to best optimize collection routes. While AI holds promise in creating sustainable cities, it also presents some challenges which need to be addressed for extracting maximum value from its implementations. One of the largest barriers will be in developing governance frameworks that exercise proper ethical rigour to keep citizens safe and protect their privacy. Greater access to enormous pools of personal data creates new worries about privacy and algorithmic bias for the AI systems that cities are embracing. Policymaker should define transparent and accountable rules on the frontline of AI, as a first deployment, in public sector use cases. Smart city projects cannot be successful without strong collaboration involving all types of stakeholders—city governments, technology vendors and local residents alike. In the engagement of citizens in decisions this engendered a greater trust and helped ensure that technology solutions were designed at the local level to meet need rather than imposing top-down approaches likely to miss community nuance. The nature of smart city governance is changing rapidly as cities across the globe journey down the path to intelligent, sustainable urban futures. The trend across policy making is to move away from silo based approaches towards interdisciplinary solutions that draw from concepts born in urban planning, environmental science, public administration and technology development alike. Cities that create pathways for collaboration between these disciplines can craft holistic approaches that leverage

AI, and meet the real-world challenges of urban governance. In conclusion, the intertwining of artificial intelligence with urban sustainability offers cities a novel way to rebuild their city infrastructure and services that can uplift the life of its PGs as citizens (Halilaj et al., 2021). This research paper aims at understanding the role of Artificial Intelligence and its applications across various sectors in smart cities) as we get into further detail about this topic, through the pursuit of analysing a few instances from different areas involved in it. In addition, we will discuss the governance frameworks required for successful implementation and consider the ethical questions associated with deploying such technologies in cities.

2. THEORETICAL FRAMEWORK

Urban governance frameworks are being redefined with the infusion of Artificial Intelligence (AI) in smart cities more so, in sustainability. The next part of the report goes on to detail important definitions and related concepts established for AI and urban sustainability, outlines different governance models appropriate to smart cities, discusses the relationship between AI with policy and sustainability (Singh, 2024). These are big, important challenges to address and understanding them matters a lot to designing policies that leverage AI as it can solve sustainability problems in the urban realm.

2.1 Definition of Concepts

Artificial Intelligence, Join Free Artificial intelligence is the simulation of human intelligence processes by machines, especially computer systems (Singh, 2023). These functions include learning (the acquisition of information and rules for using it), reasoning (using rules to reach approximate or definite conclusions) and self-correction. AI in Urban Sustainability AI includes many different technologies such as machine learning, natural language processing, and robotics that enable computers to process massive amounts of information to determine best methods to JSON-output protein for cellular function (Halilaj et al., 2023).

Urban Sustainability Urban sustainability refers to the ability of an urban area to reduce its leakage of resources. This concept seeks to marry economic development, environmental protection and social equity. United Nations Secretary-General Ban Ki-moon published a report supporting the Sustainable Development Goals (SDGs) including Goal 11, which reads "making cities and human settlements inclusive, safe, resilient and sustainable. AI can be very helpful in accomplishing these targets, by providing insights which might drive at better use of resources, better public service as well as a better quality of life for urban residents. Smart cities technology improves the efficiency of city government and services, such as transportation, energy management, waste management, and public safety. IoT devices and AI analytics are used by smart cities to gather data from various locations, thereby enabling near real-time monitoring and management of the city operations. This connectivity contributes to smarter decision making, promoting ecological structures (Xu et al., 2022).

2.2 Governance Models Smart Cities

For an AI-enabled smart city initiative to be successful, leadership and direction are vital. There are several governance models which have been proposed to handle the complexities of managing smart cities.

Collaborative Governance: This model focuses on partnerships between government agencies, private sector stakeholders, non-profit organizations, and citizens. The approach of collaborative governance means AI solutions are implemented in an open and inclusive decision-making process whereby a broader vibrant mix of perspectives can be enlisted. For example, public engagement campaigns can be used to collect feedback from residents on the deployment of AI technologies in their neighborhoods (Elahi et al., 2020).

Flexible strategies: Adaptive governance reflects the fast pace of technological change and allows cities to be flexible in how they develop and apply policy. In turn, this model is creating a culture of perpetual learning and piloting of new technologies that are evolving to meet urban sustainability challenges. Cities, for example, can test out AI applications at the sector-level before taking them to scale after receiving feedback and lessons learnt (Lecue, 2020).

Data-Driven Governance: using data analytics to influence policy decision-making across the spectrum of governance. The use of AI tools enables the extraction of invaluable insights from the gold mine, that is big data sets originating from urban systems for city officials to delivery trends through forecasting needs and consumption of resources effectively. With data-driven governance, accountability can be increased with trackable results over time (Dong et al., 2020).

Integrated Governance: This is the governance approach which attempts to bust silos within different city administration sectors through promoting collaborations among departments. Such integrated governance would preempt the development of a suite of AI applications to address transportation challenges, each siloed off from the next—in this case those related to energy use or environmental quality. For instance, a road-wise transport system might connect with local and regional energy management systems to highway-energy-efficiently prioritize EV charging according to the real-time traffic in the area (Kim et al., 2021).

2.3 Intersection of AI, Policy and Sustainability

The integration of AI technology into policy frameworks at the intersection of these objectives allows us to better leverage policies towards achieving urban sustainability goals. Policy-makers must now traverse a tricky terrain in which technological change has the potential to greatly alter social forms as well as environmental results (Singh, 2023). Most importantly, AI algorithms that can efficiently and accurately process large datasets create one major area of opportunity. As an example: A city planner could leverage predictive analytics to predict when traffic will be at it heaviest or how varying the energy demands of a community are given trends from historical data. This functionality empowers operators to take proactive load relief and energy distribution optimization before congestion starts. In addition, AI makes resource optimization easier; in the hope that cities can better use their resources and not waste too many things. By deploying AI algorithms to analyze collection patterns and predict disposal requirements with high accuracy, municipalities can optimize their operations to save costs while also encouraging recycling campaigns. At the same time, but recognizing that integrating AI with urban policy raises serious ethical concerns regarding data privacy, algorithmic bias**, and transparency in decision-making mechanisms. With AI technologies driving more data-driven strategies in cities, we are starving for the solid framework that can align about how data is collected and operated upon while safeguarding citizens right (Sharma & Singh, 2022). Local policy-makers may also address the digital divide prevailing in urban populations between socially differentiated groups that might have unequal access to the technology. It is important to develop smart city technologies in such a way that they genuinely contribute to more inclusive growth and do not serve to simply nurture further existing social inequalities.

Ultimately, a conceptual grounding in artificial intelligence and urban sustainability is necessary to build sound governance approaches that capitalize upon new technology. This research paper aims to use policy considerations in the context of smart cities and urban sustainability to provide insights into how key elements—including defining concepts, governance structures (collaborative governance, adaptive governance, data-driven governance and integrated governance) meet with some problems AI brings. Over the upcoming years as cities become more smart and digital through artificial intelligence-driven intelligent systems, sustainability in urban environments will be critical to existing necessities and incoming strains (Nickel et al., 2015).

3. AI APPLICATIONS IN URBAN SUSTAINABILITY

One of today's most challenging areas for cities is urban sustainability, and with AI (Artificial Intelligence), cities get one of the most advanced tools to tackle this problem. In a world of rapidly urbanizing areas, the incorporation of AI in just about every sector like transportation management, energy management and environmental quality monitoring has become quite indispensable towards making cities greener. This section focuses on AI applications in the built environment, transport systems, and the infrastructure of cities to underlie urban sustainability by increasing efficiency, minimizing resource consumption, and improving quality of life for city dwellers (Buchner et al., 2023).

3.1 AI Applications in Transport Management

Urban infrastructure focuses largely on the mobility and transportation within a city that has a significant part to play when it comes to sustainability. Lastly, cities are using AI technologies to not merely make their transport facilities smarter but also greener. One of its most famous uses is smart traffic management where it deploys AI algorithms to read real-time sensor and camera-derived data to control the flow. Cities use this information to streamline movement of people around the city by optimizing traffic signals for pedestrians and cyclists. Take adaptive signals, which can adjust their timing in real time based on how many vehicles are approaching an intersection — this speeds traffic flow and reduces pollution from idling engines at red lights (Yu et al., 2019).

On the other hand, AI contributes to improve public transportation systems using predictive analytics. Utilizing a similar process found in dynamic pricing models, AI can make predictions regarding how many passengers to expect on buses and trains by analyzing historical ridership data and ensuring that cities optimally allocate resources (Hu et al., 2023). This will mean buses and trains are used where the demand is, reducing overcrowding and improved service punctuality. More over, AI technology can offer people real-time information about the transit schedules fixing for the time and encourage them to use public transit more and reduce private vehicle usage (Liu et al., 2023).

A major application again is smart parking solutions. These AI-powered systems evaluate parking availability in real time, and direct drivers to open spots more quickly than they would otherwise find them on their own. This makes not only the lives of drivers easy, but it also reduces traffic congestion because less time spent stuck in the line means lesser emissions. In addition to this, some cities are considering using the technology of automated vehicles that use Artificial intelligence to travel, making them much more profitable and efficient. These vehicles can revolutionize urban mobility because they offer shared modes of transportation that cut down on the number of cars on the road (Liu et al., 2023).

3.2 AI Applications In Energy Management

AI also has a strong influence on urban sustainability in energy management. Given the dual mission of cities to lower their carbon footprints and shift to renewable energy, AI is being leveraged in order to minimize consumption, in all sectors. Major applications for smart meters, providing real-time detailed energy usage patterns. These devices allow consumers to monitor the amount of energy they are using instead of just receive a bill so that they would be more conscious of how much energy they are consuming and seek ways to use less, find said devices could help open doors to saving opportunities (Singh, 2024).

Artificial intelligence as a Behaviour of Smart Grids Safeguarding the tendency through questionable web-get together meters require sparkly new requiring structures for securing ones points joined to grand smart grid — a development electrical structure that gives utilizing notwithstanding essentialness, amazing charge convey and hopeful asset segregation Input Stream (Plonnings et al., 2022). Append("AI perceives expanding utilization in overseeing brilliant grids — imaginative electrical systems that solidify pattern vitality sources like sun arranged what's more effect standing"). AI can use predictive analytics to predict energy demand depending on historical data as well as current conditions. This helps electric utilities respond to the grid changing dynamics of renewable energy, using more clean energy and less natural gas generation at periods of highest demand. On top of this, cutting-edge AI technologies similarly benefit microgrids—small-scale energy systems that are able to work independently or in conjunction with the main grid. Artificial Intelligence also helps in improving the functioning of microgrids by controlling energy storage systems and synchronizing generation from different sources. The future of the power network will be made up from interconnected microgrids which increase the resiliency to outages and offer sustainability benefits through localised energy production (Kraft et al., 2023).

The urban landscape is additionally seeing the proliferation of AI-based smart street lighting systems. These systems adapt lighting levels in response to real-time pedestrian counts and existing light levels. The cities can cut energy usage by dimming or turning off lights that aren't needed during quiet times of the night, which also improves safety for residents.

3.3 AI Applications in Environmental Quality Monitoring

The detection of Environmental Pollution is very important to keep the equilibrium in Urban Ecosystems on the earth. Assessment of public health issues (e.g., air and water quality, other environmental factors) that are integrated directly with AI technologies. An important application is predicting air quality, which uses multiple sources of data like weather conditions, traffic information, and industrial discharge to predict how polluted the environment will be. With this predictive ability, officials can deploy apparent initiatives — such as issuing health advisories or changing traffic management to reduce pollution during dangerous time frames (Li et al., 2023).

Thermoresponsive Cellulose Nanocrystals-Based Sensor System for Real-time Water Quality Monitoring with Artificial Intelligence These systems use, sensors with AI algorithms to detect real-time water quality data anomalies. Correcting contamination problems quickly allows cities to provide safe drinking water to their citizens and with a minimal environmental footprint. Flood Detection and Management: AI plays a significant role in flood detection and management which is again one of the major category under Environmental Monitoring. Cities are able to better predict where, when and to what extremity flooding will occur through the power of machine learning models trained on historical weather patterns and topographical data. Such information can be used to take action, mitigating flood risk through deploying emergency services or upgrading infrastructure. In addition, AI solutions enable better disposal methods with automated waste sorting mechanisms integrated by means of image recognition algorithms that are able to make proper distinctions between materials that should be recycled. Cities can also encourage a circular economy in which recycling rates are increased and landfill waste is reduced (Schier et al., 2023).

4. POLICY GOVERNANCE AND CHALLENGES

The use of artificial intelligence in the process helps determine how to line up laws and standards that need to govern efforts by policymakers, governance organizations. The pace of technological innovation is such that, combined with the sprawling, mesoscopic and interdependent weave of elements at play in any smart city initiative, we need to meet policy challenges both proactively and collaboratively. This section examines the major policy and governance challenges of AI implementation in urban sustainability in order to determine whether an appropriate legal regime would be sufficient to address those issues, while also addressing the ethical considerations involved (Mahon et al., 2020).

4.1 Legal Frameworks for AI Implementation

The integration of AI in smart city projects demands a well-defined and detailed legal framework that will draw boundaries across realms like data ownership, privacy, liability and algorithmic accountability. Rule makers need to prevent companies from stifling new AI applications while also making sure that these technologies remain in compliance with current laws. Coming up with ways to fit these new things — AI systems that may have no humanlike characteristics at all, or only reminiscent of a small subset — into old legal categories is extremely difficult. As with the use of AI in decision-making, this dependence raises issues as to liability from mistakes or unforeseen outcomes. In the event that an AI system causes harm or loss, then who is responsible: the developer, the depolyer (or trainer), or even the user? Ultimately, policymakers will need to set out the rules of the road and must have appropriate guardrails in place to protect against certain risks. In addition, the smart city systems involve huge collection and use of data, thus strict laws on data management should be in place for privacy and security controlling (Xiao et al., 2023). Citizens need to feel that their information is safe and they also need to know that due process is being respected. Policymakers should take inspiration from data protection regulations (e.g. GDPR in the European Union) that introduces norms on how to collect, store and use data around individual users. Algorithmic transparency and accountability are another very important dimension in legal framework for AI deployment. The output decision logic of AI systems is also getting too complex and opaque to understand. Policymakers must make it their priority to ensure transparency in AI systems since only by combining them with auditing and oversight can bias consciousness and discrimination prevent. This might include obligating AI developers to submit detailed documentation on their algorithms, and their data sources (Liu et al., 2019).

4.2 Ethical Considerations In AI Use

In addition to the legal frameworks, policymakers must also address the ethical considerations of AI in smart city settings. Ironically, AI technologies are able to help cities reinvent themselves in ways previously unconceived, but with that great power comes concerns regarding bias and privacy as well as displacement of human labour (Liu et al., 2020). Tempered enthusiasm is the order of the day. Addressing these important ethical challenges is critical for creating trust with the public and using AI in a way that serves all of society, not only those that develop and deploy it. One of the most pressing ethical concerns about AI is that we train these systems to learn from reality, and then we let loose imperfect products in a socially unfair world, which can end up reinforcing it full-blown. So we have to be careful not to inadvertently train our AI algorithms with similar discriminations of the past. Therefore, policymakers will have to collaborate with machine developers and subject matter experts to detect such biases and make them AI fair and inclusive. The effects of AI on employment and the workforce remain another chief ethical concern. As AI technologies start to automate more tasks and processes, there can be a restructuring in the labor market, putting more low-skilled workers at risk of displacements against tech majors. Policy makers should think about strategies for reskilling and upskilling workers, and ways to leverage the job-creating potential in AI. Data rights and privacy distribution of the publisher industry. Since smart city systems collect plentiful information from individuals, they could potentially allow for surveillance and invasion of personal privacy. Elected officials must ensure accountability regarding how data are collected and used and who has control over personal information. This could mean creating data rights should that will grant individuals access, correction and deletion powers over their data (Cai et al., 2022).

A systematic solution includes a combination of legal regulation, institutional procedure, ethical guidance and partnership governance to improve policy development and implementation relevant to AI in urban sustainability. To be effective, policymakers at every level and across sectors should collaborate with AI developers, domain experts like healthcare professionals, business advisors or labor union representatives, and local community stakeholders when crafting new policies to both spur innovation but also guide responsible and fair deployment. Policymakers have an important role to provide certainty and guidance for developers and deployers around AI, by laying out clear legal frameworks that also protect the rights and interests of residents. This should always be a primary concern in the formation of policies as AI technologies are designed and applied to benefit all members of society while respecting human rights. The ability of policymakers and governance bodies to manage these challenges, due to their complexity, the outcome of urban sustainability AI will depend on that. Through being proactive and inclusive, cities can leverage AI to build more sustainable, resilient, and equitable urban systems that work for all residents (Holzinger et al., 2023).

5. CASE STUDIES

Advanced solutions enabled by Artificial Intelligence (AI) are playing a significant role in creating smarter cities that are more sustainable. AI technologies can help cities to maximize resource utilization, enhance public services and promote initiatives for sustainability (Singh, 2019). This part of the series illustrates various AI implementations in smart cities across the globe, presenting several case studies and some key learnings derived from those cases around the world. The case studies demonstrate the potential of AI in tackling urban challenges to support sustainable urban development.

Case Study 1: Barcelona, Spain

Barcelona is a prime example of a smart city that has successfully integrated AI into urban management systems. Numerous AI-aligned projects were introduced the city in order to develop efficiency in transportation and lower the overall energy consumption. It is a famous use case, in applied research projects > one such project is the smart traffic management system which attracts live data through sensors and cameras to use for making traffic smooth. The system uses traffic flows to dynamically adjust traffic signals, for example alleviating congestion and improving air quality. While Barcelona also installed an AI-enabled smart parking system to assist drivers in real-time find a vacancy. This minimizes the time looking for parking and decreases the amount of emissions given out by idling vehicles. Additionally, it has also deployed smart street lighting that dims or brightens as needed given pedestrian movement and natural light conditions, leading to meaningful energy reductions.

Barcelona's experience underlines the necessity of data integration as well as real-time analytics to achieve urban sustainability goals. The success of the city has shown that more productivity can be achieved in operations without spending your money (video) In 1996, the city.

Case Study 2: Amsterdam, Netherlands

The city of Amsterdam has made AI an integral part of its strategy to bring to life a future-proof, liveable and sustainable urban harbour_REVolution. The city has followed numerous policies that reduce emissions and improve the lives of its inhabitants. An example of this is the Amsterdam Smart City initiative, a collaborative innovation program where local governments work together with companies, residents and knowledge institutions to develop innovative solutions for metropolitan challenges. AI-based platforms for managing energyAnother pillar to the approach is using AI applied to energy management. These platforms keep an eye on the energy consumption trends in both residential and commercial buildings in order to offer advice that can help users reduce their energy use. In parallel, Amsterdam has launched an AI-powered mobility platform to harness data from different transport modes (including bicycles, public transit and shared vehicles) to provide users with a live overview of the travel alternatives available to them. Its cities also use AI to monitor the environment, by placing sensors around urban areas to measure air quality and noise levels. This data is used for policy decisions that aim to enhance public health and the quality of the environment.

Key Takeaways: Amsterdam recognizes that stakeholder engagement is key to success for smart city projects. A greater diversity of actors working together, including residents, will lead to more genuinely 'community-responsive' solutions for cities. Additionally, the way in which the AI was incorporated into energy management is a shining example of how tech can promote sustainability and improve consumer satisfaction.

Case Study 3: Singapore

Singapore has earned its reputation as a pioneer in urban planning and management utilizing AI to power many of its new systems. The city-state has initiated an extensive Smart Nation initiative, reaching across multiple sectors from transportation and healthcare to energy management. Of these, as an example of what might be included in the program, is LTA's Intelligent Transport System which uses AI algorithms to analyze traffic data and improve public transport services (Bonatti et al., 2019). Using predictive analytics, the LTA model predicts future demand for public transportation by relying on historical ticket sales and current circumstances. It can lead to a more efficient use of resources and ensure buses and trains are where they need to be. Meanwhile, Singapore introduced an AI-informed smart waste management system that features sensors in city-wide recyclable bins to measure waste level. This optimizes collection routes, which not only saves on operational costs but also protects the environment.

Takeaway: Singapore demonstrates the importance of exploring smart city development innovations and solutions in a more comprehensive manner. Artificial intelligence applied horizontally across multiple domains can achieve synergies and enhance both efficiency and sustainability within cities. Similarly, this helps to highlight the important of data-informed decision-making in some of these urban design profession by stressing on predictive analytics (Zheng et al., 2022).

6. CONCLUSION

Urban sustainability is complicated, and embedding Artificial Intelligence (AI) in such an ecosystem serves as a disruptive means to resolving their conundrums. The continual expansion of urban populations increases the need for innovation in order to manage resources more effectively, minimize environmental impacts and improve the quality of life for residents. The paper identified several dimensions of AI in contributing to sustainable city development, and studied the implications for smart cities. In this conclusion we will recap the main findings, discuss what AI means for urban governance and outline experience – science areas in which later research and policy-making can build upon.

In this paper, we have discussed the three main applications of AI in smart cities among others: transportation management, energy management and environmental quality monitoring. AI is able to analyze data, make predictions and act in real-time and so each of these areas benefits greatly from the power of artificial intelligence. AI-enabled traffic management system reduces road congestion, thus enabling better fuel efficiency and cleaner air as exemplified by AI-led results in optimized traffic flow. Likewise, AI helps energy management further by making it easier to put renewable sources on urban grids, which reduces the demands and enhances sustainability. The paper also picked out examples from fellow cities such as Barcelona, Singapore, Amsterdam, Los Angeles and Toronto. These examples serve as models for cities throughout the world that adopt AI solutions to solve urban problems and to enhance sustainable development. The lessons contained in these cases underline the necessity of collaboration, data exchange and ethical questions in AI solution deployment.

The results of this study highlight the importance of robust governance regimes to navigate the integration of AI into urban systems. Cities face a complex terrain of legal, ethical and practical challenges as they experiment with AI technologics. It is such legislation which needs to be created as the foundation for regulation to tackle data protection, accountability of algorithms and liability when AI fails. Policy must also be guided by ethical considerations. It is essential also to design AI systems which impact decision making processes in urban governance to be fair and inclusive. Policymakers need to work alongside technologists and community stakeholders to address algorithmic biases and data transparency issues in AI. Final thoughts: The critical voice of the public cannot be underemphasised. Further, involving citizens in conversations with respect to the integration of AI techniques can also build trust and assure that technology solutions that are developed correspond with what communities need. Through the emphasis of public input in determining, cities could create smart city initiatives that are more alright with what community members want to see.

Given the unprecedented speed of urbanisation and impact on natural resources, we may suggest some possible further directions for research, policy and practice to help enhance AI in urban sustainability:

Interdisciplinary Research: There is necessity to explore the field with an interdisciplinary approach integrating theoretical contributions from urban planning, environmental science, data science and social sciences. Taking into account all of these aspects will lead to a better view on the deployment potential of AI within various smart cities sectors.

Evaluation Frameworks: For evaluating how well AI is being put to use in urban environs, it is critical that strong evaluation frameworks be made. These frameworks must have real-time accuracy, scale extremely well for multiple users, use as less resources as they can manage to avoid increasing the hosting cost and have a better end-user satisfaction. Establishing Performance Evaluation Metrics: Through performance measurement, cities can be confident their AI tools are consistently and reliably meeting the objectives we set forth.

Innovation in Policies: Policymakers must introduce new policy instruments that promote synergies and alliances between a public agency and the private players for pivoted AI-enabled urban sustainability solutions. Public Private Partnerships to share knowledge and support resource mobilization: Public-private partnerships offer an opportunity to drive innovation in smart city initiatives.

Equity Upfront: When cities are utilizing AI, perspectives of equity up front. Leaders can proactively promote solutions to bridge the digital divide, so that underserved areas in cities and towns do not lose out on smart city opportunities. That might mean investing in digital citizenship programs or offering material support to vulnerable neighborhoods.

In addition to Action Items derived from these streams, cities should develop and measure within the following Sustainability Metrics stream by creating clear metrics for AI-implementation-related impacts related to the achievement of a city's sustainability objectives. It is also important that these metrics capture the environmental, social and human dimensions of the communities where factories operate; so we know how local communities are benefitting from our efforts/impacts on quality of life improvements. Collaborations will also involve fostering global collaboration between cities plagued by similar issues allowing for the process of knowledge transfer on best practices regarding the deployment of AI in relation to sustainability. As such, international networks can serve a platform for the exchange of experiences and lessons learned to stimulate innovation in urban governance.

AI has the power to create a huge impact in terms of city sustainability, by optimizing the resource management and decision-making process on different domains in smart cities. Proactive governance that paid equal attention to legal clarity, ethical considerations, public engagement and also enabled interdisciplinary collaboration is needed for an effective integration of AI technologies. For cities to fully unlock the power of AI as a force for good, it is important to continue with these principles as they evolve into smarter and sustainable experiences for its citizens. These new approaches to reaching a state of sustainable development have the potential of extending and benefiting people and improving their lives globally by addressing fundamental questions about how cities are operating today using AI head-on. This effort is an ongoing journey requiring long-term dedication from government agencies, private sector partners, community organizations — and also citizens in a democratic system. However, through working together to thoughtfully use technology where necessary and practical — with focus on equity throughout and transparency in decision-making over them all — cities can lead the way for a future where tech acts as an ally in making communities resilient and thriving amidst constant innovation. In conclusion, this paper has demonstrated the opportunity for artificial intelligence to progress urban sustainability goals globally and its position as key strategy tool — and a platform for further research into how smart city policies could develop in collaboration with technological developments, to create not only livable spaces conducive to well-being today, but which protects resources for future generations.

REFERENCES

Bonatti, P. A., Decker, S., Polleres, A., & Presutti, V. (2019). Machine learning: New directions for knowledge representation on the semantic web (dagstuhl seminar 18371). [). Schloss Dagstuhl-Leibniz-Zentrum fuer Informatik.]. *Dagstuhl Reports*, 8(9), •••.

Büchner, M., Zürn, J., Todoran, I. G., Valada, A., & Burgard, W. (2023). Learning and aggregating lane MLs for urban automated driving. In *Proceedings of the IEEE/ CVF Conference on Computer Vision and Pattern Recognition* (pp. 13415-13424).

Cai, P., Wang, H., Sun, Y., & Liu, M. (2022). DQ-GAT: Towards safe and efficient autonomous driving with deep Q-learning and ML attention networks. *IEEE Transactions on Intelligent Transportation Systems*, 23(11), 21102–21112. DOI: 10.1109/ TITS.2022.3184990

Dai, S., Liang, Y., Liu, S., Wang, Y., Shao, W., Lin, X., & Feng, X. (2018, March). Learning entity and relation embeddings with entity description for AI completion. In *2018 2nd International Conference on Artificial Intelligence: Technologies and Applications (ICAITA 2018)* (pp. 194-197). Atlantis Press.

Dong, X. L., He, X., Kan, A., Li, X., Liang, Y., Ma, J., & Han, J. (2020, August). Autoknow: Self-driving knowledge collection for products of thousands of types. In *Proceedings of the 26th ACM SIGKDD International Conference on Knowledge Discovery & Data Mining* (pp. 2724-2734). DOI: 10.1145/3394486.3403323

Elahi, M. F., Luo, X., & Tian, R. (2020, July). A framework for modeling Machine learning via processing natural descriptions of vehicle-pedestrian interactions. In *International Conference on Human-Computer Interaction* (pp. 40-50). Cham: Springer International Publishing.

Guo, R., Hong, Z., & Xue, X. (2021). *Reward Function Design via Human AI and Inverse Reinforcement Learning for Intelligent Driving* (No. 2021-01-0180). SAE Technical Paper.

Halilaj, L., Dindorkar, I., Lüttin, J., & Rothermel, S. (2021). A AI-based approach for situation comprehension in driving scenarios. In *The Semantic Web: 18th International Conference, ESWC 2021, Virtual Event, June 6–10, 2021* [Springer International Publishing.]. *Proceedings*, 18, 699–716.

Halilaj, L., Luettin, J., Henson, C., & Monka, S. (2022, September). Machine learning for automated driving. In *2022 IEEE Fifth International Conference on Artificial Intelligence and Knowledge Engineering (AIKE)* (pp. 98-105). IEEE.

Halilaj, L., Luettin, J., Monka, S., Henson, C., & Schmid, S. (2023). AI-Based Integration of Autonomous Driving Datasets. *International Journal of Semantic Computing*, 17(02), 249–271. DOI: 10.1142/S1793351X23600048

Holzinger, A., Saranti, A., Hauschild, A. C., Beinecke, J., Heider, D., Roettger, R., . . . Pfeifer, B. (2023, August). Human-in-the-loop integration with domain-Machine learning for explainable federated deep learning. In *International Cross-Domain Conference for Machine Learning and Knowledge Extraction* (pp. 45-64). Cham: Springer Nature Switzerland. DOI: 10.1007/978-3-031-40837-3_4

Hu, X., Liu, Y., Tang, B., Yan, J., & Chen, L. (2023). Learning dynamic ML for overtaking strategy in autonomous driving. *IEEE Transactions on Intelligent Transportation Systems*, 24(11), 11921–11933. DOI: 10.1109/TITS.2023.3287223

Huang, H., Sun, L., Du, B., Liu, C., Lv, W., & Xiong, H. (2021, August). Representation learning on Machine learning for node importance estimation. In *Proceedings of the 27th ACM SIGKDD Conference on Knowledge Discovery & Data Mining* (pp. 646-655).

Ji, S., Pan, S., Cambria, E., Marttinen, P., & Philip, S. Y. (2021). A survey on Machine learning: Representation, acquisition, and applications. *IEEE Transactions on Neural Networks and Learning Systems*, 33(2), 494–514. DOI: 10.1109/TNNLS.2021.3070843 PMID: 33900922

Kim, J. E., Henson, C., Huang, K., Tran, T. A., & Lin, W. Y. (2021). Accelerating road sign ground truth construction with AI and machine learning. In *Intelligent Computing:Proceedings of the 2021 Computing Conference,* Volume 2 (pp. 325-340). Springer International Publishing.

Kraft, M., Bai, J., Mosbach, S., Taylor, C., Karan, D., Lee, K. F., ... & Lapkin, A. (2023). From Platform to AI: Distributed Self-Driving Laboratories.

Lecue, F. (2020). On the role of Machine learning in explainable AI. *Semantic Web*, 11(1), 41–51. DOI: 10.3233/SW-190374

Li, X., Liu, J., Li, J., Yu, W., Cao, Z., Qiu, S., . . . Jiao, X. (2023, August). ML Structure-Based Implicit Risk Reasoning for Long-Tail Scenarios of Automated Driving. In *2023 4th International Conference on Big Data, Artificial Intelligence and Internet of Things Engineering (ICBAIE)* (pp. 415-420). IEEE.

Liu, H., Li, Y., Hong, R., Li, Z., Li, M., Pan, W., Glowacz, A., & He, H. (2020). AI analysis and visualization of research trends on driver behavior. *Journal of Intelligent & Fuzzy Systems*, 38(1), 495–511. DOI: 10.3233/JIFS-179424

Liu, J., Zhang, X., Li, Y., Wang, J., & Kim, H. J. (2019). Deep learning-based reasoning with multi-ontology for IoT applications. *IEEE Access : Practical Innovations, Open Solutions*, 7, 124688–124701. DOI: 10.1109/ACCESS.2019.2937353

Liu, Q., Li, X., Tang, Y., Gao, X., Yang, F., & Li, Z. (2023). ML Reinforcement Learning-Based Decision-Making Technology for Connected and Autonomous Vehicles: Framework, Review, and Future Trends. *Sensors (Basel)*, 23(19), 8229. DOI: 10.3390/s23198229 PMID: 37837063

Luettin, J., Monka, S., Henson, C., & Halilaj, L. (2022, November). A survey on AI-based methods for automated driving. In *Iberoamerican Machine learning and Semantic Web Conference* (pp. 16–31). Springer International Publishing.

Mahon, L., Giunchiglia, E., Li, B., & Lukasiewicz, T. (2020, December). AI extraction from videos. In *2020 19th IEEE International Conference on Machine Learning and Applications (ICMLA)* (pp. 25-32). IEEE.

Nickel, M., Murphy, K., Tresp, V., & Gabrilovich, E. (2015). A review of relational machine learning for Machine learning. *Proceedings of the IEEE*, 104(1), 11–33. DOI: 10.1109/JPROC.2015.2483592

Ploennigs, J., Semertzidis, K., Lorenzi, F., & Mihindukulasooriya, N. (2022, October). Scaling Machine learning for automating AI of digital twins. In *International Semantic Web Conference* (pp. 810-826). Cham: Springer International Publishing.

Schier, M., Reinders, C., & Rosenhahn, B. (2023). Deep reinforcement learning for autonomous driving using high-level heterogeneous ML representations. In *International Conference on Robotics and Automation (ICRA)*.

Sharma, A., & Singh, B. (2022). Measuring Impact of E-commerce on Small Scale Business: A Systematic Review. *Journal of Corporate Governance and International Business Law*, 5(1).

Singh, B. (2019). Profiling Public Healthcare: A Comparative Analysis Based on the Multidimensional Healthcare Management and Legal Approach. *Indian Journal of Health and Medical Law*, 2(2), 1–5.

Singh, B. (2023). Unleashing Alternative Dispute Resolution (ADR) in Resolving Complex Legal-Technical Issues Arising in Cyberspace Lensing E-Commerce and Intellectual Property: Proliferation of E-Commerce Digital Economy. *Revista Brasileira de Alternative Dispute Resolution-Brazilian Journal of Alternative Dispute Resolution-RBADR*, 5(10), 81–105. DOI: 10.52028/rbadr.v5i10.ART04.Ind

Singh, B. (2023). Blockchain Technology in Renovating Healthcare: Legal and Future Perspectives. In *Revolutionizing Healthcare Through Artificial Intelligence and Internet of Things Applications* (pp. 177-186). IGI Global.

Singh, B. (2023). Federated Learning for Envision Future Trajectory Smart Transport System for Climate Preservation and Smart Green Planet: Insights into Global Governance and SDG-9 (Industry, Innovation and Infrastructure). *National Journal of Environmental Law*, 6(2), 6–17.

Singh, B. (2024). Cherish Data Privacy and Human Rights in the Digital Age: Harmonizing Innovation and Individual Autonomy. In Pucelj, M., & Bohinc, R. (Eds.), *Balancing Human Rights, Social Responsibility, and Digital Ethics* (pp. 199–226). IGI Global., DOI: 10.4018/979-8-3693-3334-1.ch007

Singh, B. (2024). Lensing Legal Dynamics for Examining Responsibility and Deliberation of Generative AI-Tethered Technological Privacy Concerns: Infringements and Use of Personal Data by Nefarious Actors. In Ara, A., & Ara, A. (Eds.), *Exploring the Ethical Implications of Generative AI* (pp. 146–167). IGI Global., DOI: 10.4018/979-8-3693-1565-1.ch009

Singh, B. (2024). Evolutionary Global Neuroscience for Cognition and Brain Health: Strengthening Innovation in Brain Science. In Prabhakar, P. (Ed.), *Biomedical Research Developments for Improved Healthcare* (pp. 246–272). IGI Global., DOI: 10.4018/979-8-3693-1922-2.ch012

Singh, B. (2024). Social Cognition of Incarcerated Women and Children: Addressing Exposure to Infectious Diseases and Legal Outcomes. In Reddy, K. (Ed.), *Principles and Clinical Interventions in Social Cognition* (pp. 236–251). IGI Global., DOI: 10.4018/979-8-3693-1265-0.ch014

Singh, B., & Kaunert, C. (2024). Integration of Cutting-Edge Technologies such as Internet of Things (IoT) and 5G in Health Monitoring Systems: A Comprehensive Legal Analysis and Futuristic Outcomes. *GLS Law Journal*, 6(1), 13–20. DOI: 10.69974/glslawjournal.v6i1.123

Tang, X., Sun, T., Zhu, R., & Wang, S. (2021, January). CKG: dynamic representation based on context and AI. In *2020 25th International Conference on Pattern Recognition (ICPR)* (pp. 2889-2895). IEEE.

Xiao, D., Dianati, M., Geiger, W. G., & Woodman, R. (2023). Review of ML-based hazardous event detection methods for autonomous driving systems. *IEEE Transactions on Intelligent Transportation Systems*, 24(5), 4697–4715. DOI: 10.1109/TITS.2023.3240104

Xu, W., Feng, L., & Ma, J. (2022). Understanding the domain of driving distraction with Machine learning. *PLoS One*, 17(12), e0278822. DOI: 10.1371/journal.pone.0278822 PMID: 36490240

Yu, C., Wang, X., Xu, X., Zhang, M., Ge, H., Ren, J., Sun, L., Chen, B., & Tan, G. (2019). Distributed multiagent coordinated learning for autonomous driving in highways based on dynamic coordination MLs. *IEEE Transactions on Intelligent Transportation Systems*, 21(2), 735–748. DOI: 10.1109/TITS.2019.2893683

Zheng, Z., Zhou, B., Zhou, D., Soylu, A., & Kharlamov, E. (2022, October). Executable AI for transparent machine learning in welding monitoring at bosch. In *Proceedings of the 31st ACM International Conference on Information & Knowledge Management* (pp. 5102-5103).

Chapter 3
Digital Transformation in School Education:
Machine Learning Insights Into Marketing Strategies

Purvi Makwana
Shri Ramdeobaba College of Engineering and Management, Nagpur, India

Gayathri Band
Shri Ramdeobaba College of Engineering and Management, Nagpur, India

Kanchan Naidu
Shri Ramdeobaba College of Engineering and Management, Nagpur, India

Tarun Kumar Kaushik
Sharda University, India

ABSTRACT

In today's era, school education has seen a massive transformation due to digitalization, which has changed how schools learn, administer, and engage. However, this shift amplifies the teacher-student relationship and entirely changes how educational institutions market their business by reshaping the current education system. Also, machine learning helps optimize these strategies by analyzing and interpreting very large datasets to predict trends, customize marketing campaigns, individualize student engagement, and more. By generating ML-driven insights, all the schools can spot ideal students, analyze their requirements, and provide tailor-made educational offerings. Thus, reductive analytics can monitor the success metrics of marketing campaigns, streamline communication, and increase enrollment rates. It also provides insights into student preferences, making school promotion more efficient. AI

DOI: 10.4018/979-8-3693-8684-2.ch003

has the potential to help schools align their marketing strategies with a tech-native generation and develop more sustainable educational practices in the future.

INTRODUCTION

Over the recent past, educational institutions have pressed to implement different digital technologies to improve the learning experience, promising to increase operational efficiency (Ecom et al., 2024). One of the promising areas in which machine learning (ML) can help educational institutions is incorporating ML into their marketing strategies to attract prospective students and retain current ones (Garzón & Acevedo, 2019). Also, the educational sector increasingly focuses on digital transformation and technological advancements, particularly with inventions like artificial intelligence (AI) and machine learning (ML) (Bui & Son, 2023). However, apart from making it easier to deliver education, they have changed the academic and marketing game for educational institutions. In this paper, the authors investigate the impact of digital transformation in Indian school education by studying a specific use case, machine learning, for effective marketing strategy (Garzón et al., 2020). Without a doubt, digital transformation has become an indispensable need in the contemporary education system, primarily composed of AI and ML-based technologies continuously stretching the potential scope of the school education landscape (Cavanaugh et al., 2023). In the ever-changing digital world, educational institutions are working hard to keep up with it, and integrating such technologies has led to new opportunities for improving teaching and learning experiences (Kalemkuş & Kalemkuş, 2022).

Furthermore, AI also includes machine learning, which has become an invaluable weapon that allows systems to learn from data and improve their performance without explicit programming. ML algorithms can be used at different levels of the educational process, like personalized learning, predictive analytics, operational efficiency, etc. (Kazu & Yalçin, 2022). Marketing strategies for educational institutions. Machine learning can provide meaningful insights. Thus, schools can strategically use ML to their advantage to stay ahead of the curve on enrollment trends and meet more students where they are for longer. Using ML algorithms, colleges and universities can profile a comprehensive range of data about potential students' online behaviors, preferences, and demographics to craft targeted marketing campaigns that appeal to particular groups. Also, reductive enrollment modeling is yet another fundamental area for the intervention of machine learning in educational marketing (Maor et al., 2023). When historical enrollment data is crossed with current trends, ML models can forecast the kind of students a school should expect in the future and optimize where to focus their marketing efforts and resources because better utilization of these sources ultimately leads to higher ROI. Machine learning tools can also make it easier for schools to connect with potential students by enabling more personalization in outreach efforts. For example, ML-powered chatbots can respond to inquiries in real-time and improve the overall user experience on the institutional website, thereby improving your chances of enrolling a student. Machine learning also plays a huge role in content optimization. They can also take this with the help of ML algorithms, which will analyze what kind of content is more engaging for respective target demographics. This can help institutions improve their content strategy for increased outreach and connection with potential students (Su & Yang, 2022).

However, with the numerous benefits of machine learning in educational marketing, there are also challenges to address. Indeed, data privacy is a significant problem when it comes to collecting and analyzing student data, as this impacts the ethical considerations that universities have to make regarding regulations. Moreover, integrating new ML tools with existing systems and processes is challenging and may involve many changes and decisions. Machine learning will support these points as educational organizations continue their digital transformation journey to compete more in an increasingly digital environment. Schools can utilize ML and deepen their understanding of the most successful marketing tactics to acquire and retain students amidst an ever-changing educational landscape (Han et al., 2024).

In learning, education technology uses digital hardware such as computers, smartphones, tablets, and educational software and applications (apps). Data-driven strategies are vital to how schools and universities can enhance their outreach and engagement efforts. Artificial Intelligence (AI), specifically machine learning, can analyze massive data sets to glean helpful information, forecast trends, and hyper-personalize marketing efforts (Kedi et al., 2024).

DIGITAL TRANSFORMATION IN EDUCATION

At its core, digital transformation in education is not simply about integrating new tech but revolutionizing standard pedagogical models and embracing more facilitative teaching practices that make learning an advanced experience. Moreover, the COVID-19 pandemic has made that shift more pressing than ever, instituting remote learning and digital classrooms with educational institutions globally (Han et al., 2024). Consequently, schools realized that digital systems are needed for effective teaching and internal operations, stating that the end goal of digital transformation is to produce a student-focused learning model accessible around the clock and tailored for students based on their needs. Using technology, educators can offer customized learning experiences that meet the learning needs of individual students and also continually provide them with new content, which helps improve levels of engagement in addition to retention rates. In addition, online tools provide instantaneous feedback and evaluation, enabling teachers to keep a closer eye on student development (Rêgo et al., 2024).

Enhancing Accessibility: The most crucial benefit of a digital transformation is that it democratizes education, making it available to more users. It brings forth benefits like reaching more students to whom access would have been a challenge due to remote location or physical presence of educational institute. Online learning platforms and resources enable students to access top-notch education regardless of location. This was evident in the technological advancements during the COVID-19 Pandemic when facing School Closures; few lessons were learned by many in schools; perhaps Online Devices might help to re-engineer Educational Continuity during Disruptions in traditional Classroom settings. In addition, digital transformation ensures inclusivity through the availability of personalized assistive technology that addresses various learning needs. For example, students with learning disabilities can use tools focusing on their overall educational experience, such as speech-to-text software or interactive learning applications customized to meet their needs. Such dedication to access ensures that every student has the chance to succeed in their endeavors as they further educate (Govindan, 2024).

Personalization of Learning: Data analytics and machine learning algorithms help provide personalized learning experiences, thereby facilitating the digital transformation of the industry. This technology allows teachers to drill into growth data and develop customized lesson plans. One-sided, passive learning approaches are less effective than two-sided, active methods, and online platforms capable of adaptive learning can provide more or new questions for an exercise based on the student levels to ensure everyone gets appropriately challenged (Jin et al., 2024). It is about customization in more than just academics. It also includes learning formats and desires. Digital tools allow educators to share information in other media formats, such as videos, interactive simulations, or gamified assessments, so students can interact with the material in a way that works best for them. It customizes an individual experience for a learning profile and increases student engagement, retention, and academic success (Umamaheswari, 2024).

Increasing Student Engagement: Engagement is critical to learner success, and digital transformation ensures learners are engaged in an educational environment. In an age of ever-evolving technology and increasingly shrinking attention spans, traditional teaching methods often leave students nodding off in the back of the classroom. Interactive whiteboards, virtual reality (VR), and augmented reality (AR) with digital tools present highly interactive and engaging learning experiences that can quickly draw the attention of any student. This may include placing students in the middle of historical events and scientific phenomena so they can not only read from a textbook. Gamification is also something that motivates students. By engaging students in a novel way to learn, these strategies not only take the monotony away from learning but also promote collaboration and group discussion among peers, thereby bringing a feeling of community within the classroom (Adeleye et al., 2024).

Efficiency, Administrative Process Improvement: Digital transformation digitized the administrative system in educational institutions to a large extent by making it more operable, which improved their functioning. If you struggle to manage basic admin tasks and overstretch yourself, you could find that it is not only complex but transferring beyond manually cannot be efficient (Trevisan et al., 2024). Digital solutions like learning management Systems automate all of these processes and are tools that benefit teaching even more. In addition, analyzing data can provide insights into programmatic or institutional effectiveness, which can guide leaders in their decision-making regarding the allocation of funding and the development of programs. Enrollment Trends analysis of which programs attract new students can reveal growth areas and sometimes vulnerability in an institution's portfolio. They can deliver on their operational objectives using significant data insights and enforce the optimal use of resources (Oyekunle & Boohene, 2024).

Understanding Machine Learning in Marketing: Machine learning is an area of artificial intelligence related to constructing and studying systems that can learn from data. The gaming industry uses learning to predict churn and engagement and encourage power users. ML is also used in marketing to analyze customer patterns and segments and improve campaigns. These insights are made possible through historical data and real-time analytics, arming educational marketers with the knowledge they need to craft compelling messages and outreach strategies that resonate with their students (Adesina et al., 2024).

APPLICATIONS OF MACHINE LEARNING IN EDUCATIONAL MARKETING

Predictive Analytics: One of the most valuable aspects of machine learning in educational marketing is predictive analytics. Using historical enrollment data and current trends, machine learning (ML) models can predict future enrollments with astonishing precision. This enables institutions to create an informed allocation of resources and marketing strategies. So, for example, if predictive models point to increased online degree program interest across a specific geographic region among high school grads, institutions can then use that data to target those prospects in their marketing outreach. Being proactive with your content helps drive recruitment and keeps schools in the game (Okorie et al., 2024).

Customer Segmentation: Nevertheless, effective marketing begins with knowing the person to whom you are shooting. Educational institutions can use machine learning to segment prospective students by demographics, interests, and behaviors. Marketers can also segment untargeted data with unsupervised learning techniques such as clustering algorithms to find different clusters of student segments and send a unique message for each segment. An example is an institution finding that specific demographics are more likely to participate in particular programs or courses. When marketing campaigns are tailored to reach these segments, there is an increase in the likelihood that individuals will apply and enroll, as well as an overall improvement based on engagement signals (Raghav et al., 2024).

Quality Marketing Campaigns: Personalized marketing strategies will almost always get better results, not just today but well into the future. Universities can leverage machine learning to tailor-fit marketing campaigns using student data and preferences. Through this process, ML algorithms could observe user engagements on institutional websites and social media platforms and serve up custom content that matches the interests of an inquiring student. Email marketing also reaches this level of personalization. Using engagement metrics, institutions can send automatically triggered follow-up emails to prospects who have indicated an interest in particular programs but have not applied. Talking about each applicant explicitly allows schools to speak in depth (relatively) to the interests and apprehensions of individual students, ultimately improving the chances of conversion (Gumaelius et al., 2024).

Enhanced Student Engagement: In many instances, using machine learning techniques to automate your communication strategies can improve student engagement. A chatbot backed by ML algorithms can help encourage prospective students to visit an institution site while providing instant answers to inquiries from the student end, making the whole student journey smooth on institutional websites. The chatbots are designed to analyze user queries and provide information about programs, admissions, campus life, etc. Moreover, content recommendations are another way adaptive learning technology can assist students in delivering relevant resources on prospective students recommended based on what they have been browsing. For example, if a user repeatedly visits pages on business programs, the institution can suggest articles or videos about why those programs are good. This way, machine learning will sharpen marketing strategies (Nnaji et al., 2024).

MARKETING STRATEGIES ENHANCED BY MACHINE LEARNING

Targeted Marketing Campaigns: The ability of machine learning algorithms to process large datasets from different sources, e.g., social media interactions, website visits, and application trends, is critical. This analysis shows that institutions can profile individual audience segments according to their interests and behaviors. For example, an institution may discover that some programs or courses appeal more to particular demographics. Schools can improve enrollment rates and boost overall involvement by customizing marketing messages to these segments (Aldoseri et al., 2024).

Enrollment Predictive Modeling: Predictive enrollment modeling is another critical example of machine learning in education marketing. Using past data coupled with the latest trends, ML models can predict enrollment in the future with high accuracy. Therefore, they can allocate resources efficiently and market to the volume of student demand predicted. Suppose predictive models show high school graduates in a specific region being more interested in STEM programs, for example. In that case, this is an area where institutions can focus their marketing and target those potential students. This practice helps bring students to their schools and keeps them in control over the competitive education market (Ara et al., 2024).

Enhanced Student Engagement: Leveraging machine learning tools to increase student engagement can further empower personal communication strategies. AI-enabled chatbots using ML algorithms can respond instantly to potential students' queries and reduce their friction when interacting with institutional websites. The AI-powered chatbots analyze student inquiries and deliver personalized data on programs, admissions processes, and campus life. Additionally, ML-driven email marketing campaigns can be improved by leveraging user interaction. For example, a potential student who allows the content of a program to appear many times but never applies already can receive email follow-ups encouraging him/her to complete the application process.

Optimising Content for Outreach that Works: Content is king in educational marketing strategies. Using a machine learning algorithm, you can decipher what content your target audience resonates with- blog posts, videos, or any other social media update. In this sense, knowing why candidates engage with prospective student content can inform a strategy regarding what to emphasize. One such example would be if data analysis reveals that videos of student testimonials get more engagement than written articles. They know to produce more videos when going about their campaign. Scaling with SEO best practices improves the effectiveness with which an institution can communicate externally and bridges its way in a more personal and appealing manner to the institutions concerned (Tsaples et al., 2024).

Using Machine Learning in Education: There is no doubt that AI or machine learning (ML), in particular, are the rapidly evolving technologies of our time, spreading their wings and shaping many solutions, from improving the quality of education delivered to teaching-learning and administrative processes. With pressures on higher education to provide materials that can adapt based on students' personalized needs and consolidate how resources are used for maximum efficiency, machine learning provides a powerful solution. This section discusses all the best applications of machine learning in the education sector and their advantages and overall impact on students and educators.

Tailored Learning Experiences: Personalizing learning experiences: One of the most outstanding contributions machine learning has made to education is its ability to provide personalized learning. The current teaching structure in the traditional education system is generally characterized by a one-size-fits-all model, often leading to disinterest or challenging experiences for many students. In the same class, machine learning algorithms sift through an individual student's data performance metrics, learning style, and engagement levels to customize educational content and teaching strategies for students. For example, adaptive learning platforms use machine learning to adapt the complexity of exercises to student performance (Richter & Tudora, 2024).

Predictive Analytics for Student Retention: Predictive analytics means machine learning, which is crucial for an educational setting. ML algorithms can analyze historical student attendance, behavior, and performance data to predict which students will likely become at-risk early in their academic journey. Taking a preventative approach allows educators to catch students before things unravel and offer support to get them back on track toward success. For instance, models can predict which students are at the highest risk of attrition or poor academic performance by measuring their levels of engagement and previous academic success. When data is analyzed, it can often prompt the educator into immediate action in the form of tutoring for individual students or counseling with a specific group of students who are having difficulties. Colleges and universities can use predictive analytics to improve retention rates and foster student success (Katsamakas & Saklad, 2024).

Improved Assessment and Grading: A slow assessment task for educators is grading. This is where machine learning comes to the rescue, as it automates this process and grades assignments and assessments. Grade Scope, for example, leverages ML algorithms to parse individual student submissions, cluster pattern-matching answers together, and derive scores as a function of predefined rubrics. Automating this helps take some load off the teachers and ensures consistent grading practices. In addition, AI-based assessment solutions can give students comprehensive feedback on their work, which areas went well, and where they could use the practice, suggesting personalized exam preparation recommendations. Such an instant feedback loop allows students to be active in their learning and understand where they need to change their learning process (Harrer et al., 2024).

Intelligent Tutoring Systems: Another educational use of machine learning is with intelligent tutoring systems (ITS), which help tailor lessons to individual students. These systems leverage the power of ML algorithms to accurately mimic one-on-one tutoring approaches by dynamically adjusting content delivery in response to ongoing student activity and interaction. This might take the form of a student using an ITS to demonstrate that they know how to do a specific math concept by answering some practice problems. Based on the strengths or weaknesses of given responses, instructional strategies can be altered; if students mess up in some key areas, they may receive supplementary explanations or tactics as the system directs (Jia et al., 2024).

Administrative Efficiency: Besides aiding teaching practices, machine learning also helps the education system be more efficient on the admin side. Through automated daily maintenance processes, such as running reports and statistics, keeping student enrollment intact, and managing the schedules and resources within an institution, administrators have more bandwidth to dedicate themselves to strategic initiatives that help them fulfill the goal of expansion. For example, machine learning can process data on a particular course or program and look for enrollment trends to help predict future demand. Higher education institutions can use this predictive capability to improve course planning and resource management, covering more student needs while maximizing operational efficiency. Fostering Inclusive Education: Machine learning is essential in enabling inclusive education by offering the necessary tools to fulfill the needs of all types of learners, regardless of their diversity. ML-driven adaptive learning technologies can personalize educational experiences tailored to students with disabilities and those who need accommodations. Speech recognition software or text-to-speech applications are just some examples of assistive technologies that use machine learning algorithms to improve accessibility for students who have visual and auditory deficits. These tools use personalized data for individual performances to customize the content delivery methods, maintaining an equitable availability of educational resources (Kao et al., 2024).

Drive engagement with gamification: In educational settings, gamification is one of the most successful techniques for promoting student participation. By leveraging machine learning, gamified learning activities can be optimized such that personalized games are designed to correspond with how one learner or another is interacting (or compelled to interact) with them. A good example would be an educational app that can purchase ML algorithms to follow a student through various game levels but change the challenge at each level depending on the student's capability. This dynamic method maintains learners' interest and keeps them focused on learning, resulting in achievement so that students know their achievements when completing challenging tasks (Leso et al., 2024).

CHALLENGES OF IMPLEMENTATION MACHINE LEARNING IN EDUCATIONAL MARKETING

The application of machine education in educational marketing is quite possibly latent, but there are some challenges. Some of them are as follows:

Data Quality and Availability: High-quality data is one of the essential prerequisites for effective machine learning. The target is the education sector, so obtaining clean and valuable data can sometimes be tricky. As previous blog posts have described, educational data is a complete mess in multiple ways, including format, quality, completeness, etc. This makes the creation of models that are both accurate and unbiased difficult.

Lack of Skilled Workforce: Machine learning solutions need technical hires to develop and train models and add them to production. This aids in narrowing scope and focus, limiting who might be interested in implementing Educational ML solutions, as only a few people we know have domain knowledge of both domains. This requires schools or universities to invest in internal staff training or hire outside professionals, which can be expensive and time-consuming. Ethical and Privacy Concerns: Machine learning systems are often trained on vast amounts of data, which can lead to ethical and privacy problems. These concerns are even more significant in education, where you are dealing with student data, which is very sensitive. That makes collecting and using data ethically and legally sound more critical than ever. Education institutions are expected to adopt stringent data protection practices and ensure that they require the appropriate consent in place while also fostering trust amongst students, their parents, and other stakeholders by being transparent on how their entire student data will be used (Babatunde et al., 2024).

Machine Learning Algorithms are Biased: Bias in machine learning models can often result from bias in the training data. This bias can have essential implications in education, where it may result in different opportunities for students of diverse demographics. Biased algorithms, e.g., in student performance or educational pathways, may decide to compound gender or racial stereotypes. Bias is a problem that must be corrected with ongoing measurement, data correction, and algorithm tuning to free machine learning systems from unfairness. Migration to Other Systems: Higher education often has a wealth of existing systems, including but not limited to learning management systems (LMS) and student information systems (SIS). Integrating machine learning technologies with such systems is challenging, and compatibility issues, data interoperability, and new infrastructure requirements may encumber implementation. Institutions must ensure that the integration process does not get in with teaching and learning at their end. Resistance to Change: One group continuing to resist machine learning is educators. Sometimes, the thought of change or fear of a new technology might phase out their current methods and take away teachers' autonomy. It means telling people what they can get away from machine learning, helping them learn about it, and ensuring everybody has a say in decision-making (Qiao et al., 2024).

Financial Constraints: Machine learning solutions mean financial investments for hardware, software, data infrastructure, and talent. These initiatives can be complex for educational institutions, particularly ones with constrained budgets on which to spend their resources. The main challenge during the early stages of implementation is balancing the long-term benefits machine learning might offer with immediate financial constraints. In order to address these challenges and effectively implement machine learning within their educational marketing strategies, institutions need an end-to-end strategic approach. This investment involves, in part, acquiring a robust data infrastructure, training staff and clinicians, dealing with ethical considerations regarding the use of AI (when producing or using AI models), managing bias issues, and promoting a culture of innovation and collaboration between stakeholders. By taking these burdens head-on, educational institutions can leverage the power of machine learning and enhance student engagement by streamlining systems, improving marketing effectiveness through targeting personalized messages, and ultimately learning how to teach better (Herhausen et al., 2024).

Using Machine Learning in Education Marketing - Case Study: The Head of Robert J. +Cabugao is a brief production of the University of California, Berkeley- Running on ML and Data: How UC Berkeley uses machine learning algorithms to control the new student populace. This segmentation of audiences by enrollment propensity enabled the university to create more relevant marketing strategies. This included using historical data on who had applied, demographic profiles, and engagement metrics to create signals with the most predictability of students being more likely to apply and enroll. Berkeley saw a significant 15% rise in applications due to these focused marketing efforts. This endeavor became successful as a testimony of the power of data-backed decision-making for higher education marketers. Using machine learning to understand better and predict student behaviors would allow us to make more thoughtful use of university resources by customizing outreach efforts for individual prospective students (Butt et al., 2024).

Southern New Hampshire University (SNHU): These are just case studies, but they should explain how machine learning can transform educational marketing strategies. By using data to drive their outreach efforts, UC Berkeley and SNHU improved their application rates and student retention. Data-driven Decision Making: Both organizations leverage machine learning algorithms to make sense of a vast data set to optimize their marketing. This enhances the criticality of data quality and availability for a successful machine-learning solution. Personalized Marketing Campaigns: These universities could have segmented prospective students on a continuum of "likely to enroll" and "high risk," developing targeted marketing campaigns that addressed individual needs. This type of targeting is crucial to get the best engagement and conversion. Early Intervention: A robust predictive analytics system will help institutions identify students who are at risk (before it is probably too late) and, therefore, implement interventions that can support student success. This proactive approach supports the institution as it works to continue improving outcomes in terms of retention rates and offers a significantly better educational experience. By analyzing these data, machine learning can help institutions use resources more effectively and concentrate on students most likely to succeed in better-targeted interventions. This can result in cost savings and help improve marketing ROI. Continuous improvement: A machine learning model in educational marketing must be maintained and optimized as businesses grow. Institutions must adjust their strategies based on new data insights and evolving market conditions (Ajiga et al., 2024).

These examples of UC Berkeley and Southern New Hampshire University have proved that machine learning can significantly improve educational marketing strategies. Through predictive analytics and targeted outreach initiatives, these schools have been able to increase application rates and retention and deliver a tailored experience to potential students. Educational institutions looking to mine their data with machine learning face data quality issues, privacy concerns, and integration with legacy systems. Nevertheless, the rewards span improved engagement with richer resources. The implementation of digital transformation in the education industry and machine learning at schools has modernized educational marketing strategies. Through data analytics and machine learning, schools can receive fascinating insights to more efficiently appeal to prospective students and their families (Kedi et al., 2024).

CONCLUSION

One of the most significant benefits of machine learning algorithms is that they help educational institutes analyze large quantities of data faster. This data-driven method allows schools to see trends and what their ideal "audience" behaves like and likes. This can help institutions optimize their marketing and determine how it appeals to potential students (and, perhaps more importantly, their parents). Predictive analytics allow schools to predict the wants and needs of prospective students based on past data so they can more effectively market to them in a tailored way.

This personalization enhances involvement and develops an idea that learners are valued and beneficial participants in a cutthroat educational environment.

Machine learning also gives you clear indications of the marketing channels and strategies that work for your campaign or business. With tools like A and B testing and multivariate testing (adding more variables), schools can learn what messages work on which platforms. By taking an iterative approach to marketing and treating experimentation as a strategic lever, funds are better allocated, increasing the ROI for each marketing dollar spent. The reality is that schools continually have to work within a budget, so knowing what techniques tend to be the most effective can mean the difference between successfully attracting and converting students and simply spinning your wheels. More prospective students are interacting with institutions through online platforms, so schools must be prepared to pivot their marketing strategies accordingly. Further tweaking the delivery of content with machine learning could help seal the deal and get those would-be students more engaged in moments where they are likely to pay attention most. This not only boosts the ultimate user experience but also strengthens an institutional image, which, therefore, attracts prospective families.

Future Workforce Prep for Students: The world of work is changing rapidly, with more and more of its operations being automated by technology. Digital transformation is a vital enabler of the soft skills employers crave today, such as digital literacy, critical thinking, collaboration, and adaptability (Potla & Pottla, 2024). By incorporating technology into the curriculum, students can hone their skills with various digital tools they will likely use in their future profession. Practical courses like coding boot camps or data analysis-oriented courses can be beneficial as they provide some skills to help you get a job. In addition, interacting with software wellbeing programs also helps students transition to remote learning practices, considering that working from home has become a relevant reality in the last few years.

Additionally, by using predictive analytics to form their marketing strategies, schools enjoy better audience mapping. Such segmentation allows the nature of the institution to create targeted messaging for those sub-groups and deliver the information these segments need based on their context. For global students, this could include messages on support services for your visa application; local students may be more interested in community involvement, social events, and clubs. Focusing on the needs of each audience can help schools create affinity and attachment to them.

Machine learning affects more than just marketing strategies; it improves the enrollment management process. Utilizing historical enrollment data, schools can identify trends and anticipate areas of concern fast enough to intervene before they become problems. Institutions with this foresight can adapt their marketing strategy on the fly, adjusting to changes in demand or new competition as soon as they arise. While machine learning in marketing strategies has many benefits, educational institutions must be considerate when navigating this digital transformation. All marketing must be conducted responsibly and equitably, so ethical considerations such as data privacy and the potential for algorithm bias need to come first. Being straightforward about how data is used and what it means communicates trust and reliability to students and families over the long haul.

Finally, implementing machine learning insights in school education as part of a digital transformation offers schools a fantastic opportunity to review their marketing strategies. Without a solid 2021 admissions strategy, schools will struggle to attract their new prospective audience, meaning they can use data analytics and personalize outreach, among other digital functions, to get the right students excited about attending. This ongoing transformation will see institutions that adopt these novel methodologies enhance their marketing prowess and affirm their status as leaders in the education space, effectively catering to the dynamic requirements of students and families within an evidence-based, digital ecosystem.

REFERENCES

Adeleye, R. A., Awonuga, K. F., Asuzu, O. F., Ndubuisi, N. L., & Tubokirifuruar, T. S. (2024). Digital marketing analytics: A review of strategies in the age of big data and AI. *World Journal of Advanced Research and Reviews, 21*(2), 073-084.

Adesina, A. A., Iyelolu, T. V., & Paul, P. O. (2024). Leveraging predictive analytics for strategic decision-making: Enhancing business performance through data-driven insights. *World Journal of Advanced Research and Reviews*, 22(3), 1927–1934. DOI: 10.30574/wjarr.2024.22.3.1961

Ajiga, D. I., Ndubuisi, N. L., Asuzu, O. F., Owolabi, O. R., Tubokirifuruar, T. S., & Adeleye, R. A. (2024). AI-driven predictive analytics in retail: A review of emerging trends and customer engagement strategies. *International Journal of Management & Entrepreneurship Research*, 6(2), 307–321. DOI: 10.51594/ijmer.v6i2.772

Aldoseri, A., Al-Khalifa, K. N., & Hamouda, A. M. (2024). Methodological approach to assessing the current state of organizations for AI-Based digital transformation. *Applied System Innovation*, 7(1), 14. DOI: 10.3390/asi7010014

Ara, A., Maraj, M. A. A., Rahman, M. A., & Bari, M. H. (2024). The Impact Of Machine Learning On Prescriptive Analytics For Optimized Business Decision-Making. *International Journal of Management Information Systems and Data Science*, 1(1), 7–18.

Babatunde, S. O., Odejide, O. A., Edunjobi, T. E., & Ogundipe, D. O. (2024). The role of AI in marketing personalization: A theoretical exploration of consumer engagement strategies. *International Journal of Management & Entrepreneurship Research*, 6(3), 936–949. DOI: 10.51594/ijmer.v6i3.964

Bui, T., & Son, N. T. (2023). The survey of digital transformation in education: A systematic review. *International Journal of TESOL & Education*, 3(4), 32–51. DOI: 10.54855/ijte.23343

Butt, A., Imran, F., Helo, P., & Kantola, J. (2024). Strategic design of culture for digital transformation. *Long Range Planning*, 57(2), 102415. DOI: 10.1016/j.lrp.2024.102415

Cavanaugh, C., McConney, A., Maor, D., & McCarthy, A. (2023). Digital transformation in education: Critical components for leaders of system change. *Computers in Human Behavior*, 145, 107751. Advance online publication. DOI: 10.1016/j.chb.2023.107751

Eom, T., Woo, C., & Chun, D. (2024). Predicting an ICT business process innovation as a digital transformation with machine learning techniques. *Technology Analysis and Strategic Management*, 36(9), 2271–2283. DOI: 10.1080/09537325.2022.2132927

Eom, T., Woo, C., & Chun, D. (2024). Predicting an ICT business process innovation as a digital transformation with machine learning techniques. *Technology Analysis and Strategic Management*, 36(9), 2271–2283. DOI: 10.1080/09537325.2022.2132927

Garzón, J., & Acevedo, M. (2019). Augmented reality applications in education: A systematic review of the literature. *Journal of Educational Technology & Society*, 22(1), 1–12.

Govindan, K. (2024). Unlocking the potential of quality as a core marketing strategy in remanufactured circular products: A machine learning enabled multi-theoretical perspective. *International Journal of Production Economics*, 269, 109123. DOI: 10.1016/j.ijpe.2023.109123

Gumaelius, L., Skogh, I. B., Matthíasdóttir, Á., & Pantzos, P. (2024). Engineering education in change. A case study on the impact of digital transformation on content and teaching methods in different engineering disciplines. *European Journal of Engineering Education*, 49(1), 70–93. DOI: 10.1080/03043797.2023.2285794

Han, Y., Li, Z., Feng, T., Qiu, S., Hu, J., Yadav, K. K., & Obaidullah, A. J. (2024). Unraveling the impact of digital transformation on green innovation through microdata and machine learning. *Journal of Environmental Management*, 354, 120271. DOI: 10.1016/j.jenvman.2024.120271 PMID: 38354610

Harrer, S., Menard, J., Rivers, M., Green, D. V., Karpiak, J., Jeliazkov, J. R., & Sternke, M. C. (2024). Artificial intelligence drives the digital transformation of pharma. In *Artificial intelligence in clinical practice* (pp. 345–372). Academic Press. DOI: 10.1016/B978-0-443-15688-5.00049-8

Herhausen, D., Bernritter, S. F., Ngai, E. W., Kumar, A., & Delen, D. (2024). Machine learning in marketing: Recent progress and future research directions. *Journal of Business Research*, 170, 114254. DOI: 10.1016/j.jbusres.2023.114254

Jia, J., Xu, Y., & Li, W. (2024). A study on the strategic momentum of SMEs' digital transformation: Evidence from China. *Technological Forecasting and Social Change*, 200, 123038. DOI: 10.1016/j.techfore.2023.123038

Jin, K., Zhong, Z. Z., & Zhao, E. Y. (2024). Sustainable digital marketing under big data: An AI random forest model approach. *IEEE Transactions on Engineering Management*, 71, 3566–3579. DOI: 10.1109/TEM.2023.3348991

Kalemkuş, F., & Kalemkuş, E. (2022). The effects of augmented reality on students' learning outcomes: A meta-analysis study. *Journal of Educational Technology & Society*, 25(3), 1–15.

Kao, L. J., Chiu, C. C., Lin, H. T., Hung, Y. W., & Lu, C. C. (2024). Unveiling the dimensions of digital transformation: A comprehensive taxonomy and assessment model for business. *Journal of Business Research*, 176, 114595. DOI: 10.1016/j.jbusres.2024.114595

Kaperonis, S. (2024). How artificial intelligence (AI) is transforming the user experience in digital marketing. In *The use of artificial intelligence in digital marketing: Competitive strategies and tactics* (pp. 117–141). IGI Global.

Katsamakas, E., Pavlov, O. V., & Saklad, R. (2024). Artificial intelligence and the transformation of higher education institutions. *arXiv preprint arXiv:2402.08143.*

Kazu, I., & Yalçin, M. (2022). The role of mobile devices in education: A systematic review of the literature from 2010 to 2020. *Journal of Educational Computing Research*, 60(1), 1–27.

Kedi, W. E., Ejimuda, C., Idemudia, C., & Ijomah, T. I. (2024). Machine learning software for optimizing SME social media marketing campaigns. *Computer Science & IT Research Journal*, 5(7), 1634–1647. DOI: 10.51594/csitrj.v5i7.1349

Kedi, W. E., Ejimuda, C., Idemudia, C., & Ijomah, T. I. (2024). AI software for personalized marketing automation in SMEs: Enhancing customer experience and sales. *World Journal of Advanced Research and Reviews*, 23(1), 1981–1990. DOI: 10.30574/wjarr.2024.23.1.2159

Leso, B. H., Cortimiglia, M. N., Ghezzi, A., & Minatogawa, V. (2024). Exploring digital transformation capability via a blended perspective of dynamic capabilities and digital maturity: A pattern matching approach. *Review of Managerial Science*, 18(4), 1149–1187. DOI: 10.1007/s11846-023-00692-3

Maor, D., McCarthy, A., McConney, A., & Cavanaugh, C. (2023). Leading digital transformation in education: Insights from educators and stakeholders across various contexts worldwide. *Educational Management Administration & Leadership*, 51(2), 215–234.

Nnaji, U. O., Benjamin, L. B., Eyo-Udo, N. L., & Etukudoh, E. A. (2024). A review of strategic decision-making in marketing through big data and analytics. *Magna Scientia Advanced Research and Reviews, 11*(1), 084-091.

Okorie, G. N., Egieya, Z. E., Ikwue, U., Udeh, C. A., & Adaga, E. M. (2024). Leveraging big data for personalized marketing campaigns: A review. *International Journal of Management & Entrepreneurship Research*, 6(1), 216–242. DOI: 10.51594/ijmer.v6i1.778

Oyekunle, D. O. T., & Boohene, D. (2024). Digital Transformation Potential: The Role of Artificial Intelligence in Business. *International Journal of Professional Business Review*.

Potla, R. T., & Pottla, V. K. (2024). AI-Powered Personalization in Salesforce: Enhancing Customer Engagement through Machine Learning Models. *Valley International Journal Digital Library*, 1388-1420.

Qiao, W., Ju, Y., Dong, P., & Tiong, R. L. (2024). How to realize value creation of digital transformation? A system dynamics model. *Expert Systems with Applications*, 244, 122667. DOI: 10.1016/j.eswa.2023.122667

Raghav, Y. Y., Tipu, R. K., Bhakhar, R., Gupta, T., & Sharma, K. (2024). The Future of Digital Marketing: Leveraging Artificial Intelligence for Competitive Strategies and Tactics. In *The Use of Artificial Intelligence in Digital Marketing: Competitive Strategies and Tactics* (pp. 249-274). IGI Global.

Rêgo, B. S., Lourenço, D., Moreira, F., & Pereira, C. S. (2024). Digital transformation, skills and education: A systematic literature review. *Industry and Higher Education*, 38(4), 336–349. DOI: 10.1177/09504222231208969

Richter, N. F., & Tudoran, A. A. (2024). Elevating theoretical insight and predictive accuracy in business research: Combining PLS-SEM and selected machine learning algorithms. *Journal of Business Research*, 173, 114453. DOI: 10.1016/j.jbusres.2023.114453

Su, C.-H., & Yang, Y.-T. C. (2022). Robotics and AI applications in education: Implications for teaching and learning practices in K-12 settings worldwide. *Computers & Education*, 182, 104136.

Trevisan, L. V., Eustachio, J. H. P. P., Dias, B. G., Filho, W. L., & Pedrozo, E. Á. (2024). Digital transformation towards sustainability in higher education: State-of-the-art and future research insights. *Environment, Development and Sustainability*, 26(2), 2789–2810. DOI: 10.1007/s10668-022-02874-7 PMID: 36624732

Tsaples, G., Papathanasiou, J., & Manou, D. (2024). Synergies and Challenges: Exploring Organizational Perspectives on Digital Transformation and Sustainable Development in the Context of Skills and Education. *Buildings*, 14(2), 395. DOI: 10.3390/buildings14020395

Umamaheswari, D. D. (2024). Role of Artificial Intelligence in Marketing Strategies and Performance. *Migration Letters : An International Journal of Migration Studies*, 21(S4), 1589–1599.

Chapter 4
Featuring Wave and Tidal Energy Conversion With Artificial Intelligence and Machine Learning:
Imminent of Clean Energy and Sustainability in Urban Living

Laeeq Razzak Janjua
https://orcid.org/0000-0002-1948-5859
WSB University, Poland

ABSTRACT

Artificial intelligence (AI) and machine learning (ML) are becoming indispensable tools for increasing the efficiency and sustainability of this renewable energy source and the ocean energy industry has made significant strides in recent years. The initial stages of the industry research and development are when AI and ML first started to emerge in the wave and tidal energy space. The development, management, and upkeep of maritime energy systems have all changed as a result of these innovations. An massive, unexplored renewable resource that has the potential to make an important contribution to the world's energy mix is ocean wave energy. In order to maximize the efficacy and efficiency of wave energy conversion, this chapter focuses on the incorporation of artificial intelligence (AI) and machine learning technologies. It looks at these technologies' capability to support clean energy solutions and build a sustainable energy environment particularly in the context of urban living.

DOI: 10.4018/979-8-3693-8684-2.ch004

1. INTRODUCTION

The goal of artificial intelligence (AI), a subfield of computer science, is to build intelligent computers that can carry out activities that normally require human intellect. Applications for AI and machine learning have grown throughout time as a result of several turning points and developments (Wang et al., 2024). Artificial intelligence (AI) is used in ocean energy to maximize energy extraction, keep an eye on and repair equipment, and enhance overall performance. The capacity of computer systems to learn from experience and become better at it without explicit programming is known as machine learning (ML), on the other hand. Machine learning methods are used in the ocean energy domain to gather and evaluate data, anticipate and predict energy trends, and create self-governing systems (Clemente et al., 2023).

Optimizing energy extraction in maritime energy systems requires the integration of AI and ML. These devices determine optimal conditions for energy production by analyzing real-time data. Moreover, system maintenance and monitoring use AI and ML to guarantee the seamless functioning of maritime energy systems. While using AI and ML has many advantages, such as lower costs and better efficiency, there are also potential drawbacks, such as job displacement and technical issues. The data gathering and analysis phases of ocean energy projects heavily rely on AI and ML (Bibri et al., 2024). Through the use of sophisticated data gathering methods and clever sensors, these technologies provide insightful information on the functioning of ocean energy systems. In order to find trends and maximize energy production, the acquired data is then processed using AI and ML algorithms, greatly enhancing the effectiveness and performance of these systems (Abedinia et al., 2023).

Figure 1. The landscapes of Introduction Split Sections (Source- Original)

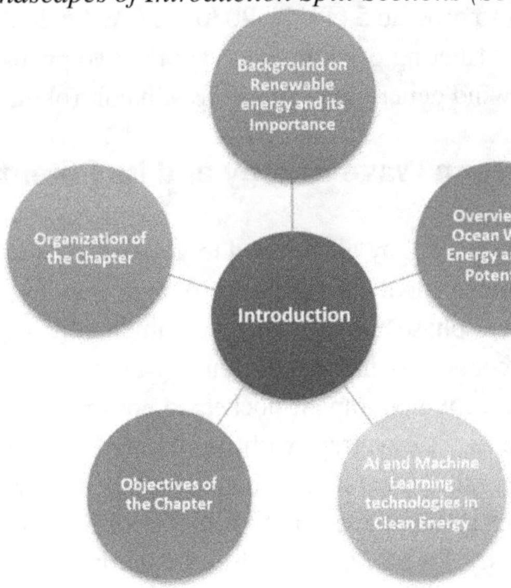

1.1 Background on Renewable energy and its Importance

Autonomous systems for ocean energy have been developed as a result of the fusion of AI and ML. These technologies improve operating efficiency and safety by making judgments in real-time using AI algorithms. Additionally essential to the upkeep and repair of maritime energy infrastructure are robotics and artificial intelligence. Although autonomous systems have advantages like higher output and lower danger for human operators, issues like legal frameworks and possible environmental effects must be resolved. In the ocean energy sector, AI and ML approaches are used for forecasting and predictive modeling. These technologies help optimize energy production by enabling precise forecasts of wave and tidal patterns via the analysis of historical data and integration of real-time data. In this regard, AI and ML techniques like neural networks and genetic algorithms are utilized as examples. Precise forecasting is essential for maximizing energy production and reducing idle time (Alahi et al., 2023). There is potential for offshore renewable energy to contribute significantly to the global supply of sustainable and renewable energy. Ireland has a substantial offshore renewable energy (ORE) potential that may help decarbonize its economy because of its favorable wind and wave conditions and seabed area, which is seven times larger than its mainland. With ORE's capacity to produce more energy than Ireland consumes, the nation will be able to export substantial volumes of low-carbon electricity to customers across Europe.

Ireland's decarbonized electrical grid will rely heavily on offshore wind energy, with the government hoping to provide 5 GW by 2030, 20 GW by 2040, and 37 GW by 2050. Fixed-bottom and floating offshore wind are the two primary technological types used in offshore wind generation (Samaei & Ghahfarrokhi, 2023).

1.2 Overview of Ocean Wave Energy and its Potential

Currents, waves, and tides may all be used to generate power and these ocean technologies show promise for the future even though they are currently in the research and development phase and are not yet commercially accessible. In order to produce electricity, converters harness the energy contained in ocean waves. - Oscillating water columns, which trap air pockets to power a turbine, are one kind of converter. These are as - Converters with oscillating bodies that make use of wave motion; Converters that overtake, taking advantage of variations in height. Tidal-range technologies generate electricity from the difference in tides between high and low tides by use of a barrage, which is a dam or barrier (Kumar, 2023). Technologies that use tidal current or stream technology harness the energy of flowing water and Hybrid apps integrate many techniques. Salinity gradient energy produced at the interface of a river and the ocean, where there are variations in the concentration of salt. The following are some examples of demonstration projects "Pressure Retardant Osmosis" in which freshwater passes through a membrane to raise the pressure in a saltwater tank (Almalki et al., 2023).

1.3 AI and Machine Learning technologies in Clean Energy

Modern artificial intelligence (AI), the Internet of Things (Machine Learning), and renewable energy sources are coming together to provide a paradigm-shifting change in environmental sustainability and energy efficiency. This examines the ways in which various renewable energy sources, AI, microgrids, machine learning, and the Internet of Things are transforming the energy industry, improving financial results, and significantly lowering greenhouse gas emissions. Within machine learning, artificial neural networks are particularly good at processing large and complex datasets. They excel in forecasting the erratic and unexpected patterns of solar and wind energy in the field of renewable energy. To maximize the amount of energy produced by solar and wind power, this skill is essential (Salama & Al-Turjman, 2023). Energy producers may streamline their processes, reduce waste, and guarantee a steady supply of energy by precisely predicting when and how much renewable energy will be available. Artificial intelligence algorithms make optimal use of these insights to distribute electricity produced by renewable sources. This permits storage or conservation during times of low demand while also guaranteeing that energy is distributed effectively to where it is most required. With optimizing the use of renewable energy sources and supporting the stability and dependability of power supply systems, this kind of optimization increases the overall efficiency of microgrid operations.

1.4 Objectives of the Chapter

This chapter has the following objectives to:

- examine the efficiency, scalability, and integration potential of current ocean wave energy conversion technologies using AI and Machine Learning in smart city settings.
- scrutinize how deploying ocean wave energy technologies improved by AI and Machine Learning would affect the environment, taking into account advantages for sustainability, the effect on marine life, and the reduction of carbon emissions.
- inspect how ocean wave energy systems might benefit from the integration of AI and Machine Learning to maximize energy output, improve operational effectiveness, and guarantee dependability under a range of environmental circumstances.
- study the long-term financial sustainability, capital investment, and possible cost-effectiveness of using AI and Machine Learning technologies in ocean wave energy projects.
- encourage innovation, reduce risks, and advance sustainable development in smart cities, make recommendations for policy frameworks and regulatory measures that will facilitate the adoption and deployment of AI and Machine Learning technologies in ocean wave energy.

Figure 2. The Objectives of the Chapter (Source- Original)

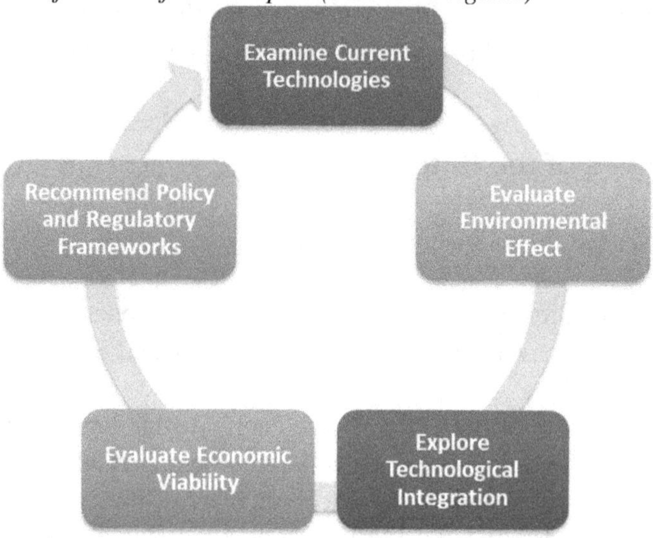

1.5 Organization of the Chapter

This chapter comprehensively explores the diverse arena of the Abundance Ocean Wave Energy to Electricity with Artificial Intelligence and Machine Learning Solutions: Future of Clean Energy and Sustainability in Smart Cities. Section 2 discusses the Ocean Wave Energy: Potential and Benefits. Section 3 specifies the Wave Energy Conversion Technologies. Section 4 lays down the Role of AI and Machine Learning Solutions for in Maximizing Ocean Wave Energy Solutions. Section 5 conveys the Impact on Smart Cities. Section 6 highlights the Policy and Regulatory Considerations. Section 7 Conclude the Chapter with Future Scope.

Figure 3. The Flow of this Chapter (Source- Original)

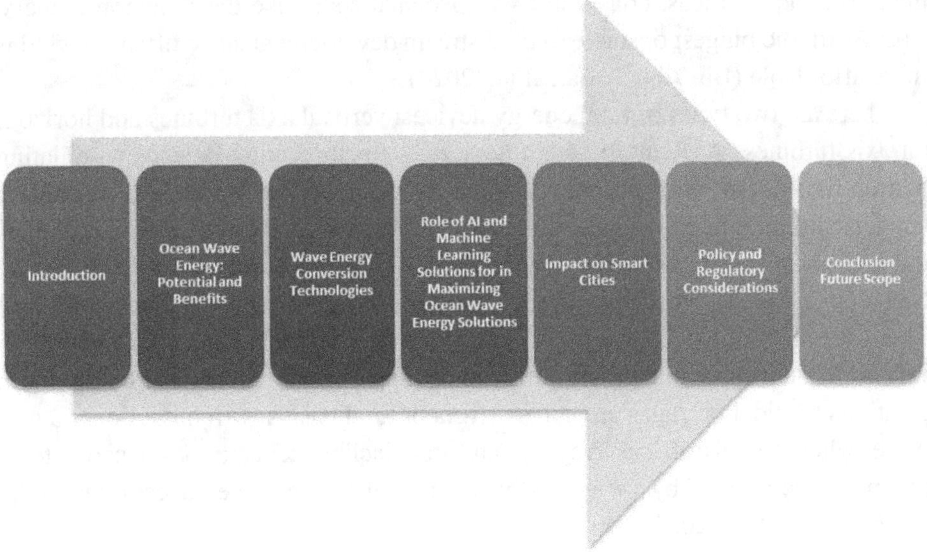

2. OCEAN WAVE ENERGY: POTENTIAL AND BENEFITS

Utilizing the kinetic energy of the water's inherent tide-driven movement, tidal energy produces electricity. Because of the sun, moon, and earth's relative motions, tides may be forecast years in advance. The biggest and lowest tidal heights are separated by the tidal range, which is largest during spring tides and smallest during neap tides. The two primary approaches for capturing tidal energy are tidal barrage and tidal stream. To harness the kinetic energy of flowing water, tidal stream energy converters (TECs) are installed in certain locations. TECs use water movement as a source of energy instead of air, working similarly to wind turbines in this regard. The majority of TECs are found in places like tidal lagoons, estuaries, or channels that have significant flow rates because of their appropriate bathymetry. Some recent designs, nevertheless, are tailored for slower flow rates (Mohammed et al., 2023). The richest tidal stream resources are usually found in places with large tidal ranges, where local seafloor and shoreline features, such as channels between islands, narrow straits, and inlets, enhance current activity. TECs may be fixed or floating, and they are positioned straight into the tidal stream. Generally speaking, tidal energy devices are found in tidal streams, which include waterways that connect islands, headlands, and small straits. Although some of these devices have surface components, most of them are beneath. Within a tidal stream, one or more devices may be positioned. The finest places to find Ireland's tidal stream resource are on the north and northeast coastlines, with isolated spots like the Shannon Estuary. Cost is still the biggest obstacle to tidal stream development since offshore wind is more affordable (Bin Abu Sofian et al., 2024).

There are two types of tidal energy devices: vertical axis turbines and horizontal axis turbines. As Venturi effect devices; - Reciprocating devices (oscillating hydrofoils); Archimedes fasteners - Tidal inflatables Tidal Barrage as building a barrier with integrated turbines that catch water flow as the tide rises and falls is the process of creating a tidal barrage (Fan et al., 2023). With a far lower pressure head than hydroelectric dams, water flowing across the barrage powers the turbines. The production of tidal barrages is not expected to be a major source of energy in Ireland in the future since there aren't many appropriate locations. The optimum locations for tidal barrages are across rivers or tidal basins with large tidal ranges, places where vessel traffic is irregular, and areas lacking delicate coastal ecosystems that might be harmed by variations in water level and restricted access due to the barrage (Jian et al., 2023).

Figure 4. The Points on Wave and Tidal Energy (Source- Original)

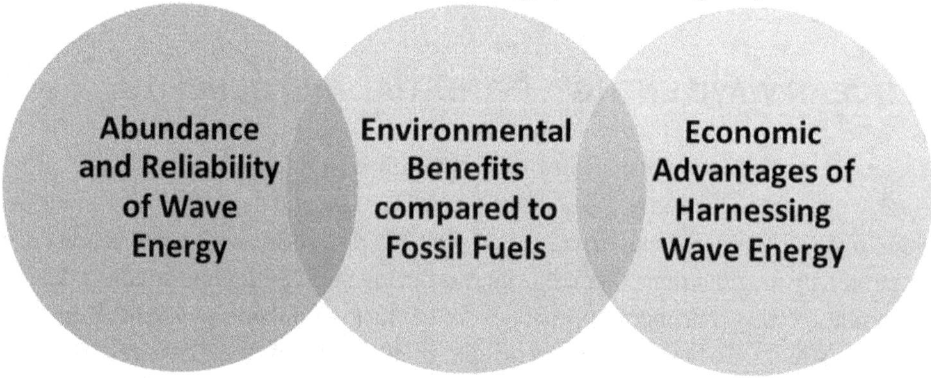

2.1 Abundance and Reliability of Wave Energy

In a world where sustainability is becoming more and more important, wave power has emerged as a viable form of renewable energy. Wave power technology, which captures the kinetic energy of ocean waves, provides a plentiful and clean source of electricity for usage worldwide. This in-depth manual paves the path for a future driven by the ocean's unceasing motion by examining the physics, benefits, difficulties, and application of wave energy (Mirshafiee et al., 2023). Wave power produces energy by harnessing the oscillatory motion of ocean waves. Many methods, including buoys with mechanical or hydraulic systems or floating turbine platforms, are used to harvest this energy. These devices harness the energy of moving waves to create mechanical power, which is then utilized to link generators to produce electricity (Yousef et al., 2023).

2.1.1 Principal Benefits of Wave Power

Wave power has a number of special benefits as-

Plenty and Predictable Resource: Because ocean waves are always moving, wave energy is steady and predictable, in contrast to solar and wind power.

Environmentally Friendly: Wave power production is a sustainable and environmentally friendly energy source that doesn't add to air pollution or greenhouse gas emissions since it doesn't generate any harmful emissions or pollutants.

High Energy Density: Compared to other renewable energy sources, wave power systems may produce a considerable amount of electricity per unit area since waves have a high energy density.

Versatility: Adapting to changing circumstances and optimizing energy collection, wave power systems may be installed in a variety of maritime situations, including offshore and close to coasts (Zhou & Liu, 2024).

2.1.2 Wave Power's Challenges

Wave power technique is not without its obstacles, though:

High Initial Costs: Developing and installing specialized energy collectors is an expensive initial investment in wave power infrastructure, often surpassing those of other renewable energy sources.

Harsh Ocean Environment: Wave energy devices placed offshore have durability and maintenance issues due to the forceful and unpredictable nature of ocean waves. Regular upkeep is thus necessary to guarantee operating efficiency.

Environmental Impact: Although wave power doesn't harm the environment directly, installing and maintaining wave energy systems could have an impact on nearby marine ecosystems. The use of mitigation techniques is essential to reducing any negative environmental effects (Nizetic et al., 2023).

2.2 Environmental Benefits compared to Fossil Fuels

Ocean wave energy is an ecologically favorable substitute for conventional fossil fuels since it produces a notably less amount of carbon emissions. It is dependent on elements that influence the motion of ocean waves, such as winds, tides, moon cycles, and weather. When waves break the surface of the water, they release kinetic energy that may be used to drive turbines to create electricity. Attenuators, absorbers, and other devices are among the technologies used to effectively collect wave energy. In particular, wave power is the electrical energy that comes from the surface water movement in the ocean (Hui et al., 2023). Turbine platforms and floating buoys are used to harvest this energy, which rises and falls in tandem with ocean waves. It takes use of variations in wind pressure inside of chambers or by directly capturing wave pressure at the ocean's bottom. Although ocean waves have enormous power potential, successfully using this energy will need overcoming technological and technical obstacles (Alshmeel et al., 2024).

Applications for wave power include desalination, water pumping, and electricity generating. It is used in many nations throughout the globe and is regarded as a renewable energy source. Wave farms have been developed in nations including Portugal, the UK, Australia, and the US, and they produce a significant amount of wave power. For example, wave farms in the UK have the capacity to power over 7,500 houses and are predicted to reduce $CO2$ emissions by over 300,000 tons over the course of the next 25 years.

2.3 Economic Advantages of Harnessing Wave Energy

Wave energy technology is still in its infancy, yet it can already transfer wave energy from the ocean into electrical power. Kilowatts per unit meter of wave crest length are the unit of measurement for the worldwide wave energy resource. In addition to being fascinating for surfers and beachgoers, the force of ocean waves is also revered because of its capacity for devastation, as boaters, fishers, and seafarers have long understood. But these waves may also be converted into useful energy, providing fossil fuels with a sustainable substitute. We may lessen pollution and our dependency on finite fossil fuel supplies by using the plentiful and clean energy of ocean waves.

3. WAVE ENERGY CONVERSION TECHNOLOGIES

Because AI-driven predictive maintenance may identify equipment faults before they happen, it is essential to the renewable energy industry. This proactive strategy lowers maintenance costs, saves downtime, and increases the lifetime of equipment like solar panels and wind turbines. Machine Learning technology's capacity to gather and send real-time data has made it essential to contemporary renewable energy systems. Wind turbines and solar panels, for example, are examples of renewable energy facilities with integrated sensors that collect vital data on wind speed, sun irradiance, temperature, and equipment performance. This information is sent to centralized monitoring systems, which provide a thorough picture of the infrastructure supporting renewable energy sources (Vijayalakshmi et al., 2023). It makes it possible to analyze things in-depth and make wise decisions that will maximize productivity and efficiency. Real-time data makes it easier to spot irregularities or inefficiencies in the system right away. For example, it is possible to identify and fix mechanical problems that are causing a wind turbine's decreased efficiency quickly (Genish & Boopathiraja, 2023). Reducing downtime and increasing equipment lifetime, Machine Learning-enabled predictive maintenance examines data patterns to anticipate equipment breakdowns or maintenance requirements. Electric vehicle (EV) charging stations are being revolutionized by Internet of Things (Machine Learning) technology, which automatically manages the charging process depending on variables including current demand and the availability of renewable energy. EV charging stations with Machine Learning capabilities may dynamically change to favor charging at times when renewable energy output is highest, as during prime solar or wind hours. Through this integration, the use of non-renewable energy sources is reduced, and EV charging procedures are in line with sustainability objectives (Vijayalakshmi et al., 2023).

An important step toward intelligent, effective, and sustainable energy management has been made with the integration of Machine Learning in renewable energy systems. Machine Learning technology will play a bigger part in assisting the development of EV infrastructure and maximizing renewable energy sources as it continues to advance. This technical advancement increases environmental conservation objectives, improves operating efficiency, and supports energy sustainability (Shukla et al., 2024). For communities and stakeholders who may potentially use marine energy as a power source there is several advantages to be gained. Marine energy may provide vital electricity to even the most isolated villages by using wave, tidal, current, and temperature variations in the sea (Al Hajri et al., 2024). This energy source has a number of benefits that demonstrate how crucial it is to meeting US clean energy targets. The following are some of the main advantages of marine energy that may aid in the US's transition to net-zero emissions by 2050 and 100% clean power by 2035. These are multifarious benefits of Ocean Energy as-

Renewable and Clean Energy: Marine energy, which depends only on the flow of water naturally, is a clean and renewable energy source.

Plenty of Resources: The United States has a vast amount of river and ocean shoreline, which presents a significant opportunity for the use of marine energy.

High Predictability: Marine energy is quite predictable since tides, waves, and currents are cyclical phenomena.

Complementary to Other Sources: Because marine energy is predictable, it may be used in conjunction with solar and wind energy, which often produce less electricity during high tides and waves.

Resilient and Reliable: By placing marine energy technologies near to the source of electricity, they minimize the need for long transmission lines, which improves grid resilience and reliability.

Power for Remote Communities: With using the nearby water to power microgrids for residences and businesses, marine energy may provide electricity to isolated, island, and coastal communities.

Innovative Opportunities: Technologies related to marine energy may support additional developments in the maritime domain, such desalination for potable water and energy-based sea and ocean research.

Job Creation: Because local workers are required for operation and maintenance, the implementation of marine energy technology might lead to the creation of employment in isolated island or coastal communities.

Diverse Professional Opportunities; There are many different career options in environmental science, ecology, engineering, outreach, policy, research, and environmental science available in the marine energy industry.

Blue Economy Experience: Gaining practical knowledge of the multitrillion dollar blue economy which is expanding twice as quickly as the rest of the world economy can be achieved by working in the marine energy sector (Issa Zadeh et al., 2023).

Figure 5. The Wave Energy Conversion Technologies (Source- Original)

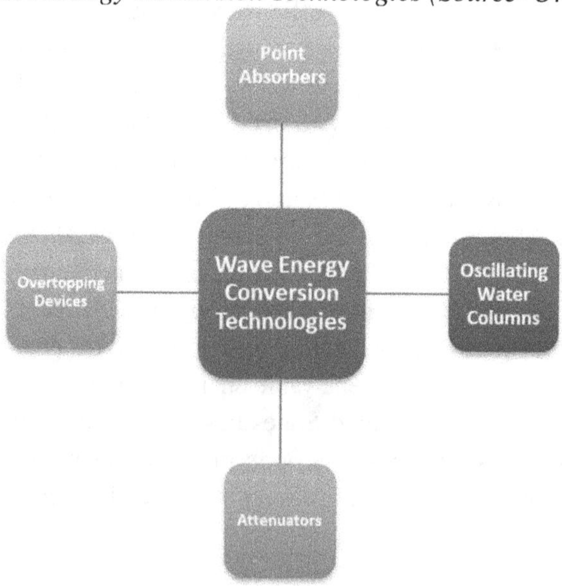

4. ROLE OF AI AND MACHINE LEARNING SOLUTIONS FOR IN MAXIMIZING OCEAN WAVE ENERGY SOLUTIONS

Osmotic energy uses the concentration of salt in saltwater to produce power. This technology's key component is a double-sided, semi-permeable membrane that traps mineral salts while allowing water to flow through. This membrane is situated between saltwater and freshwater on opposite sides. Osmosis is the mechanism by which freshwater migrates toward the saltwater compartment as a result of the salt molecules drawing it in. Power is generated by a turbine that is driven by the flow of water. High-performance membranes that demonstrate robust durability and the capacity to draw in enough water to enable optimum turbine operation and power production efficiency are critical components of the efficacy of osmotic energy technology. These membranes also need to be reasonably priced. The plant has to be located close to fresh and saltwater sources; it cannot be built randomly along the coast. As a result, river mouths are where these plants are usually found. Osmotic power plants function by use of a system of pipelines that move saltwater and freshwater into distinct chambers separated by membranes (Li & Zhou, 2023).

Pierre-Simon Girard, a French inventor, received a patent for the first wave energy converter in 1799. His idea was simple: a raft that floated at sea and was attached to the land by wires and pulleys (Orumwense & Abo-Al-Ez, 2023). The raft was dragged by the waves, which created kinetic energy that could be used for a variety of purposes. But Girard's creation never made it beyond the drawing board, and wave power remained mostly unrealized for more than 200 years. Modern technological developments are producing wave energy converters that are more durable and effective, which might open the door for widespread commercialization (Raj, 2023). The basic idea behind wave energy converters, despite their very different designs, is to transform the sea's inherent motion into mechanical energy, which is then often used to power turbines that produce electricity. The most popular designs, which are similar to the invention of Girard, use the rise and fall of buoys to move hydraulic arms or cables. Other systems use big oscillating wings that move with the waves to function continuously below the surface (Camacho et al., 2024).

The energy production of "wave farms" is predicted to be three times higher per square kilometer than that of floating offshore wind farms. In comparison to solar and wind energy, waves are also far more predictable and readily foreseen; wave energy is accessible for use around 90% of the time, while solar and wind energy only do so 20–30% of the time. Significant advancements have been made in commercial-scale pilot programs in recent years (Kannammal & Chandia, 2023). For example, in February 2024, the Swedish company Using mechanical rods within the buoys, the system, which consists of buoy arrays, transforms wave motion into rotational energy. With "digital twin" technology, the C4 seeks to counteract storm damage and optimize energy production in typical circumstances by synchronizing with incoming waves (Zearah et al., 2023). This is another creative technique. Onboard controllers in the x1 converter allow for remote control and self-orientation for optimal efficiency. The x1 was able to minimize visual effect on the surrounding environment, resist strong storms, and achieve high system uptime of 99% while operating autonomously 80% of the time throughout its trial program (Awogbemi et al., 2024). Designers want to create lighter, more efficient autonomous gadgets that can predict and avoid storms as artificial intelligence and processing capacity grow. The progression of wave energy converters is anticipated to augment their dependability and efficacy, hence expediting their assimilation into worldwide renewable energy portfolios.

4.1 Machine Learning Algorithms: Enhancing Efficiency of Wave Energy Converters

The current state of the globe has made it more urgent than ever to combat global warming, which is driving countries to look for ecologically acceptable alternatives to conventional fossil fuels (Verma et al., 2024). These days, there's a big push to investigate alternate renewable energy sources, including solar, wind, and ocean wave energy, which have gotten a lot of traction. Although wind and solar energy have reached a mature commercial stage, ocean wave energy is still a viable field for further research (Khalid, 2024). There are a number of obstacles to overcome in order to harvest energy from waves, such as low efficiency and expensive deployment costs for wave energy devices. Thus, there are presently no completely implemented commercial wave energy structures. However, a lot of experts think that ocean wave energy is a very promising field for environmentally friendly, sustainable energy sources (Issa Zadeh et al., 2023). Wave Swing, Wave Star, Oyster and Wave Dragon, using data analysis and numerical simulations for effeciency. Based on variables including capacity factor, rated capacity factor, and operating duration, their analysis determined the best WEC designs.

With converting the kinetic and potential energy of waves into rotational mechanical energy, this asymmetric WEC improves energy conversion efficiency using a pitch-type mechanism where the lee side generates a circular pattern and the fore side displaces fluid particles. Numerical research and experiments have improved this technology over time in a variety of sea conditions (Obaideen et al., 2024). Nevertheless, difficulties still exist, especially in complicated wave circumstances that may drastically lower efficiency, such as very steep waves, 3D nonlinear effects, and close proximity to breaking waves. In order to tackle these obstacles, scholars have investigated inventive approaches, such as integrating negative stiffness mechanisms, which exhibit potential in augmenting WEC efficacy in both linear and nonlinear wave situations (El-Hussieny, 2023). However, the current models of asymmetric wave energy converters have not undergone a comprehensive geometric design optimization, which results in differences in performance across various wave environments. In order to close this performance gap, machine learning (ML) has been the focus of current research efforts. The goal of using machine learning (ML) techniques is to increase predictability and flexibility, which will raise the efficiency and consistency of energy production from ocean waves (Salman & Hasar, 2023).

5. SMART CITIES AND OCEAN ENERGY INTO ELECTRICITY: FUTURE OF SUSTAINABILITY

Energy harvesting is the process of turning energy that would otherwise be squandered into energy that can power autonomous devices (EH). By physically or chemically obtaining ambient energy from natural or artificial environmental sources, it solves the present energy problems faced by Internet of Things (Machine Learning) networks. The term "smart city" refers to the degree of urbanization and technical sophistication of a certain city (Van Hoang, 2024). Its definition varies across nations and localities, often including certain application sectors that cater to the everyday requirements of people and attract the attention of smart city developers (Huda et al., 2024). Some see it as crucial for resiliency in the face of natural disasters like fires, floods, or earthquakes; others see it in relation to security and terrorism. With linking people and things online, smart cities hope to improve people's quality of life overall. City planners may obtain real-time data and automate complicated municipal challenges thanks to this connectedness (Ali, 2023).

Web technologies are essential to smart cities because they allow machine-to-machine communication, human-machine interaction, and control over physical assets that are linked to the internet. Smart cities rely on the Internet of Things (Machine Learning), which is made up of networks where devices are linked to each other and use sensors to send data wirelessly or via the cloud. This link improves everyday living circumstances and human sustainability while facilitating real-time decision-making for urban planning (Nezhad et al., 2024). A progressive approach to urban administration, smart cities aspire to improve sustainability and the quality of life for their citizens. Via producing real value, digitization must be included into smart city initiatives in order to achieve ecological and economic sustainability. To support the financial outlays, the advantages of these initiatives must be made abundantly evident (Moemen et al., 2023). The creating of infrastructure with a low environmental effect while using contemporary, energy-efficient techniques is a major problem for smart cities. Innovations like smart buildings and effective transit systems are needed to address environmental problems and climate change as well.

The implementation of automation systems that possess the ability to independently transform electrical energy into functional goods is crucial for attaining a balanced energy exchange in smart cities. Smart cities are putting in place integrated systems that combine disparate energy, heat, gas, and water networks with telecommunications infrastructure in order to maximize energy production, consumption, and generating capacity while affecting larger energy markets. The transition to electricity as the main energy source for civilization is symbolized by electrification, which is seen to be necessary for the energy sector's long-term survival (Nasserddine et al., 2023).

6. POLICY AND REGULATORY CONSIDERATIONS: REGULATORY FRAMEWORKS FOR WAVE ENERGY

A rising emphasis on marine resource exploitation often referred to as a "industrial revolution" of the ocean has coincided with the pressing need to cut carbon emissions in the energy sector, which has inspired major interest in renewable energy technology. The growing ocean energy sector, which seeks to use the force of waves and tides to create electricity, is located where these trends converge. The European Union's (EU) Blue Growth policy has highlighted ocean energy as one of the major industries with promise for sustainable "Blue Growth". Significant attention and investment have been generated worldwide by this (Lopez et al., 2023). Because of its extensive maritime knowledge, financial and political supports, and continuing legal and regulatory changes, the United Kingdom, and especially Scotland, have showed a significant interest in ocean energy. On the other hand, a more crowded and contentious maritime environment is being entered by ocean energy technology (Logeswaran et al., 2023). As a result, in order to protect the increasingly industrialized and fragile ocean environment, policymakers, developers, and academics are investigating how changing marine governance frameworks may successfully support the sustainable growth of these cutting-edge enterprises.

6.1 Promoting Innovation and Investment

For governments to encourage investment in sustainable energy, policies and regulations are essential. Investors evaluate a nation's ability to plan for the growth of its main electrical infrastructure and to accommodate independent and mini-grid solutions. Investment choices are also influenced by the availability of financial incentives for clean energy projects, especially those pertaining to power access and clean cooking in rural and isolated locations (Li et al., 2024). Environmental standards, consumer energy subsidies, and other regulatory frameworks are important additional variables. The World Bank's Regulatory Indicators for Sustainable Energy

(RISE) show that the adoption of complete policy frameworks for sustainable energy is now only about halfway done globally. Countries must put in place strong laws and regulations that attract investment if they are to reach Sustainable Development Goal 7 and attain universal energy access by 2030. But doing so is contingent upon the ability of supporting agencies and governments, many of which need technical help (Xie et al., 2023). There are among the initiatives are:

- Creating an online information center where policymakers may get a variety of policy instruments, templates, and tools for decision-making.
- Providing assistance in tailoring laws and rules to enhance national capacity building, strategy creation, program execution, and adoption of best practices.
- Promoting the development of regional value chains for solar energy in order to increase off-grid energy solutions' affordability and uptake, optimizing financial gains.
- Making use of and supporting the Mini-Grids Partnership (MGP), an alliance including more than 300 stakeholders, in order to facilitate cooperation, information sharing, and group problem-solving in order to accelerate the deployment and progress of clean energy mini-grids (Khan et al., 2024).

7. CONCLUSION AND FUTURE SCOPE

There is a lot of potential for efficiency and sustainability gains when AI and ML are integrated into maritime energy systems. The ocean energy industry may enhance its efficiency and have a positive impact on the environment by using these technologies for energy extraction, data analysis, prediction, and the creation of autonomous systems. To completely achieve the promise of AI and ML in ocean energy, further study and development are necessary. It is anticipated that AI and ML will have a significant influence on the sustainability and efficiency of maritime energy. With the use of these technologies, energy output may be boosted and environmental impact can be decreased via real-time optimization, predictive maintenance, and better decision-making. AI and ML in ocean energy have both benefits and difficulties in store for the future. Overcoming technological constraints and reducing possible dangers are challenges. Possibilities include expanding R&D, encouraging innovation, and improving stakeholder engagement. Wave power technology has the potential to become a major part of the world's energy mix as it develops. It is anticipated that more research, development, and funding will increase deployment, lower prices, and improve efficiency. Government regulations and incentives are essential in hastening the assimilation of wave power into the global energy framework and guaranteeing a clean, sustainable energy future fueled by the bounty of the ocean.

REFERENCES

Abedinia, O., Ghasemi-Marzbali, A., Gouran-Orimi, S., & Bagheri, M. (2023). Presence of renewable resources in a smart city for supplying clean and sustainable energy. In *Decision making using AI in energy and sustainability: methods and models for policy and practice* (pp. 233–251). Springer International Publishing. DOI: 10.1007/978-3-031-38387-8_14

Al Hajri, N. H., Al Harthi, R. N., Pasam, G. K., & Natarajan, R. (2024). Machine Learning and Machine Learning based Green Energy Generation using Hybrid Renewable Energy Sources of Solar, Wind and Hydrogen Fuel Cells. In *E3S Web of Conferences* (Vol. 472, p. 01008). EDP Sciences. DOI: 10.1051/e3sconf/202447201008

Alahi, M. E. E., Sukkuea, A., Tina, F. W., Nag, A., Kurdthongmee, W., Suwannarat, K., & Mukhopadhyay, S. C. (2023). Integration of Machine Learning-enabled technologies and artificial intelligence (AI) for smart city scenario: Recent advancements and future trends. *Sensors (Basel)*, 23(11), 5206. DOI: 10.3390/s23115206 PMID: 37299934

Ali, M. (2023). Synergizing Solar Cell Innovation, Radio Wave Technology, and AI for Sustainable Business Growth: A Comprehensive Review. *International Journal of Social. Humanities and Life Sciences*, 1(1), 19–28.

Almalki, F. A., Alsamhi, S. H., Sahal, R., Hassan, J., Hawbani, A., Rajput, N. S., Saif, A., Morgan, J., & Breslin, J. (2023). Green Machine Learning for eco-friendly and sustainable smart cities: Future directions and opportunities. *Mobile Networks and Applications*, 28(1), 178–202. DOI: 10.1007/s11036-021-01790-w

Alshmeel, G. H. A., Al-Doori, A. S. B., Ahmed, S. R., Abrahim, Z. A., Ghaffoori, A. J., & Hussain, A. S. T. (2024, May). Self-Sustaining Buoy System: Harnessing Water Wave Energy for Smart, Wireless Sensing and Data Transmission. In *Proceedings of the Cognitive Models and Artificial Intelligence Conference* (pp. 349-356). DOI: 10.1145/3660853.3660940

Asaduz-Zaman, M., Ongsakul, W., & Hossain, M. J. (2023). Microgrid Energy Management for Smart City Planning on Saint Martin's Island in Bangladesh. *Energies*, 16(10), 4088. DOI: 10.3390/en16104088

Awogbemi, O., Von Kallon, D. V., & Kumar, K. S. (2024). Contributions of Artificial Intelligence and Digitization in Achieving Clean and Affordable Energy. *Intelligent Systems with Applications*, 22, 200389. DOI: 10.1016/j.iswa.2024.200389

Bibri, S. E., Krogstie, J., Kaboli, A., & Alahi, A. (2024). Smarter eco-cities and their leading-edge artificial intelligence of things solutions for environmental sustainability: A comprehensive systematic review. *Environmental Science and Ecotechnology*, 19, 100330. DOI: 10.1016/j.ese.2023.100330 PMID: 38021367

Bin Abu Sofian, A. D. A., Lim, H. R., Siti Halimatul Munawaroh, H., Ma, Z., Chew, K. W., & Show, P. L. (2024). Machine learning and the renewable energy revolution: Exploring solar and wind energy solutions for a sustainable future including innovations in energy storage. *Sustainable Development (Bradford)*, 32(4), 3953–3978. DOI: 10.1002/sd.2885

Camacho, J. D. J., Aguirre, B., Ponce, P., Anthony, B., & Molina, A. (2024). Leveraging Artificial Intelligence to Bolster the Energy Sector in Smart Cities: A Literature Review. *Energies*, 17(2), 353. DOI: 10.3390/en17020353

Clemente, D., Cabral, T., Rosa-Santos, P., & Taveira-Pinto, F. (2023). Blue seaports: The smart, sustainable and electrified ports of the future. *Smart Cities*, 6(3), 1560–1588. DOI: 10.3390/smartcities6030074

El-Hussieny, E. (2023). Emerging the concept of Sustainable Architecture in forming a Smart City. *Resourceedings*, 3(1), 22–32. DOI: 10.21625/resourceedings.v3i1.948

Fan, Z., Yan, Z., & Wen, S. (2023). Deep learning and artificial intelligence in sustainability: A review of SDGs, renewable energy, and environmental health. *Sustainability (Basel)*, 15(18), 13493. DOI: 10.3390/su151813493

Genish, T., & Boopathiraja, S. (2023). Biomass Renewable Energy: Introduction and Application of AI and Machine Learning. In *AI-Powered Machine Learning in the Energy Industry: Digital Technology and Sustainable Energy Systems* (pp. 165–187). Springer International Publishing.

Huda, N. U., Ahmed, I., Adnan, M., Ali, M., & Naeem, F. (2024). Experts and intelligent systems for smart homes' Transformation to Sustainable Smart Cities: A comprehensive review. *Expert Systems with Applications*, 238, 122380. DOI: 10.1016/j.eswa.2023.122380

Hui, C. X., Dan, G., Alamri, S., & Toghraie, D. (2023). Greening smart cities: An investigation of the integration of urban natural resources and smart city technologies for promoting environmental sustainability. *Sustainable Cities and Society*, 99, 104985. DOI: 10.1016/j.scs.2023.104985

Issa Zadeh, S. B., Esteban Perez, M. D., López-Gutiérrez, J. S., & Fernández-Sánchez, G. (2023). Optimizing Smart Energy Infrastructure in Smart Ports: A Systematic Scoping Review of Carbon Footprint Reduction. *Journal of Marine Science and Engineering*, 11(10), 1921. DOI: 10.3390/jmse11101921

Issa Zadeh, S. B., & Perez, E. (2023). MD; López-Gutiérrez, J.-S.; Fernández-Sánchez, G. Optimizing Smart Energy Infrastructure in Smart Ports: A Systematic Scoping Review of Carbon Footprint Reduction. *Journal of Marine Science and Engineering*, 11(10), 1921. DOI: 10.3390/jmse11101921

Jain, A., Sharma, A., Jately, V., & Azzopardi, B. (Eds.). (2023). *Sustainable Energy Solutions with Artificial Intelligence, Blockchain Technology, and Internet of Things*. CRC Press. DOI: 10.1201/9781003356639

. Kannammal, A., & Chandia, S. (2023). Applications of AI and Machine Learning for Smart Cities. *Research Trends in Artificial Intelligence: Internet of Things*, 186.

Khalid, M. (2024). Energy 4.0: AI-enabled digital transformation for sustainable power networks. *Computers & Industrial Engineering*, 193, 110253. DOI: 10.1016/j.cie.2024.110253

Khan, M. W., Saad, S., Ammad, S., Rasheed, K., & Jamal, Q. (2024). Smart Infrastructure and AI. In *AI in Material Science* (pp. 193–215). CRC Press. DOI: 10.1201/9781003438489-9

Kumar, D. (2023). Concept of Sustainable Energy System for Smart Cities. In *Renewable Energy Scenarios in Future Indian Smart Cities: A Geospatial Technology Perspective* (pp. 1–20). Springer Nature Singapore. DOI: 10.1007/978-981-19-8456-3_1

Li, J., & Zhou, T. (2023). Multiagent deep meta reinforcement learning for sea computing-based energy management of interconnected grids considering renewable energy sources in sustainable cities. *Sustainable Cities and Society*, 99, 104917. DOI: 10.1016/j.scs.2023.104917

Li, X., Zhao, H., Feng, Y., Li, J., Zhao, Y., & Wang, X. (2024). Research on key technologies of high energy efficiency and low power consumption of new data acquisition equipment of power Internet of Things based on artificial intelligence. *International Journal of Thermofluids*, 21, 100575. DOI: 10.1016/j.ijft.2024.100575

. Logeswaran, K., Suresh, P., Ponselvakumar, A. P., Savitha, S., Sentamilselvan, K., & Adhithyaa, N. (2023). Exploration of Assorted Modernizations in Forecasting Renewable Energy Using Low Power Wireless Technologies for Machine LearningSG. *Smart Grids and Internet of Things: An Energy Perspective*, 157-191.

López, O. L., Rosabal, O. M., Ruiz-Guirola, D. E., Raghuwanshi, P., Mikhaylov, K., Lovén, L., & Iyer, S. (2023). Energy-sustainable Machine Learning connectivity: Vision, technological enablers, challenges, and future directions. *IEEE Open Journal of the Communications Society*, 4, 2609–2666. DOI: 10.1109/OJCOMS.2023.3323832

Mirshafiee, F., Shahbazi, E., Safi, M., & Rituraj, R. (2023). Predicting power and hydrogen generation of a renewable energy converter utilizing data-driven methods: A sustainable smart grid case study. *Energies*, 16(1), 502. DOI: 10.3390/en16010502

Moemen, Y. S., Alshater, H., & El-Sayed, I. E. T. (2023). Clean Energy Management Based on Internet of Things and Sensor Networks for Climate Change Problems. In *The Power of Data: Driving Climate Change with Data Science and Artificial Intelligence Innovations* (pp. 117–136). Springer Nature Switzerland.

Mohammed, M. N., Aljibori, H. S. S., Al-Tamimi, A., Alani, Z. N., Hameed, M. S., Ibrahim, F. M., . . . Alfiras, M. (2023, October). Toward Sustainable Smart Cities in Bahrain: A Groundbreaking Approach to Marine Renewable Energy Harnessing Sea Tides and Waves for a Greener Energy Future. In *2023 IEEE 8th International Conference on Engineering Technologies and Applied Sciences (ICETAS)* (pp. 1-5). IEEE. DOI: 10.1109/ICETAS59148.2023.10346346

Nasserddine, G., Nassereddine, M., & El Arid, A. A. (2023). Internet of things integration in renewable energy Systems. In *Handbook of Research on Applications of AI, Digital Twin, and Internet of Things for Sustainable Development* (pp. 159-185). IGI Global. DOI: 10.4018/978-1-6684-6821-0.ch010

Nezhad, M. M., Neshat, M., Sylaios, G., & Garcia, D. A. (2024). Marine energy digitalization digital twin's approaches. *Renewable & Sustainable Energy Reviews*, 191, 114065. DOI: 10.1016/j.rser.2023.114065

Nizetic, S., Arıcı, M., & Hoang, A. T. (2023). Smart and Sustainable Technologies in energy transition. *Journal of Cleaner Production*, 389, 135944. DOI: 10.1016/j.jclepro.2023.135944

Obaideen, K., Albasha, L., Iqbal, U., & Mir, H. (2024). Wireless power transfer: Applications, challenges, barriers, and the role of AI in achieving sustainable development goals-A bibliometric analysis. *Energy Strategy Reviews*, 53, 101376. DOI: 10.1016/j.esr.2024.101376

Orumwense, E. F., & Abo-Al-Ez, K. (2023). Internet of Things for smart energy systems: A review on its applications, challenges and future trends. *AIMS Electronics & Electrical Engineering, 7*(1).

Raj, E. F. I. (2023). A Detailed Review on Wind and Solar Hybrid Green Energy Technologies for Sustainable Smart Cities. *Polytechnica*, 6(1), 1. DOI: 10.1007/s41050-023-00040-0

Salama, R., & Al-Turjman, F. (2023). Sustainable energy production in smart cities. *Sustainability (Basel)*, 15(22), 16052. DOI: 10.3390/su152216052

Salman, M. Y., & Hasar, H. (2023). Review on environmental aspects in smart city concept: Water, waste, air pollution and transportation smart applications using Machine Learning techniques. *Sustainable Cities and Society*, 94, 104567. DOI: 10.1016/j.scs.2023.104567

Samaei, S. R., & Ghahfarrokhi, M. A. (2023). Optimizing Coastal Hydro Turbines Integrating Artificial Intelligence for Sustainable Energy Conversion. In *2th International Conference on Creative achievements of architecture, urban planning, civil engineering and environment in the sustainable development of the Middle East.*

Shukla, S. S., Jain, M., Jain, A., Bajpai, S., Mishra, A. K., Gupta, A., & Pramanik, S. (2024). Industry 5.0: Blue Wind Energy for Sustainable Urbanization and Intelligent Control of Energy. In *Convergence of Human Resources Technologies and Industry 5.0* (pp. 170-187). IGI Global.

Singh, B., & Kaunert, C. (2024). Future of Digital Marketing: Hyper-Personalized Customer Dynamic Experience with AI-Based Predictive Models. *Revolutionizing the AI-Digital Landscape: A Guide to Sustainable Emerging Technologies for Marketing Professionals*, 189.

. Van Hoang, T. (2024). Impact of integrated artificial intelligence and internet of things technologies on smart city transformation. *Journal of Technical Education Science, 19*(Special Issue 01), 64-73.

Verma, J., Sandys, L., Matthews, A., & Goel, S. (2024). Readiness of artificial intelligence technology for managing energy demands from renewable sources. *Engineering Applications of Artificial Intelligence*, 135, 108831. DOI: 10.1016/j.engappai.2024.108831

Vijayalakshmi, S. Savita, Genish, T., & George, J. P. (2023). The role of artificial intelligence in renewable energy. In *AI-Powered Machine Learning in the Energy Industry: Digital Technology and Sustainable Energy Systems* (pp. 253-269). Cham: Springer International Publishing.

Vijayalakshmi, S. Savita, & Durgadevi, P. (2023). AI and Machine Learning in Improving Resilience of Smart Energy Infrastructure. In *AI-Powered Machine Learning in the Energy Industry: Digital Technology and Sustainable Energy Systems* (pp. 189-213). Cham: Springer International Publishing.

Wang, H., Wen, C., Duan, L., Li, X., Liu, D., & Guo, W. (2024). Sustainable energy transition in cities: A deep statistical prediction model for renewable energy sources management for low-carbon urban development. *Sustainable Cities and Society*, 107, 105434. DOI: 10.1016/j.scs.2024.105434

Xie, H., Huang, R., Sun, H., Han, Z., Jiang, M., Zhang, D., Goh, H. H., Kurniawan, T. A., Han, F., Liu, H., & Wu, T. (2023). Wireless energy: Paving the way for smart cities and a greener future. *Energy and Building*, 297, 113469. DOI: 10.1016/j.enbuild.2023.113469

Yousef, L. A., Yousef, H., & Rocha-Meneses, L. (2023). Artificial intelligence for management of variable renewable energy systems: A review of current status and future directions. *Energies*, 16(24), 8057. DOI: 10.3390/en16248057

Zearah, S. A., Al-Tahee, M., Mohammed, A., Shnawa, A. H., Mahdi, H. H. J., & Majeed, Z. (2023, December). Smart City Application based on Internet-of-Things and Artificial Intelligence for Energy Management Dynamic Memory Modelling. In *2023 International Conference on Emerging Research in Computational Science (ICERCS)* (pp. 1-6). IEEE. DOI: 10.1109/ICERCS57948.2023.10434196

Zhou, Y., & Liu, J. (2024). Advances in emerging digital technologies for energy efficiency and energy integration in smart cities. *Energy and Building*, 315, 114289. DOI: 10.1016/j.enbuild.2024.114289

Chapter 5
Heatwave Prediction on Era5 Data Using Deep Learning Techniques

Abhishek Basak
VIT-AP University, India

Soumyajit Pal
https://orcid.org/0009-0006-9902-8653
St. Xavier's College, Kolkata, India

ABSTRACT

This chapter seeks to establish the usefulness of deep learning methods in heatwave prediction with the help of the ERA5 dataset. Heat waves, which are defined as periods of elevated temperature and humidity, are a critical concern because they affect the health of people, farming, water supply, and energy consumption. This study relies on state-of-the-art deep learning models such as Convolutional Neural Networks (CNNs), Recurrent Neural Networks (RNNs), and Long Short-Term Memory (LSTM) to predict heatwave cases. The data set utilized in this work is the ERA5-HEAT dataset; this is a high-resolution atmospheric reanalysis data set. In this paper, the process of data cleaning, temporal alignment, and spatial aggregation in handling large climate data sets has been described. More importantly, the idea of using deep learning models in capturing the temporal and spatial dependencies from meteorological data is well supported through real-life applications. Deep learning methods provide a higher level of accuracy and flexibility than traditional meteorological models.

DOI: 10.4018/979-8-3693-8684-2.ch005

1 INTRODUCTION TO HEATWAVES AND THEIR IMPACT

A heatwave is usually a succession of unusually hot days with extended periods of high humidity, commonly experienced in countries of oceanic climate type. Such periods of high temperature far exceed the usual for a particular area and season and may last for days to weeks.

The definition of a heatwave is determined by region because of its dependence on the climate of the area being considered. Generally, a heat wave is proclaimed when the day's top temperature exceeds the customary maximum temperature by approximately 5 degrees Celsius (9 degrees Fahrenheit) or higher for several consecutive days (Center for Climate and Energy Solutions, 2023; Westerling et al., 2006; Seidl et al., 2014).

Heat waves are dangerous meteorological events because of their huge impact on human health, agriculture, water resources, and energy demand ("Heat wave," 2024; Lobell et al., 2011; Stone et al., 2010). Constant staying for long at high temperatures causes heat-related illnesses, for example, heat exhaustion, heatstroke, or even death, with the oldest and the young being most affected, as well as those with previous medical conditions (Ebi & Meehl, 2007; Schlesinger et al., 1990; Dai, 2013). Also, heat waves can worsen drought impact on crops and so might increase the risk of wildfires (Sun et al., 2019; Zuo et al., 2015; Al Mahri & Bouchama, 2018).

In a scientific sense, heat waves are usually connected with high-pressure systems that are stationary over an area (Koppe et al., 2004; Thomas et al., 2004; Hoegh-Guldberg et al., 2007). These weather patterns, known as "heat domes", can hinder the formation of clouds, and keep cool air levels at bay, which thus prevents the warm air from escaping to other areas (Zuo et al., 2015). Climate change as the world is getting warmer is the cause of the increasing intervals of heat waves (Lopez-Gomez et al., 2023; Kenny et al., 2010; Knowlton et al., 2009).

Understanding and predicting heat waves are essential for preparing and mitigating their effects (Campbell et al., 2018; Parmesan, 2006; Ciais et al., 2005). This can be a combination of meteorological monitoring, climate modeling, public health planning, and the establishment of heat action plans that include public alerting mechanisms and reduced exposure strategies (Costello et al., 2009; Watts et al., 2015; Caissie, 2006). Heat waves are detrimental to physical health in various ways such as high mortality rates due to heat-related illnesses and also chronic diseases, wildfires which are disastrous and result in habitat loss hence air pollution (Whitman et al., 1997) (Kosatsky, 2005). Also, heat waves impact agriculture systems where yields are reduced and food security is compromised (Nitschke et al., 2011; Kovats & Koppe, 2005; Parmesan & Yohe, 2003). Figure 1 and Figure 2 elucidate them.

Figure 1. Environmental impact of heatwaves: A systematic diagram

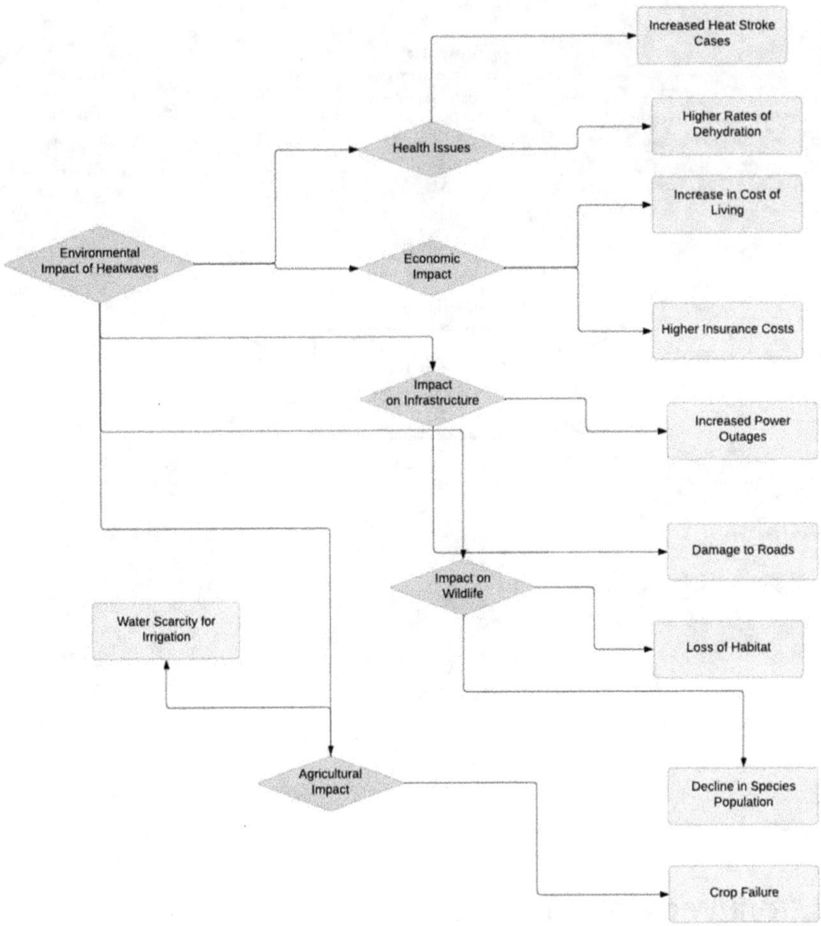

Figure 2. Impact of heatwaves on Health

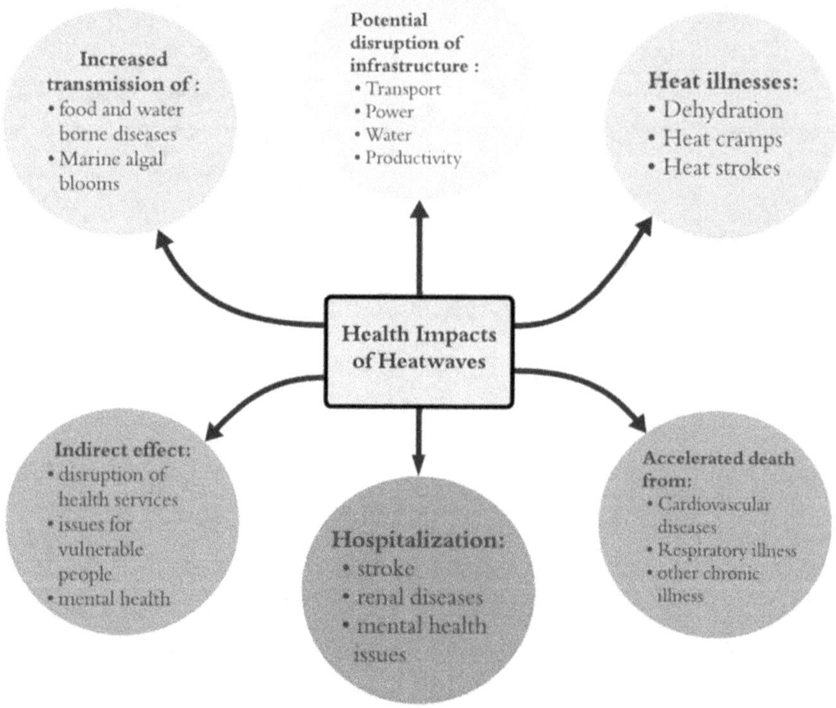

2 OVERVIEW OF ERA5 DATA

The ERA5-HEAT (Human thErmAl comforT) dataset is a historical dataset that offers selected indices of human thermal stress and discomfort under outdoor situations. It uses the datasets from the European Centre for Medium-Range Forecasts (ECMWF) ERA5 reanalysis which is a global, high-resolution, atmospheric reanalysis, based on the ECMWF interim reanalysis project started in 1991, that blends model data with observations from around the world to provide a second-by-second description of the climate across the planet and its changes over the past few decades (Espinosa et al., 2023). Due to this, ERA5 is a good proxy for observed atmospheric conditions because it incorporates all the components above.

Main variables

Mean Radiant Temperature (MRT)

Units: Kelvin (K)

Description: Mean radiant temperature, or MRT, is one of the primary parameters used to determine thermal comfort. It measures the constant temperature of an abstract black body radiation cavity (ε = 1) that one would predict the gross phenomena of radiation heat transfer to or from a person, as does the actual and rather complex environment. Essentially, MRT captures the combined effects of all radiant heat sources surrounding a person. Essentially, MRT captures the combined effects of all radiant heat sources surrounding a person, including:

Direct Solar Radiation: Real heat obtained from the sun-exposed areas depends on time, weather conditions, and shading.

Diffuse Sky Radiation: Diffuse radiation: is a form of solar radiation that hits the Earth's surface from every direction of the sky and on any type of day, including snowy ones.

Reflected Radiation: Energy that bounces back from grounds, structures, and other frills like buildings and structures, etc.. This could vary depending on the type of surfaces and the degree of color that is seen on them.

Long-wave Radiation: This is released by surrounding objects and surfaces inclusive of buildings vegetation ground and other objects present in the environment, as they have accumulated solar energy and have the potential to emit heat energy.

The application of MRT is even more justified in the context of bioclimatology and thermal comfort investigations since this method provides a single calculational parameter that quantitatively describes the conditions of heat radiation in the immediate proximity of the human body. It is thus worthwhile to get acquainted with MRT to grasp the extent of the overall thermal load placed upon people. At the same time, they engage themselves in various endeavors across different weather conditions.

Universal Thermal Climate Index (UTCI)

Units: Kelvin (K)

Description: The UTCI on the other hand is a complicated index formulated to accurately depict the thermal comfort of people under the open sky in any weather condition as discussed below. The temperature within the reference environment provides similar thermal stress to our body like sweat rate, shivering pace, skin wetness, blood circulation, and deep body temperatures as in the real environment. UTCI takes into account multiple factors:

Air Temperature: The first one is the ambient temperature based on the feel outside or possibly inside a car or any other enclosed space.

Wind Speed: About convectional heat loss, they have the following implications: They imply that when speed increases the convection, heating, or cooling, of an environment is enhanced and may be useful in keeping a climate cool.

Humidity: Affects evaporative cooling. High temperatures also increase the emission of metabolic heat, while high humidity slows the rate of heat loss through the vaporization of sweat.

Radiation: Comprises both solar and thermal radiation, whose effects determine heat in or out of the building. The combination of these factors defines the thermal injury endured by people.

Combined, these elements give us UTCI, which represents one number that tells the overall thermal strain on people. Thus, UTCI is a highly helpful resource for public health organizations, commune and city designers, and other investigators; It can be used to learn more about the detrimental impacts of different climates and to devise ways how to minimize these consequences. Fig. 3 describes the UTCI for July 1981-2010 (as an example) (Espinosa et al., 2023).

Figure 3. UTCI for July 1981-2010

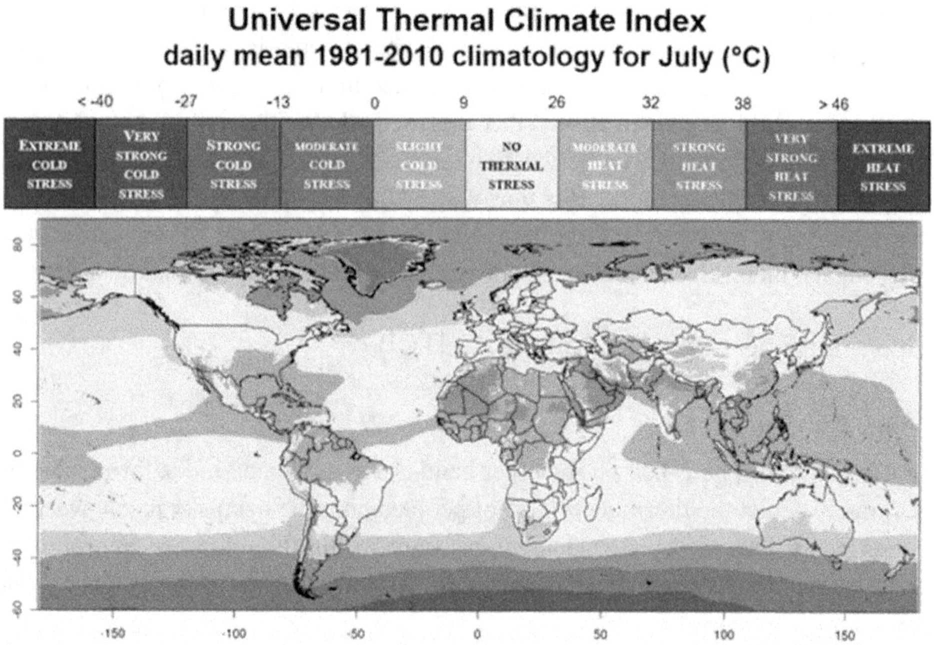

Description of the Dataset

Data type: Gridded

Projection: Regular latitude-longitude grid

Horizontal coverage: Global except for Antarctica (90N-60S, 180W-180E)

Horizontal resolution: 0.25° x 0.25°

Vertical resolution: Surface level

Temporal coverage: From January 1940 to near real-time for the most recent version.

Temporal resolution: Hourly data

File format: NetCDF

Conventions: Climate and Forecast (CF) Metadata Convention v1.6

Versions: v1.1 (latest). This new version uses the new thermofeel library. The data is available in GRIB format.

Update frequency: Intermediate dataset updated daily in near real-time. Consolidated dataset monthly updates with a 2–3-month delay behind real-time.

3 DATA PREPROCESSING AND FEATURE ENGINEERING

Handling datasets of different magnitudes of large climatic variations

1. Data Cleaning

Missing Data: Weather records and climate data are often incomplete due to various factors like failures in the sensors or the instruments used, lack of data transmission, or lack of observation. Remembering that these missing values have appeared during the data collection can be a major advantage since they can later create biases or inaccurate results. Methods to handle missing data include:

Imputation: Imputing the samples based on those values that have been omitted but those that have the same characteristics as the samples. Some of the most frequently used approaches are simple linear interpolation that approximates the unspecified data values by averaging the neighbor values or employing more sophisticated scenarios such as employing a machine learning algorithm to approximate the missing values by analyzing the given data.

Forward or Backward Fill: Using the last known value to fill in missing data points (forward fill) or using the next known value (backward fill). This is simple but can be flawed if the period of missing data is big.

Outliers: Outliers are observed values that greatly differ from other observations and must be treated with utmost caution because they can distort results. The first step that always has to be taken is that of dealing with outliers. Methods include:

Capping or Flooring: Restriction of output values to a certain upper or lower bound (maximum or minimum threshold).

Robust Statistical Methods: This normally can be done with methods that are less likely to be skewed by outliers such as median analysis instead of mean, or even calculated standard deviation.

2. Temporal Alignment

Resampling: Regarding temporal data resolutions, climate data could be on an hourly, daily, or even monthly basis depending on the observation made. For a more consistent method of data analysis, the resampling techniques can resample the data to a standard time scale. For example:

Upsampling: Pros: It could also be useful to convert daily data to hourly, in which case, one has to interpolate the values.

Downsampling: Averaging or summing data on an hourly basis to present data on an assessment of a day.

Time Zones: The time zone is important if climate data is obtained at different times from different geographical regions, this is to make sure that all data sets have a common temporal reference frame. This becomes important to analyze the time series more adequately.

3. Spatial Aggregation

Grid Aggregation: Climate information is usually in gridded forms and the grids may have fine resolutions (for instance $0.25° \times 0.25°$). To reduce the complexity or focus on specific areas, spatial aggregation can be performed:

Averaging or Summarizing: Integrating and aggregating information derived from multiple geographical grid cells into a single unified larger cell or group of distinct cells.

Region-Based Analysis: Subsetting for small tracts including deltas, coastal lowlands, great plains, and large river basins, where local effects prevail.

NORMALIZATION AND STANDARDIZATION SCHEMES FOR TIME SERIES DATA

Normalization is a procedure used in data preparation to change all data records so that their absolute values fall within a predefined level; the most common level is the interval [0, 1]. Consequently, this method makes a certain sense as it scales the data within the range of zero and a specified value which can be very helpful for machine learning methods that can be very sensitive to the detail and the scale of the data being taken.

Min-Max Scaling

Formula I:

$$x' = \frac{x - min\ (X)}{max\ (X) - min\ (X)}$$

where:
x is the original value.
x' is the normalized value.
min(X) is the minimum value in the dataset.
max(X) is the maximum value in the dataset.
Application:

This is required if the data needs to be manipulated and restricted to form a certain range. Besides, it proved effective for those algorithms, such as neural networks or gradient descent optimization methods which work better when input is normalized.

Standardization

Standardization rescales the data such that the mean of the data is mean = 0, and its standard deviation = 1. It is applied for the normally distributed data or when the comparative values of features are important.

Z-score Normalization:

Formula II:

$$x' = \frac{x - \mu}{\sigma}$$

where:
x is the original value
x' is the standardized value

μ is the mean of the dataset

σ is the standard deviation of the dataset

Application:

This technique is most suitable when algorithms assume that the input data follows a normal distribution like linear, logistic regression, principal component analysis (PCA), etc.

Handling Seasonality and Trends

Detrending:

The purpose is to remove trends from the time series data allowing for the analysis of fluctuations around the trend. Differencing is the technique followed, where each data point is subtracted from the previous one.

Formula III:

$$y_t = x_t - x_{t-1}$$

Application: This method is useful for converting non-stationary data into stationary one as it is a requirement for many time series models.

Seasonal adjustment

The purpose of this is that seasonality can be either filtered out or removed to allow for other underlying patterns in the data to be observed. Time-series decomposition, for instance, Seasonal Decomposition of Time Series (STL), enables the breakdown of the series into various components such as seasonal, trend, and residuals.

Application: This is especially useful when trying to disentangle the seasonal patterns from other overlying fluctuations that occur in the time series; this assists in enhancing the precision of forecasting.

Time Series-Specific Scaling

Rolling Window Scaling:

The purpose is to normalize or standardize data within a rolling window to capture local trends and variations. We apply scaling methods within a window that moves over the data. For instance, a rolling mean normalization:

Formula IV:

$$x'_t = \frac{x_t - mean\left(x_{t-n+1:t}\right)}{std\left(x_{t-n+1:t}\right)}$$

where: n is the window size

Application: This method is useful for non-stationary time series as it easily adapts to changes over time.

Feature Scaling

Log Transformation:

The purpose is to handle skewed data and reduce the impact of large outliers. We can use logarithmic transformation to transform the data:

Formula V:

$$y = log(x + 1)$$

1 is added to avoid log 0 as it is undefined.

Application: It is useful for data spanning several orders of magnitude, as it compresses the range of values and reduces skewness.

For details refer to Figure. 4 where an example related to data preprocessing has been displayed for better understanding. In this case, an example of applying the data preprocessing steps is given using a synthetic time series data set containing temperature data. The dataset includes a set of 100 randomly generated daily temperatures.

Figure 4. This is an example that aims at outlining the preprocessing of time series data in detail concerning the temperature data to help in creating models for analysis. This entails practicing essential pre-processing techniques like normalization, standardization, as well as transformation to deal with issues like scale, trends, seasonality, and outliers.

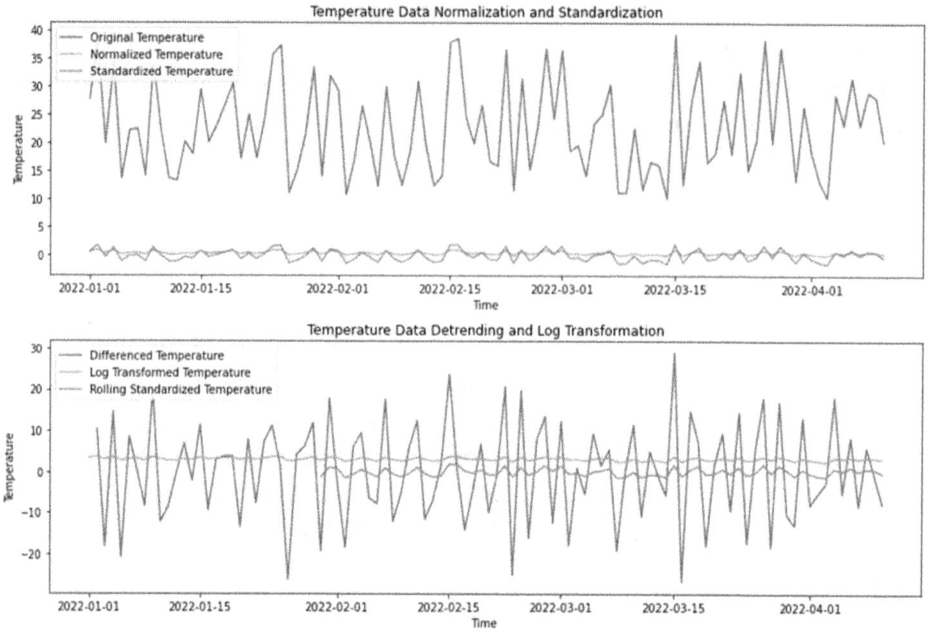

4 DEEP LEARNING FOR PREDICTION OF WEATHER CONDITIONS

Machine learning involves a specialized branch called deep learning that has greatly transformed various disciplines such as meteorology in terms of enormous dataset pattern recognition. Contrary to standard approaches that require engineering features and creating physical models, deep learning models are capable of self-learned representations from given data. This capability makes them very suitable for high-dimensional and non-linear spaces, which translate to weather information.

Deep Learning Neural Networks in Weather Predictions

It shows that deep learning Neural networks have become essential tools for weather prediction since they can address temporal and spatial data elaboratively. Here's a detailed exploration of the three most notable types of deep learning neural networks used in this field: from a family of CNNs, RNNs, and LSTM networks.

1. Convolutional Neural Networks (CNNs)

Description:

Architecture: There are several layers in CNNs such as convolution layers, pooling layers, and the output fully connected layers. The convolutional layers help in performing the filtering of the data that was inputted to the network then move to the pooling layer to minimize the dimension of the map and then the fully connected layer to predict the data.

Learning Mechanism: Backpropagation uses the error for training CNNs through which the spatial hierarchies of features are learned. During the training phase, they at some point reconfigure the weights attached to the filters to achieve the smallest possible error.

Application in Meteorology:

Satellite Image Classification: Based on experiments, CNNs are effective in the classification of distributed cloud formations, actual land surfaces, and atmospheric phenomena in satellite images. This is essential and significant in comprehending and outlining the climate conditions and situations.

Precipitation Forecasting: Hence, through the use of image radars, the CNNs can predict the occurrence and intensity of precipitation expected. This is useful in predicting rainfall, snowfall, and other relevant occurrences on the ground.

Identifying Weather Patterns: Other demonstrations include cyclones and certain fronts of weather from the satellite and radar are captured and other demonstrations from the satellite and radar are detected by CNNs. These patterns are valuable for learning known as short-range forecasts of the climate.

Example:

For example, if we have to build a system that can distinguish between various forms of clouds from satellite images, we can train a CNN for this purpose. For this reason, the system can predict the formation of conditions such as storms or hurricanes and provide valuable information for constructing warnings.

2. Recurrent Neural Networks (RNNs)

Description:

Architecture: The actual architecture of RNN is recursive given the fact that out(t) that is produced at one-time step is turned into input for the next time step. It entails a loop that could allow the networking to store information on the earlier inputs.

Learning Mechanism: Changes in the weights in RNNs use backpropagation of errors through time (BPTT). This empowers them to detect temporal relations about the data that is usually in chronological order.

Application in Meteorology:

Time Series Forecasting: Examining the ability of RNN as one of the feed-forward neural networks; this capacity of the network can be understood by training it with some previous data to predict variables such as temperature, humidity, wind speed, and pressure.

Sequential Data Analysis: They can consecutively take a set of weather information and provide its forecast information concerning the actual condition. For example, the activity entails an endeavor to predict which team would be victorious in the subsequent football match bases with the other one being the database of the previous football match results.

Example:

For the RNN to be used the following are possibilities: making predictions concerning changes in temperature over some time that are marked by certain temperature recordings. This capability is useful in situations where there is a more frequent requirement for real-time and/or more frequent updates of the common climate in the environment.

3. Long Short-Term Memory (LSTM):

Description:

Architecture: LSTMs are a further development of the basic RNN; they have a more complex architecture encompassing gates to regulate information processing: input, forget, and output gates. In this way, the LSTMs can remember and use information over long periods, with the help of this design.

Learning Mechanism: The short memories appear by maintaining a constant error flow, which makes LSTMs able to learn long-term dependencies as opposed to standard RNNs which suffer from the vanishing gradient problem.

Application in Meteorology:

Long-Term Weather Forecasting: There are benefits to LSTMs such as the ability to forecast the weather over large periods like seasonal changes or even fluctuations over a long term regarding temperature or rainfall.

Extreme Weather Event Prediction: They ensure that they predict disasters that may occur such as hurricanes, droughts, or heat waves within the weeks, months, or years to come based on previous data patterns.

Example:

LSTM network can be applied for analyzing the seasonal changes in temperature and for this. For instance, it can predict summer temperatures in previous years which is quite helpful for those planning during the heat period.

Below is a case study where the objective is as follows:

The purpose of this assessment is to utilize various deep-learning neural network structures including CNNs, RNNs, and LSTM for the modeling of the weather parameters based on the time series data. This includes calibration and testing of these models based on the synthetic temperature data to evaluate their capabilities of representing the temporal and spatial features within the meteorological data.

Figure 5 describes the arbitrary Synthetic daily temperature data for a specific period. Conceptually, the contribution can be illustrated with an example of the generation of synthetic weather data and, in particular, daily temperature values for the year. The temperature data was created using a combination of random noise and a sinusoidal function to simulate daily variations:

Random Component: Thus, it became possible to include certain randomness into the data, while keeping the overall distribution of the temperature values relatively stable: The temperature values were randomly generated from a normal distribution with a mean of 20°C and a standard deviation of 5°C.

Seasonal Component: To replicate annual temperature changes, a second harmonic was incorporated in the form of a sinusoidal function. There are little variations in temperatures and this function simulates the effects of temperatures rising and falling that can be attributed to the different seasons.

The conjoining of these aspects results in a dataset that reproduces random fluctuations in temperature and the fluctuation that results from daily and seasonal changes. This synthetic dataset is useful for the illustration of each of the time series forecasting methods without the nuisance of real data anomalies.

Figure 5. Daily Temperature Data (1st January 2023 - 1st January 2024)

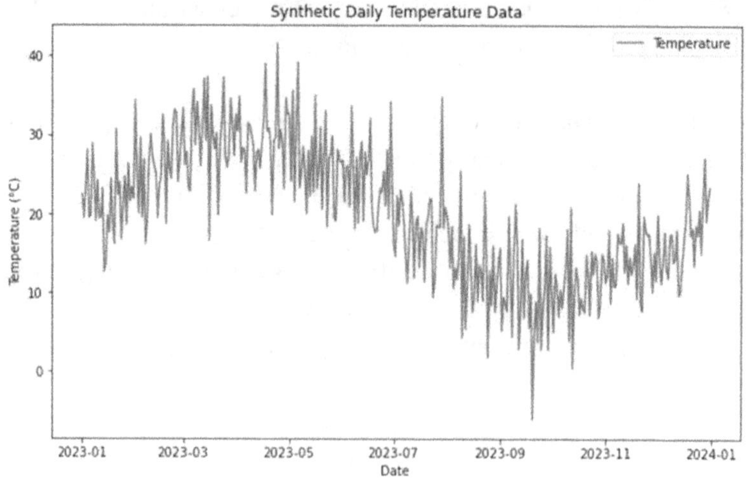

We then implement CNN, RNN, and LSTM to make future predictions. It is then run over 100 epochs. Figures 6, 7, and 8 describe the predictions of the above Deep Learning techniques.

Figure 6. CNN predictions of the Synthetic time series

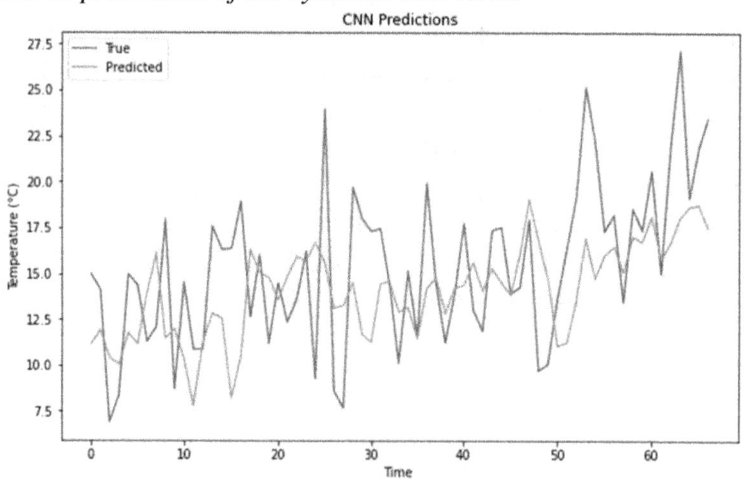

Figure 7. RNN predictions of the Synthetic time series

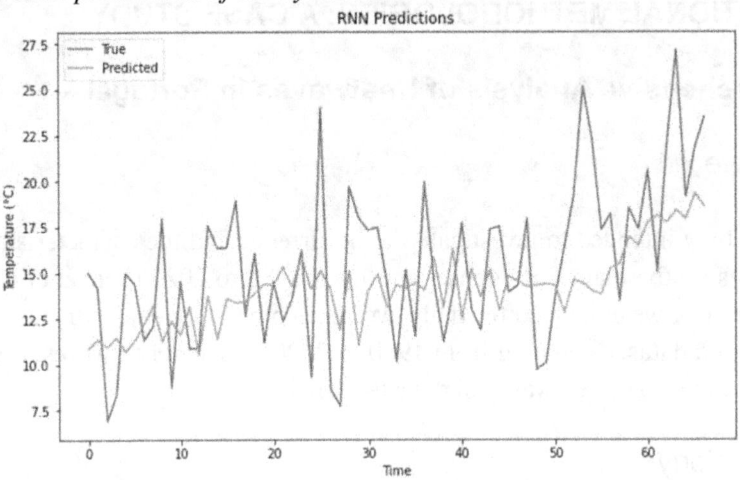

Figure 8. LSTM predictions of the Synthetic time series

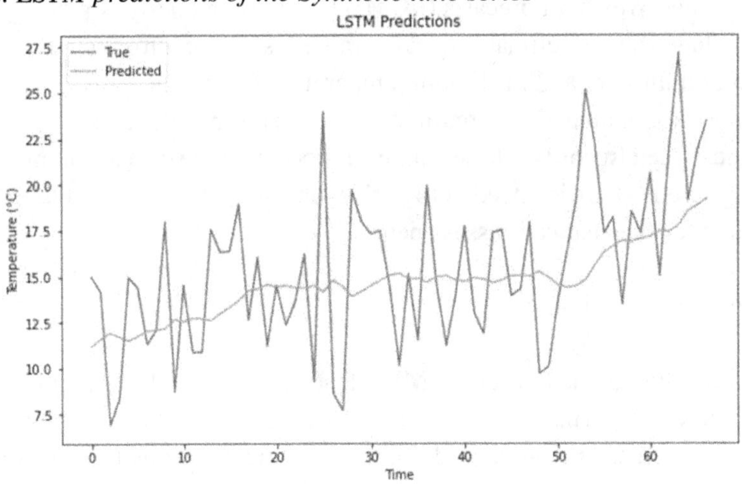

5 HEATWAVE PREDICTION ON ERA5 DATA USING TRADITIONAL METHODOLOGIES: A CASE STUDY

Comprehensive Analysis of Heatwaves in Portugal

Objective

The study intended to investigate the occurrence and the characteristics of the heatwaves on the main Portuguese land from 1980 to 2021 (Sehgal et al., 2023). For this study, we employed the daily average temperature data extracted from the ERA5-Land dataset spanning from 1980 to 2020 to detect heatwaves, trends, and spatial frequency of heatwaves in the study area.

Methodology

1. Data Source:

ERA5-Land Dataset: This dataset was extracted from the European Centre for Medium-Range Weather Forecasts (ECMWF) and is an hourly climatological data set that includes a lot of climatological parameters. The control variables included meant daily minimum and maximum temperature (Tmin and Tmax) were derived Mean daily maximum and minimum temperatures were calculated.

Ground-Based Records: These studies indicate that the data accomplished from the Portuguese Met Office needed to verify the ERA5-Land dataset to ensure that it is reliable for the heatwave assessment.

2. Heatwave Definition:

Heatwave Magnitude Index (HWMI): Heatwave Meteorological Index (HWMI) was applied to categorize the heatwave into two groups and for operationalization of heatwave events. The baseline definition of a heat wave is three or more days of temperatures that are higher than the normal average temperatures for the given place and period.

3. Validation and Analysis:

Validation: This paper used the ERA5-Land Monthly Mean of Temperature and attempted to assess its accuracy based on ground measurements for nine grid cells over Portugal.

Trend Analysis: Because of temporal data, statistical methods were applied to establish the heat-wave pattern from frequency or severity. Kernel rate estimation was also used in addition to exploring the numbers of heatwave events realized by Tmin and Tmax within the year.

4. Spatial Analysis:

Geographical Focus: The study was carried out in a number of regions in the main body of Portugal, some of the most developed regions with cosmopolitan features were subjected to this study like Lisbon and some of the northern and Core regions of Portugal were also used in the study. This in turn provided a general picture of the variability in heat waves relative to geographic locality.

Findings

1. Trend Analysis:

Increasing Heatwave Days: It was observed that for the same region, the number of heat wave days does enhance progressively as the period of observation advanced. In particular, the investigation of the tendencies of the increase in heat waves was conducted on the climates of the southern interior and coastal zones, where a significantly higher rise in such indicators was identified.

Regional Variations: Hence, comparing the number of heatwaves in Laeti with other regions of Portugal and mainland Europe, it is seen that Northern and interior regions exhibited the maximum number of days correlated with 'D Max ' heatwaves, whereas Lisbon displayed a high number of days related to 'D Min ' heatwaves due to the impact of UHI.

2. Temperature Changes:

Temperature data disclosed in the research pointed to the fact that temperature changes over the twenty years of the heatwave days are more. For instance, the figures observed earlier by using the annual Tmax heatwave had it that these parts had risen by 10-30 in the year 2019 compared to the year 2018.

3. Spatial Distribution:

In the current study, the geographic information system was used to spatially identify the heatwave days distribution, and found out that there is a geography variation. In like manner, in the comparison of the heat waves of the regions, it, too, increases the maximum number of heat waves in the days for the southern interior and the coastal points of the grid.

Significance

This study is significant for several reasons:

Validation of ERA5 Data: Accordingly, the present work validated the feasibility of employing the aforementioned ERA5-Land datasets concerning heatwaves through their comparison with observational data.

Long-Term Trends: The heat wave research conducted in the study assessed forty years and from the analysis, it can be seen that heat waves are getting more successive and severe because of global warming.

Regional Insights: Ultimately, Regional specificity enhances awareness of how heat waves affect regions and may be employed in developing regional-based adaptive measures.

Policy and Planning: Thus, this research can be useful for policymakers and Urban planners to develop the necessary countermeasures depending on the specific sector to combat the impact of heat waves, primarily in high-risk zones.

Techniques

Statistical Analysis: Various statistical analyses were used to test the number of heatwave days, including the rising and fluctuation analysis.

Kernel Rate Estimation: The following are the annual frequency of heat wave risks and it was useful to know heat wave risks and & get details about each one of them.

Comparative Analysis: Measurement of yielded data against ground records made it correct and provided an assurance of the study outcomes.

Retrospection of Heatwave and Heat Index

Objective

This study aimed at studying the extremely severe heat waves in Europe of 2003 chiefly through recreating temperature trends and the heat index (HI): in Marmara, Turkey, and Naples, Italy regions (Jiao et al., 2021). The heat sensitization study was conducted to identify the variations in the temperature during the heat waves and to further establish the relationship between changes in temperature and the health effects documented.

Methodology

1. Data source:

ERA5 Dataset: The study utilized the ECMWF's ERA5 reanalysis dataset to capture the various climate parameters at a high temporal and spatial resolution. This dataset contains statistics of atmospheric, ocean-wave, and land-surface, statistics that have been calculated for each hour. Among all the factors, temperature and relative humidity the most attention was paid to in this work.

2. Heat Index Calculation:

Heat Index (HI): They include; the Steadman equation which helps in estimating the HI when taken in conjunction with the prevalence points of temperature and relative humidity standard when used in establishing the Apparent temperature of a place.

3. Region-Specific Analysis:

Marmara Region: Thus, the observations of fourteen weather stations in the present study and all of them are located in the Marmara region of Turkey. The sources of information included data in HI for the given period which ranges between 2007, and 2016 by stressing averages and peaks.

Naples Region: In the case of the work under discussion, the author aims to describe certain aspects of two strong heat waves that affected Naples, Italy in 2003. Temp, rel humidity & wind speed, are essential motifs when working out the HI and assessing on health impact.

4. Health Impact Correlation:

The study also investigated the correlation between HI and health risks or health risks related to events like hospitalization and mortality to determine the impact of heat on people's health.

Findings

Temperature Trends:

Marmara Region: Looking at the above data analysis, it could be realized that the temperatures have progressively risen with an excellent appreciation of about 1. Produce 19 percent less carbon dioxide emission below the average rate estimated in the 1980s – that is 5°C below the average global temperature in the 1980s. On the same note, and regarding the rigidity of the climate, compared to the initial status, the regional relative humidity decreased by approximately 4% for the same duration.

Naples Region: However, by generally higher in Neapolis by a margin of about 2 C degrees in the present climate compared to the 1980s. This resulted in higher temperatures within a year especially during summer because of African and Azores anticyclones.

Heat Index Trends:

Marmara Region: HI revealed that when the temperatures were slightly above twenty-one Celsius, the climates were regarded as semi-comfortable while climates with percentages above twenty-seven were seen as very uncomfortable. Thus in the case of the HI value the values obtained were observed to be increasing progressively over the period under consideration for the region.

Naples Region: Referring to HI the degrees of heat in the summer of 2003 were higher as it identifies more impressions on health such as heat injuries and some fatal deaths. To reiterate, Subsection HI dealt with attitude to extreme conditions and their effects on human health.

Health Impacts:

These findings were corroborated by the HI above the HI admission rates and total deaths, which further confirmed the model. This was even more so during heatwaves such as the heatwaves that were witnessed during the summer of 2003 whereby many people suffered very much; this was occasioned when temperature and humidity hit the roofs.

Significance

Use of ERA5 Data:

Therefore it becomes relevant to use the atmospheric data set of ERA5 to determine the temperature and HI. To advance our knowledge of heatwave impacts and apply high-resolution satellite imagery.

Health Impact Analysis:

Hence, the present research provides important data on the effects of heat waves on public health, having matched HI to the resultant health impacts. Consequently, such briefs are of great importance in the enhancement of risk and adaptation of frameworks relating to population health.

Region-Specific Insights:

This is done based on the regional analysis of the specific impacts of heat waves, as well as the possibilities of the prevention and treatment of heat-related diseases in the given geographic region. These findings apply to the regions that commonly undergo high heat and will assist in planning and executing preventive measures against future effects of heat streaks.

Dataset Features and Techniques

Dataset Features:

ERA5 Dataset: Provides information on the current weather of the places of the user's interest in regards to the temperature, humidity and wind rate in the hour-space format. Due to the high density of the spatial and temporal resolution, ERA5 is suitable to be used in climatology applications.

Techniques:

Statistical Analysis: Calculating different means of temperature and humidity variations over time, and applying different statistic tests, the HI was estimated.

Heat Index Formula: This HI average was computed with the help of Steadman's formula taking into consideration the HI data collection of temperature and humidity based on the ERA5 dataset.

Validation: Performing the validation: The use of weather observation station data to some extent made to verify ERA5 data which in turn shows the accuracy of the study.

Analysis of Heatwave Events in Central Asia

Objective

In this study, the authors aimed to assess the increasing trend of heatwaves and their frequency and duration in the Central Asia region from 1981 to 2020 based on the ERA5-Land data (Copernicus Climate Data Store, n.d.-b). Such changes in the context of vital interest necessitated the study's goal of monitoring alterations in heatwaves' incidence and time course and determining climatic factors' role in intensifying these conditions.

Methodology

Data Source:

ERA5-Land Dataset: First of all the dataset entitled as ERA5-Land presents the real time as well as hourly average of the climate parameters and climate factors and it is made available by the European Centre for Medium-Range Weather Forecasts (ECMWF). In this connection, the parameters of temperature and humidity were used for this work more frequently.

Heatwave Identification:

Definition: L1: Heatwave: the occurrence of three more consecutive days where the maximum temperature recorded that day is above the 90th percentile of the annual temperature scale.

Data Extraction: When analyzing the climatic factors in the study area, the climatic characteristics of the Central Asian countries, including Kazakhstan, Uzbekistan, Turkmenistan, Kyrgyzstan, and Tajikistan, the two factors compared were the temperature and humidity conditions.

Analysis Techniques:

Trend Analysis: Statistical procedures were employed in analyzing heatwave frequency and duration data to decipher the developmental characteristics of the heatwaves spanning the study period.

Correlation with Climatic Variables: Temperature profiles as well as humidity and pressure changes were also taken into consideration by the study when searching for possible instances of heatwaves.

Findings

Increase in Heatwave Frequency and Duration:

In detail, the study disproportionally exposed heat-wave frequency enhancing four folds and heat-wave duration enhancing four points during 1981-2020 in Central Asia. This trend was established throughout several countries within the region However, a recent review by Wabiri-Math et al (2014) showed that some countries in the area were registering an increased trend in the rate of this disease.

Climatic Variables:

Temperature Anomalies: Heatwave was understood to be days with above average temperatures for a given location and the result indicated that positive temperature departure involves heatwave. New data shows that communities where temperature fluctuations from the average are drastically higher have been subject to longer durations of heat waves.

Humidity Levels: Another factor that also contributed to worsening heat wave situations was the reduction in relative humidity as the level reduced. Lower humidity levels indicated in radiation balance less frequently, but the heatwaves were more significant coincidingly: perhaps, there is some kind of evaporative cooling in operation in such a location.

Atmospheric Pressure: Some pressure distributions had been observed to increase the length of heat waves while others acted as seen by the significant effect of pressure oscillations in differentiating heatwave periods.

Spatial Variations:

He concentrated his analysis on the variations of anticipated heat waves across the regions and in terms of heatwave magnitude. For example, the southern countries of Central Asia – Turkmenistan and Uzbekistan – had a significantly higher HWS score than the countries in Northern Central Asia.

Significance

Geographical Insights:

This paper builds on prior studies about the tendencies in heatwaves by adding geographical location data and using ERA5 data in Regions efficiently. Besides, it is crucial to consider any alterations in the characteristics of heatwaves by adopting a broader and more diverse territory of the CAS region.

Climate Change Indicators:

Ascribed to this category, the study contributes to the underlying general trend of higher heatwave intensity and several occurrences globally. Changes have also been highlighted in climatic variables and the study focuses on global warming as a cause of change in climatic systems of a specific region.

Policy and Planning:

According to the findings of the thesis, the chance to approve public policy and planning, in general and specifically within Central Asia, is important to grasp. Simulating the heatwave source and the reasons behind it would help in the determination of measures to prevent the heatwaves from getting worse in the future and the measures put in place to protect vulnerable populations and properties.

Dataset Features and Techniques

Dataset Features:

ERA5-Land Dataset: Functions with detailed data sets representing temperature, humidity, wind speed, and some other climatic variables. Hourly data is important for temporal heatwave hotspot identification; therefore, the offered dataset provides valuable information in this regard.

Techniques:

Statistical Analysis: A descriptive type of analysis was utilized where trends of the variables and correlation coefficients were employed in the analysis of data.

Heatwave Definition and Detection: Thus, adopting a rather stringent set of criteria for defining heatwaves while relying on historical temperatures and their percentile distribution helped to provide accurate identification of events.

Correlation Analysis: Concerning the heatwave analysis and the climatic parameters, the following findings were developed to determine the drivers of heatwaves.

6 HEATWAVE PREDICTION ON ERA5 DATA USING DEEP LEARNING TECHNIQUES: A CASE STUDY

A deep learning model for forecasting global monthly mean sea surface temperature anomalies

The scholarly study titled 'A Deep Learning Model for Forecasting Global Monthly Mean Sea Surface Temperature Anomalies' presents and utilizes an Unet-LSTM model to predict sea surface temperatures (SST) using ERA5 data (He et al., 2024). This model is a dual conception of CNN and LSTM network topologies together to incorporate spatial and temporal features of the data.

Dataset and Features

The climate data used in this study comes from an ERA5 reanalysis dataset containing higher time-resolution data. Key features extracted from ERA5 for this study include:

Sea Surface Temperature (SST): This forms the core parameter of interest and is essentially the 'sea surface temperature'.

2-meter Atmospheric Temperature (t2m): This variable complements the SST forecast by expressing temperature at a two-meter height above the ground level.

Also of the class ERA5, the class contains data from January 1950 to May 2021 and has a temporal resolution of one month. The data has been collected worldwide but the analysis has been done specifically in the equatorial and subtropical Pacific Island region where the model has better skill.

Data Preprocessing

The data preprocessing steps involve:

Grid Preparation: SST and t2m are combined into a global coordinate system with Latitude: 720 points & Longitude: 1440 points. This grid is then averaged to a 1×1 degree grid [180, 360] and then to [64, 128] grid to estimate the interpolated values.

Normalization: The data is prepared so that the deep learning model behaves as expected for an input with a mean of 0 and a standard deviation of 1. Normalization is particularly effective in smoothening the scaling of input features, which can be useful in training neural networks, as they normally converge during their training.

Model Architecture

The Unet-LSTM model consists of:

Unet Architecture: This is a variant of CNN called the Unet, and it is particularly tailored for image segmentation problems. These architectures have an encoder-decoder design, which is enhanced by skip connections that facilitate an understanding of the details as well as the context.

LSTM Layers: LSTM or Long Short-Term Memory is a specific kind of RNN structure considered one of the best solutions for predicting sequences. They aid in capturing the temporal interactions as available in the SST data.

The model accepted the latitude and longitude as inputs in the two-dimensional grids for each time step and used the SST anomalies for the future time steps as output. The training procedure is changing weights within a chosen object for the minimum possible prediction deviation.

Results

The evaluation of the Unet-LSTM model shows that the proposed model offers a good prediction accuracy, with an even stronger performance in the regions of interacting and sophisticated ocean-atmosphere phenomena such as throughout the equatorial and subtropical areas of the Pacific Ocean. Model performance is assessed through measures like Root Mean Squared Error (RMSE) and coefficients of determination. Our findings indicate that the proposed model is well capable of predicting SST anomalies with a reasonable degree of accuracy to adequate specification of both seasonal as also inter-annual variations.

Techniques

Key techniques used in this study include:

2D Convolutions: For this reason, the top three features namely; globe, population density, and land cover were included in the model to capture spatial dependencies in the SST data.

LSTM Networks: The above is about living dynamic relationships to capture temporal dependencies and/or give long-horizon forecasts.

Normalization and Interpolation: As a form of data pre-processing, we wanted to ensure that the data were in the best possible state for the model to deliver its best performance.

Encoder-Decoder Architecture: To make the model aware of both formant frequencies that are local and over the signal window as a whole.

These aspects of the model and preprocessing steps are suitable for emerging climate data which has higher data dimensionality and more complex features to perform SST prediction with deep learning algorithms.

An Interpretable Deep Learning Approach for Detecting Marine Heatwave Patterns

The study "An Interpretable Deep Learning Approach for Detecting Marine Heatwave Patterns" aims to adopt LSTM models to predict patterns of marine heatwaves by using SST data of ERA5. Unlike most studies that focus on predicting heat waves, this study also focuses on the ease of understanding the deep learning models as well as their decision-making processes (Awasthi et al., 2021).

Dataset and Features

The data employed in the current research is based on the ERA5 reanalysis database, which is an extensive source of historical climate data. Key features extracted from ERA5 for this study include:

Sea Surface Temperature (SST): The first is a measure of the temperature of the ocean's surface – the primary variable of interest.

Atmospheric Variables: It is clear that there may be additional independent variables such as air temperature, wind speed, and humidity that might also be added to the input layer to capture the maximum variation in the atmosphere and enhance the forecasting power of the model.

Data Preprocessing

Data preprocessing steps include:

Temporal and Spatial Averaging: Real SST data is taken at certain temporal intervals (such as daily or monthly, according to the data availability) and geographic areas, which are often some ocean basins.

Normalization: The normalizing of data which will enable all the features to have equal contribution in the learning process of the model.

Feature Engineering: Introducing new variables that could capture a temporal framework or an interaction between variables.

Model Architecture

The main model chosen for the research of this paper is the LSTM (Long Short-Term Memory) network, which is a type of recurrent neural network that is used for stationary sequences and is capable of handling both short-term and long-term relations among the data. As a result, the LSTM model stands as one of the more appropriate models when it comes to time series prediction due to its long-term memory capabilities. See Fig. 9 for details

Figure 9. Basic LSTM unit structure explanation

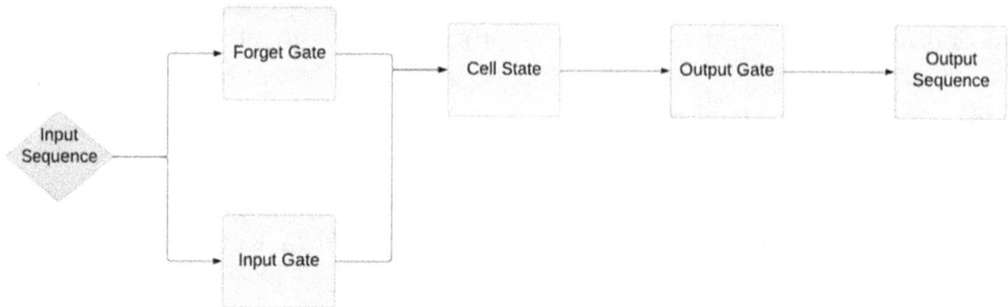

Techniques

The following are the key methods and techniques employed in the study:

LSTM Networks: To predict Future SST values from Past SST Observations and capture Temporal Dependencies in SST data.

Explainable AI (XAI): Methods such as SHAP (SHapley Additive explanations) or LIME (Local Interpretable Model-agnostic Explanations) are employed to provide explanations of the predictions of the LSTM model, and the importance of different features in predicting the patterns is highlighted.

Simple thresholding: The identification of marine heatwaves often entails identifying SST events that exceed a certain threshold (say, SST in the 90th percentile of all SST observations at a given location over a particular period).

Regions of Focus

The study is limited to certain regions and ocean basins, including regions where marine heatwave frequency is increasing. Examples include:

- Pacific Ocean

- Indian Ocean
- Atlantic Ocean

Results

The predictive ability of the LSTM model for spotting marine heatwave patterns was clear. Key findings include:

Excellent Forecast Skill: The model gave successful predictions mainly because it was able to represent the non-linear relations of the SST temperature as well as marine heatwave occurrences.

Interpretability: The transparent nature of the model provided an understanding of which features were most important in predicting heat waves, thereby making the model more interpretable and trustable.

This study demonstrates the feasibility of using deep learning models, in particular LSTM networks, for predicting marine heatwaves in SST fields based on only SST data from ERA5. And it emphasizes model interpretability, making sure the predictions are not only accurate but also human-interpretable. This is fundamental for decision-making in climate change and marine management tiers (Jacques-Dumas et al., 2022).

Deep Learning-Based Extreme Heatwave Forecast

The study 'Deep Learning-Based Extreme Heatwave Forecast' uses the Convolutional Neural Network (CNN) model to predict heatwaves with data captured from the mid-latitude regions of the Northern Hemisphere. The model learns from Fourier-transformed data to work effectively with the geospatial data since the model is required to operate on spherical data of the Earth. This work has employed a supervised learning model where the Convolutional Neural Network (CNN) model gives the chances of a heatwave occurrence given the previous temperature data (Taylor & Feng, 2022).

Dataset and Features

The data employed in this research is based on climate models and observations with an emphasis on the mid-latitudinal region of the Northern Hemisphere, particularly above 30° latitude. Key features include:

Temperature Data: Surface and atmospheric temperature history data temperature measurements, historical records, and temperature data.

Fourier-Transformed Data: To control the spherical nature of the Earth and set boundary conditions the data is projected into the frequency domain through the use of the Fourier transform.

The data covers the period from the 1st of June to the 31st of August over a thousand simulated years and offers a large corpus for training and testing purposes.

Data Preprocessing

Data preprocessing involves several steps:

Fourier Transformation: The temperature data is then transformed from a spatial domain to the frequency domain using a 64×64 grid which is more helpful when working with spherical geometry of the Earth.

Normalization: Data is normalized so that every feature has the same importance during the process of model learning.

Labeling: The prepared data is assigned binary labels that signify the occurrence (1) or lack (0) of heatwave conditions.

Model Architecture

The base model that has been employed in this study is the CNN as this model is well suited for handling high dimensional data and is also good at capturing spatial relationships. The architecture includes:

Convolutional Layers: Both initial layers use filters of size 12x12 and 9x9 and ReLU activation function for learning spatial features of the input data.

Max Pooling and Dropout: The max-pooling layers are used for the input size and dropout layers are used to eliminate overfitting in the process of training where some features are set to zero value.

Fully Connected Layers: Thus, the feature maps are flattened and then passed through other fully connected layers and then finally passed through the sigmoid layer to produce the probability of a heatwave.

Training and Testing

The overall list of number series was further separated into two groups namely the training list and the testing list, The training list contained 900 numerical years while the testing list contained 100 numerical years. Here also the training set is sub-sampled so as not to be dominated by outliers which is common with big data that is imbalanced. Key training parameters include:

Batch Size: The case will be one thousand samples per batch.

Learning Rate: The learning rate is set to 2e-4 mainly due to AMSGrad than other optimizers which helps in faster and better convergence of the model.

Epochs: The number of NFA for one was to be fixed at 10 while for the remaining ones, it was fixed at 5, as observed to boost performance.

Results

This study found that the CNN model produced high accuracy in predicting heat waves. Key findings include:

High Accuracy: It was observed that the model had high prediction accuracy for the occurrence of heat waves based on performance measures like precision, recall, and F1-score.

Spatial and Temporal Precision: To incorporate the spatiotemporal dependencies, the model used Fourier-transformed data and was able to provide accurate predictions of heat waves.

Techniques

The study employs several advanced techniques:

Fourier Transform: To cope with spherical topology and to avoid problems related to the transformation of boundary conditions.

Convolutional Neural Networks: To include the spatial correlation structure present in the data.

Supervised Classification: With the help of historical data, it tries to estimate the likelihood of a heatwave to happen.

7 CONCLUSION AND FUTURE WORK

Deep learning has proven to be a better approach to the prediction of heatwaves than conventional methods due to certain remarkable advantages. Climate analysis is characterized by high dimensionality and non-stationary, so the analysis or modeling of climate variables, such as temperature or humidity, and their correlations with other atmospheric and oceanic variables, cannot be simplified. Other deep learning structures such as the LSTM and CNN are preferable to tackle such problems as they are good at handling long dependencies and spatial characteristics respectively. For example, the Unet LSTM model to extract spatial features employs convolution and LSTM for the temporal characteristic of sea surface temperature abnormalities in the equatorial and subtropical Pacific.

About the above-stated models, the following benefits apply. First of all, the appeal of deep learning methods is in the fact that they can learn features by themselves from the raw data, thus eliminating the need to predefine features by hand, which is a time-consuming and error-prone process. This is especially useful when analyzing big and highly detailed data such as ERA5 which incorporates a large amount of temporal and spatial data. Second, the deep learning models have easier ways to update the information they have learned; this means that they can always have the most updated information. This makes them suitable to use in areas where it is believed that the data patterns may shift due to climate changes.

Also, as it was mentioned earlier deep learning models are well tailored to work with high-dimensions data. For example, the CNN-based model used to forecast heatwaves employs the Fourier transform of data to transform the earth to a spherical structure with easier boundary conditions hence improving the model's predictive performance. This in turn not only increases the accuracy of the data but also allows us to get a deeper insight into the interconnection between the field features, which are characteristic of the climate.

This approach not only raises the level of correctness of the data but also enables a deeper insight into the relationships between the field characteristics that are characteristic of the climate.

Applying the XAI methods to explain the results from the deep learning models is also helpful in climate science. By using recently proposed techniques like SHAP (SHapley Additive exPlanations) and LIME (Local Interpretable Model-agnostic Explanations), experts can reveal how these models make decisions thereby improving the models' interpretability and credibility. This is because the number of people willing to believe and rely on the models to get things right in a policy and disaster preparedness context will be very high.

In the future, further research in this field should be directed toward the following: First of all, some issues are still considered not solved, and one of them is the interpretability of the deep learning models. Thus, it is essential to develop new strategies for presenting these, and for the most part, other complex models, including LSTM and CNN, to further audiences and explain their working mechanisms to non-professionals. However, to improve the model, some suggestions can be made like using data from more than one source like satellite data and also many other reanalysis data sets. This would help to achieve a somewhat better understanding of the conditions that lead to heat waves as identified in this study by Frontiers.

Also, there will be enhancement of computing power and algorithms in an area that will facilitate the real-time prediction and early warning of heat waves. These modifications will assist in handling the scaling of deep learning models to possibly a greater number of users as well as in scenarios that are resource-scarce to foster climate resilience and disaster preparedness. In conclusion, deep learning for heat wave forecasting is not only an innovation in climate science but also opens up certain possibilities for development and research on combating the impact of such climatic events.

8 REFERENCES

Al Mahri, S., & Bouchama, A. (2018). Heatstroke. *Handbook of Clinical Neurology*, 157, 531–545. DOI: 10.1016/B978-0-444-64074-1.00032-X PMID: 30459024

Awasthi, A., Vishwakarma, K., & Pattnayak, K. C. (2021). Retrospection of heatwave and heat index: a case study using ERA5 dataset.

Caissie, D. (2006). The thermal regime of rivers: A review. *Freshwater Biology*, 51(8), 1389–1406. DOI: 10.1111/j.1365-2427.2006.01597.x

Campbell, S., Remenyi, T. A., White, C. J., & Johnston, F. H. (2018). Heatwave and health impact research: A global review. *Health & Place*, 53, 210–218. DOI: 10.1016/j.healthplace.2018.08.017 PMID: 30189362

Ciais, P., Reichstein, M., Viovy, N., Granier, A., Ogée, J., Allard, V., Aubinet, M., Buchmann, N., Bernhofer, C., Carrara, A., Chevallier, F., De Noblet, N., Friend, A. D., Friedlingstein, P., Grünwald, T., Heinesch, B., Keronen, P., Knohl, A., Krinner, G., & Valentini, R. (2005). Europe-wide reduction in primary productivity caused by the heat and drought in 2003. *Nature*, 437(7058), 529–533. DOI: 10.1038/nature03972 PMID: 16177786

Copernicus Climate Data Store | Copernicus Climate Data Store. (n.d.-b). https://cds.climate.copernicus.eu/cdsapp#!/dataset/derived-utci-historical?tab=overview

Costello, A., Abbas, M. M. E., Allen, A., Ball, S., Bell, S., Bellamy, R., Friel, S., Groce, N., Johnson, A. M., Kett, M., Lee, M., Lévy, C., Maslin, M., McCoy, D., McGuire, B., Montgomery, H., Napier, D., Pagel, C., Patel, J., & Patterson, C. (2009, May 1). Managing the health effects of climate change. *Lancet*, 373(9676), 1693–1733. Advance online publication. DOI: 10.1016/S0140-6736(09)60935-1 PMID: 19447250

Dai, A. (2013). Increasing drought under global warming in observations and models. *Nature Climate Change*, 3(1), 52–58. DOI: 10.1038/nclimate1633

Ebi, K. L., & Meehl, G. A. (2007). The Heat is On: Climate Change & Heatwaves in the Midwest. In Regional Impacts of Climate Change: Four Case Studies in the United States. Pew Center on Global Climate Change.

Espinosa, L. A., Portela, M. M., Moreira Freitas, L. M., & Gharbia, S. (2023). Addressing the Spatiotemporal Patterns of Heatwaves in Portugal with a Validated ERA5-Land Dataset (1980–2021). *Water (Basel)*, 15(17), 3102. DOI: 10.3390/w15173102

He, Q., Zhu, Z., Zhao, D., Song, W., & Huang, D. (2024). An Interpretable Deep Learning Approach for Detecting Marine Heatwaves Patterns. *Applied Sciences (Basel, Switzerland)*, 14(2), 601. DOI: 10.3390/app14020601

Heat wave. (2024, May 7). Wikipedia. https://en.wikipedia.org/wiki/Heat_wave

Heat Waves and Climate Change - Center for Climate and Energy Solutions. (2023, July 14). Center for Climate and Energy Solutions. https://www.c2es.org/content/heat-waves-and-climate-change/

Hoegh-Guldberg, O., Mumby, P. J., Hooten, A. J., Steneck, R. S., Greenfield, P., Gomez, E., ... & Hatziolos, M. (2007). Coral reefs under rapid climate change and ocean acidification. science, 318(5857), 1737-1742.

Jacques-Dumas, V., Ragone, F., Borgnat, P., Abry, P., & Bouchet, F. (2022). Deep learning-based extreme heatwave forecast. *Frontiers in Climate*, •••, 4.

Jiao, D., Xu, N., Yang, F., & Xu, K. (2021). Evaluation of spatial-temporal variation performance of ERA5 precipitation data in China. *Scientific Reports*, 11(1), 17956. DOI: 10.1038/s41598-021-97432-y PMID: 34504211

Kenny, G. P., Yardley, J., Brown, C., Sigal, R. J., & Jay, O. (2010). Heat stress in older individuals and patients with common chronic diseases. *Canadian Medical Association Journal*, 182(10), 1053–1060. DOI: 10.1503/cmaj.081050 PMID: 19703915

Knowlton, K., Rotkin-Ellman, M., King, G., Margolis, H. G., Smith, D., Solomon, G., Trent, R., & English, P. (2009). The 2006 California heat wave: Impacts on hospitalizations and emergency department visits. *Environmental Health Perspectives*, 117(1), 61–67. DOI: 10.1289/ehp.11594 PMID: 19165388

Koppe, C., Kovats, S., Jendritzky, G., & Menne, B. (2004). *Heat-waves: risks and responses (No. EUR/03/5036810).* World Health Organization. Regional Office for Europe.

Kosatsky, T. (2005). The 2003 European heat waves. *Eurosurveillance*, 10(7), 3–4. DOI: 10.2807/esm.10.07.00552-en PMID: 29208081

Kovats, R. S., & Koppe, C. (2005). Heatwaves: past and future impacts on health. Integration of public health with adaptation to climate change: lessons learned and new directions, 136-160.

Lobell, D. B., Schlenker, W., & Costa-Roberts, J. (2011). Climate trends and global crop production since 1980. *Science*, 333(6042), 616–620. DOI: 10.1126/science.1204531 PMID: 21551030

Lopez-Gomez, I., McGovern, A., Agrawal, S., & Hickey, J. (2023). Global Extreme Heat Forecasting Using Neural Weather Models. *Artificial Intelligence for the Earth Systems*, 2(1), e220035. DOI: 10.1175/AIES-D-22-0035.1

Nitschke, M., Tucker, G. R., Hansen, A. L., Williams, S., Zhang, Y., & Bi, P. (2011). Impact of two recent extreme heat episodes on morbidity and mortality in Adelaide, South Australia: A case-series analysis. *Environmental Health*, 10(1), 1–9. DOI: 10.1186/1476-069X-10-42 PMID: 21592410

Parmesan, C. (2006). Ecological and evolutionary responses to recent climate change. *Annual Review of Ecology, Evolution, and Systematics*, 37(1), 637–669. DOI: 10.1146/annurev.ecolsys.37.091305.110100

Parmesan, C., & Yohe, G. (2003). A globally coherent fingerprint of climate change impacts across natural systems. nature, 421(6918), 37-42.

Schlesinger, W. H., Reynolds, J. F., Cunningham, G. L., Huenneke, L. F., Jarrell, W. M., Virginia, R. A., & Whitford, W. G. (1990). Biological feedbacks in global desertification. *Science*, 247(4946), 1043–1048. DOI: 10.1126/science.247.4946.1043 PMID: 17800060

Sehgal, V., Gaur, N., & Mohanty, B. P. Global Flash Drought Monitoring using Surface Soil.

Seidl, R., Schelhaas, M. J., Rammer, W., & Verkerk, P. J. (2014). Increasing forest disturbances in Europe and their impact on carbon storage. *Nature Climate Change*, 4(9), 806–810. DOI: 10.1038/nclimate2318 PMID: 25737744

Stone, B., Hess, J. J., & Frumkin, H. (2010). Urban form and extreme heat events: Are sprawling cities more vulnerable to climate change than compact cities? *Environmental Health Perspectives*, 118(10), 1425–1428. DOI: 10.1289/ehp.0901879 PMID: 21114000

Sun, Q., Miao, C., Hanel, M., Borthwick, A. G., Duan, Q., Ji, D., & Li, H. (2019). Global heat stress on health, wildfires, and agricultural crops under different levels of climate warming. *Environment International*, 128, 125–136. DOI: 10.1016/j.envint.2019.04.025 PMID: 31048130

Taylor, J., & Feng, M. (2022). A deep learning model for forecasting global monthly mean sea surface temperature anomalies. *Frontiers in Climate*, 4, 932932. DOI: 10.3389/fclim.2022.932932

Thomas, C. D., Cameron, A., Green, R. E., Bakkenes, M., Beaumont, L. J., Collingham, Y. C., Erasmus, B. F. N., de Siqueira, M. F., Grainger, A., Hannah, L., Hughes, L., Huntley, B., van Jaarsveld, A. S., Midgley, G. F., Miles, L., Ortega-Huerta, M. A., Townsend Peterson, A., Phillips, O. L., & Williams, S. E. (2004). Extinction risk from climate change. *Nature*, 427(6970), 145–148. DOI: 10.1038/nature02121 PMID: 14712274

Watts, N., Adger, W. N., Agnolucci, P., Blackstock, J., Byass, P., Cai, W., Chaytor, S., Colbourn, T., Collins, M., Cooper, A., Cox, P. M., Depledge, J., Drummond, P., Ekins, P., Galaz, V., Grace, D., Graham, H., Grubb, M., Haines, A., & Costello, A. (2015). Health and climate change: Policy responses to protect public health. *Lancet*, 386(10006), 1861–1914. DOI: 10.1016/S0140-6736(15)60854-6 PMID: 26111439

Westerling, A. L., Hidalgo, H. G., Cayan, D. R., & Swetnam, T. W. (2006). Warming and earlier spring increase western US forest wildfire activity. science, 313(5789), 940-943.

Whitman, S., Good, G., Donoghue, E. R., Benbow, N., Shou, W., & Mou, S. (1997). Mortality in Chicago attributed to the July 1995 heat wave. *American Journal of Public Health*, 87(9), 1515–1518. DOI: 10.2105/AJPH.87.9.1515 PMID: 9314806

Zuo, J., Pullen, S., Palmer, J., Bennetts, H., Chileshe, N., & Ma, T. (2015). Impacts of heat waves and corresponding measures: A review. *Journal of Cleaner Production*, 92, 1–12. DOI: 10.1016/j.jclepro.2014.12.078

Zuo, J., Pullen, S., Palmer, J., Bennetts, H., Chileshe, N., & Ma, T. (2015). Impacts of heat waves and corresponding measures: A review. *Journal of Cleaner Production*, 92, 1–12. DOI: 10.1016/j.jclepro.2014.12.078

Chapter 6
How Can Artificial Intelligence Contribute to Mitigating Climate Change?
Harnessing Technology for Environmental Resilience and Sustainability

Pratiti Nayak
https://orcid.org/0009-0003-0959-0334
Kalinga Institute of Industrial Technology, India

Aakash Bag
https://orcid.org/0000-0001-8820-171X
Adamas University, India

Anant Vijay Maria
Supreme Court of India, India

ABSTRACT

This paper argues that only artificial intelligence can confront climate change, as it offers unique methods for adaptation and mitigation. Climate change threatens the protection of the environment and the future of society, making the use of AI crucial in supporting sustainable development. This paper seeks to discuss the use of AI, particularly in understanding climate change and the various tasks in this area. In order to promote ecological conservation, AI improves weather forecasting and regulates the environment. Legal systems like the UN Climate Change

DOI: 10.4018/979-8-3693-8684-2.ch006

Agreements incorporate AI to reduce emissions. However, ethical questions and regulatory issues are crucial for ensuring a more sensible use of AI that does not pose a threat. Technological change alone is not enough to harness AI for climate strategies without organizational change, and policy adjustment. Therefore, this paper advocates for the proactive integration of AI into climate action, adhering to legal and ethical guidelines, to prevent the emergence of an AI-driven catastrophe.

INTRODUCTION

The pandemic's guru Bill Gates has called climate change the biggest danger to civilization after COVID-19 and has suggested perhaps elevated mortality levels (Gates, 2020). The United Nations Intergovernmental Panel on Climate Change (IPCC) in a special report of October 2018 pointed to the catastrophic effect of global warming. These effects include extreme and frequent hurricanes, floods, and more frequent and severe droughts and storms sea-level rise, melting glaciers, and flooding of coastal communities and islands. Also, the IPCC (2018) asserts that desertification of land and a decline in agricultural productivity due to depletion of water supply are likely to worsen tensions and conflicts at regional levels. The IPCC prescribes a cut of greenhouse gas emissions by 45% in the year 2030 and total elimination in the year 2050 to avert the more drastic impacts of a 2°C increase in the global temperature. They can only be obtained from further exceptional reforms in every sphere of the socioeconomic production system. The major greenhouse gas sources in the US per the Environmental Protection Agency (2020) breakdown is transportation (29%), energy (28%), industry (22%), buildings (12%), and agriculture (9%). Therefore, specialists and decision-makers have come up with several interventions to reduce such negative effects and facilitate adaptation (2019 Mulvaney).

Combatting and preparing for climate change entails knowledge of the gravity of the problem and climate change's difficulties. That is why the recommendations underscore the need for global actions and shifts regarding the norms in society to address this pressing problem. Through CU and EP laws, international agreements have been developed to counter climate change situations. Climate change was discussed formally for the first time in 1992 when the United Nations Framework Convention on Climate Change or UNFCCC was formed. The emission control of greenhouse gases was the objective of the 1997 Kyoto Protocol (United Nations, 1997). The Paris Agreement made in 2015 as a legally binding document outlined a global action plan to limit global warming to as far as 1. 5°C to start with in the year 2020 (United Nations, 2015). The global pledge to reduce emissions might be in place, however, the WMO (2019) affirmed that climate change is advancing at a faster pace than estimated. The information society must further search for climate solutions Raban, Gordon, and Geifman, 2011; Duff 2015; Martins et al. 2019; Filippova 2021. Here, therefore, artificial intelligence (AI) is most vital. The EU High-Level Expert Group on Artificial Intelligence defines artificial intelligence as human-created systems capable of perceiving the environment and acquiring data, understanding structured or unstructured data, reasoning, and choosing the most appropriate action to achieve their objectives. The definition of the above term involves application in issues of ecology, environmental law, and policy. AI needs to put down cognitive reference points and systemic solutions to address environmental issues.

Being digital systems, AI can learn numerical models or use symbolic rules to change their functioning depending on what they glimpse. Artificial intelligence shows intelligent action in that it assesses the environment on its own and responds to meet planned goals. These systems can be incorporated into robots and cars for example or in voice assistants and image analysis software among others (European Commission, 2019). AI's legal status needs to be established so that it can be effectively governed and utilized – especially because the concept has a prominent role in solving critical global issues. Complete social domination and constant technology and law adjustments to the dynamic virtual environment are essential for creating a new legal subject out of AI. Because AI has a rather large impact on social life, it must be credible and its application must be made based on fundamental human rights and freedoms. While the use of AI in climate-related matters has a vast another potential, this discussion implies moral, political, or even ethical issues. The need to address the proliferation, emergence, experimentation, implementation, and cessation of artificial intelligence projects by employing a coherent system of laws is a significant problem. This should include a common conceptual approach to such a concept as well as the set of core values and regulations that pertain to this system; legal responsibility for the undesirable outcomes should also be embedded herein. For AI to be used appropriately in the fight against climate change, some things need to be solved.

LITERATURE REVIEW

Climate change is another area where the use of AI has emerged as a useful tool in the struggle due to its inventive solutions. There has been an attempt to adopt deep learning for the presentation of the subgrid process in climate modeling (Rasp, Kaiser, & Gentine, 2018). Machine learning, a subfield of artificial intelligence, has been implemented in earth system research through the following management of several environmental problems that have exploited data availability; (Reichstein et al., 2019). Vinuesa et al. (2020) have pointed out that AI can contribute towards the

SDGs and that measuring its impact on the sustainable development agenda is therefore important. In the light of carbon management, Wang et al. (2023) argue that AI can facilitate the overall decrease of carbon emissions and the fight against climate change. And AI and machine learning showed a potential for spatially separated climate change mitigation in cities (Milojevic-Dupont & Creutzig, 2021). Applying AI to enhance information facilities, optimally changing structures in industries, and developing eco-friendly technologies can play a specific role in decreasing CO2 emissions in manufacturing companies (Chen, 2023). Such a characteristic of machine learning to mimic human preference suggests that the concept can help in addressing environmental challenges and the climate crisis (Rolnick et al., 2022). In addition, a large number of interdisciplinary fields have attempted to incorporate AI to address and respond to climate change (Efremova et al., 2022). In the use of AI in the fight against climate change, there are opportunities and challenges but there are ideas on how to best harness the potential of AI in environmentalism (Cowls et al., 2021).

Moreover, the research done on AI for net-zero carbon emissions in the network systems has established that the recommended solutions make a large impact on carbon emissions (Zhang, 2024). Although the incorporation of AI within climate change predictions draws more and more attention, its application in the sphere of climate change adaptation is understudied (Cheong et al., 2022). About attempts in climate change, AI integration enhances climate action and deletes hurdles (Kaack et al., 2022). AI has particularly enhanced the experimental phases with reduced carbon footprint and enhanced processes (Wang 2024). Environmental monitoring is gradually employing Artificial Intelligence (AI) for multiple affairs concerning the environment. In conservation, resource management, and climate change management, AI technology has been integrated (2024). This paper argues that the application of AI boosts environmental data-based approaches and it is quite efficient in monitoring, analyzing, diagnosing, predicting, and controlling environmental problems (Oprea & Iliadis, 2011). Due to artificial intelligence, the important object of sustainable development is the development of smart technologies in environmental control and monitoring.

Thus, AI was also applied to study the impacts of climate change on the marine environment; IoT was employed to optimize data acquisition and analysis (Guobao et al., 2019). They have pointed out the development of AI in environmental toxicology for air pollution monitoring by IoT-enabled system in Asa et al. 2022. Some AI systems that have been used to monitor and enhance the IEQ include AI-based environmental assessment systems for indoor environmental quality development showing how versatile AI is in a wide range of environmental practices (Hou & Xu, 2021). AI has also been useful in environmental justice concerns, for example, in the monitoring of urban air pollution, where machine learning models and AI are applied in predicting the concentration of pollutants and the associated health risks of exposure to them (Krupnova et al., 2022). There are examples of how scholars have employed artificial intelligence to minimize pollution in marine environments, evidencing that AI is crucial in addressing the issues of environmental protection (Fazri et al., 2023). Ensuring climate justice and ecosystem monitoring therefore require AI to be used ethically and responsibly (McGovern, 2021).

METHODOLOGY

Based on the study, there are ways in which artificial intelligence (AI) is battling climate change. AI works towards ensuring that the positive impacts of human nurture are achieved supplemented by the eradication of its negative impacts. Success stories reveal specifically how on many climate-relevant issues AI operates well. However, applying AI for environmental protection challenges have arisen especially concerning areas that directly affect the climate posing ethical and political questions. This research work was undertaken based on general principles and methods of science such as analysis, systems approach, synergetics, and modeling. In the paper, we focus on the legal regulation of AI in the European Union and Ukraine with an emphasis on the latest legislation and possible amendments. Legal data was collected and legal analysis was done with specific sociological and comparative legal research practices. This strategy aims at enhancing the laws governing the tasks that artificial intelligence can handle in the natural environment. Understanding these interactions is crucial to realizing the potential of AI while managing the very nature of AI and the issues it presents in terms of better living. In light of this, the inclusion of AI in the formulation of environmental plans is only possible after making sure that those plans will only involve suitable technologies and can be backed by ironclad legal frameworks and regulations.

ANALYSIS

Artificial Intelligence Usage in Climate Change Problem

Across the globe, polity and the economy regard artificial intelligence as a disruptive innovation. The UN Artificial Intelligence Summit that was held in Geneva that was held in 2017 showed how AI can be used to enhance almost every aspect of human life. In previous concerns, people have tried to apply AI technology used in smartphones and self-driving cars to general applications such as food shortage and environmental conservation (Muraleedharan, 2021). AI can predict hurricanes, floods, and other difficult calamities, enhance climate prediction, and generate more reliable weather alerts. As for the strengths, it can mimic the historical weather conditions and the impacts they can have on society. The study by Rasp, Pritchard, & Gentine (2018) establishes that AI and artificial neural networks effectively control complex varieties of climate behavior like the process of convective cloud formation though current climate regimes often exclude it. Large-scale computations of data from several climate models are made possible because of AI. Monteleoni et al. (2011) and McQuade and Monteleoni (2012) performed climate model forecast combinations using the AMNC from about thirty climate models that are endorsed by the IPCC. In their study, McGovern et al. (2017) anticipate that natural catastrophes are likely to become more frequent and severe and it is these AI algorithms that enable the equilibrium of global climate simulation to be controlled thereby enhancing accuracy.

Table 1. AI Applications in Climate Change Mitigation

Application Area Affected	Description of the Area	Technologies Applied	Positive Climate Change Impact	Peer-Examples
Climate Prediction	AI improves the predictions of climate and weather phenomena.	Artificial Intelligence, Natural Language Processing, Predictive Modeling, Algorithm Click-Through Rate, Artificial Neural Networks	Enhances the formulation of better disaster preparedness and response plans.	IPCC climate simulation models, Microsoft Conscient AI Labs GAN models.
Renewable Energy Management	AI is applied in the management of renewable energy generation, storage, and power distribution.	Smart Grids using Artificial Intelligence Sensors	Helps to enhance the effectiveness in energy delivery and diminish reliance on fossil fuels.	Google's DeepMind to increase efficiency of data center energy consumption.
Environmental Monitoring	Real-time monitoring and analysis of environmental conditions are made possible by AI.	IoT, AI-based Satellite Imagery Analysis	Enhances better policing of emissions and environmental crimes.	Spatial monitoring of CO_2 emission levels, Integrated Collections of eBirds and iNaturalist.

continued on following page

Table 1. Continued

Application Area Affected	Description of the Area	Technologies Applied	Positive Climate Change Impact	Peer-Examples
Sustainable Agriculture	AI is applied in agriculture to cut emissions, improve production, and implement sustainable practices.	AI-Led Crop Practices, Robotics	Facilitates food production and security.	Plantix application for soil analysis, FarmShots for pest and disease identification.
Carbon Footprint Reduction	AI locates and manages potential sources of greenhouse gas emissions.	AI Algorithms, Carbon Tracker Tools	Reduces industrial emissions and boosts the adoption of sustainable measures.	Carbon Tracker's AI-based emissions monitoring; underwater, self-sufficient data centers.

Accurate prognosis matters in making appropriate climate policy because it allows the government to adapt to changing conditions and note that such impacts can be mitigated. AI algorithms enhance the readiness for environmental risk by helping provide timely and accurate decisions. We draw them to both local and worldwide factors: we can, for instance, predict the implications of a 2°C rise in the global temperature. In the same year, Ise and Oba also proved that a neural network using monthly global temperature data for 30 years to forecast the change in the subsequent 10 years had an accuracy level of 97 percent. AI is also useful in establishing the causes of climate change. For instance, satellite imagery can reveal and locate huge emissions of CO2 in countries that are not obliged to report their emission levels. This skill is crucial in pursuing and combating the globe's climate alteration. But that's where artificial intelligence comes in; if countries are to control droughts and other hydrological hazards, then artificial intelligence is essential. The current precipitation values that UNESCO's G-WADI Geoserver uses are estimated using an artificial neural network. This instrument is known as G-WADI PERSIANN-CCS and assists in the analysis of hydrological risks and planning for disasters. By applying the crowdsourced data, the iRain mobile application monitors extreme rainfall events, as well as real-time satellite rainfall measurements. It also allows the users to input local rainfall data for use in monitoring rainfall at the global level (UNESCO, 2019).

AI can also depict the impact of severe climate change conditions. In this project, Microsoft, Conscient AI Labs, and MILA came up with a model that allows them to estimate how homes that may potentially be reduced to mere wrecks by extreme storms and sea level rise will look in the future. The purpose of this app is to present to users how climate change may affect their areas in the future years (Snow, 2019).

It will be seen with the current development of AI that it is completely capable of enhancing the CO2 emissions measurement and the management of the systems that lower the emissions. Carbon Tracker looks at the satellite data and tracks sources of emissions in the form of coal-fired power stations as well as demonstrating their financial unproductiveness. This system's goal is to accumulate data about emissions from 4,000–5,000 power plants; whereas the Wildfire system can track emissions all over the world without permission. This database might be critical in identifying the key sources of emissions and setting up carbon prices. Microsoft has developed underwater AI-powered self-sustaining data centers that are powered by wave energy and opt for ocean water as a cooling system, (Roach, 2020). AI also accelerates the building of nuclear fusion reactors, which offer stable, clean, and renewable energy. Smart-built environments utilize artificial intelligence in controlling the operation of the sensors to adjust energy consumption depending on climatic conditions and the surrounding environment. These structures are important in metropolitan environments where resources are much required because they might adjust their power consumption when low-carbon electricity is hardly available.

The AI used by Google's Deep Mind reduced energy consumption in data centers by 40% and in the process thereby cutting greenhouse emissions. AI has a complex technique of learning and the first level is to understand standards and laws and the second level is to recognize any deviation in industries and pollution and management of waste. IoT technology keeps cool the industrial plants and collects information about species populations and ecosystems in eBird and iNaturalist important for the preservation of freshwater and marine habitats. LEDs of such AgTech, for example, the Plantix app developed by Berlin-based PEAT business, diagnose problems with the soil. With the help of satellites and self-learning software, American businesses like AWhere and FarmShots can predict the weather, assess the resistance of crops, and detect diseases and pests on farms. These are some of the aspects that have led to the introduction of these smart farming technologies to increase resilience and crops. Currently, AI has a lot of potential in the advancement of clean energy. Most hydroelectric dam builders in the Amazon Basin undertake the process sequentially in that they build one dam at a stretch and have no blueprints. To determine which of the dams offered the least GHG emissions we created an AI simulation and we were able to come up with more complex and realistic solutions than necessarily initially conceived (Cornell University, 2019).

AI can make use of zero-carbon electrification using isolated microgrids as over a billion people still have no access to electricity (Ritchie, 2019). The 2019 document "Tackling Climate Change with Machine Learning" (Rolnick et al., 2019) lists the following 13 areas: Typologies of solution areas include power and electricity; transport; built environment and settlements, manufacturing industries; agriculture, silviculture, and forestry; carbon capture and storage; weather and climate; social implications; solar engineering; doing ones' part; decision making; knowledge enhancement; and funding/financing. The authors of the study mentioned that the opinions of prominent specialists were sought for (Hao, 2019). The transition to sustainable energy provision calls for the collection of infrastructure data, and improvements in energy estimation. AI is to look at satellite imagery to detect which buildings need to be rehabilitated in terms of energy efficiency and to detect energy use for a municipal region. Firstly, novel chemical structures reveal to machine learning tools, that advancements in materials with enhanced energy storage, collection, and utilization rates are developed more quickly. For instance, Airbus introduced a new aircraft part that is made from 3D printing technology and comes with higher strength, lesser weight, and lesser CO2 emission during flight (Autodesk, 2016). It then combines the cargoes and tends to minimize travel and emission; artificial intelligence (AI)—improves chains and cargo routes. Thus, analysis of current and future supply and demand leads to minimization of waste in delivery and production. Advancement in artificial intelligence (AI) sees to it that there is a decrease in carbon footprints in electric vehicles by enhancing the transport systems while enhancing the batteries. A range of other applications of AI is yet manageable and worth mentioning; one of them is the use of AI in monitoring forest loss. Satellite pictures and programming instruments gather information on the loss of tree cover at the macro level. The application of chainsaw sound detection algorithms can better perform conservation tasks by immediately notifying the police to act and suppress the illicit work. These developments show that revolutionary AI is in the making in providing a stop to global warming and promoting sustainability in a plethora of industries. Issues of Artificial Intelligence Usage in Environmental Protection

Cryptocurrency mining, particularly inefficient mining, has continued to expand the amounts of data and traffic captured; the initiation of the Internet of Things through 5G networks has also led to data and traffic collection. Moreover, electronic devices in particular consume a lot of energy during their production, do not spare raw materials, and are in a great part made of plastic, as well as being packed in plastic. Such factors raise the call for the evaluation of the environmental effects of AI and the development of comprehensive strategies to minimize the negative impacts. The advancement in technology has seen AI being combined with the fossil fuel industry a development that is creating havoc to the environment. Some of the world's most significant tech firms have been pushing for the use of carbon-intensive AI services to enhance the production of oil and resources. For example, Amazon offers services such as "Predicting the Next Oil Field in Seconds with Machine Learning." Google Cloud deals with fossil fuel industries, while Microsoft hosts events such as "Empowering Oil & Gas with AI" (Microsoft News Center, 2018). Notably, C3 IoT today aids large oil and gas organizations to optimize the extraction of hydrocarbon after it has earlier aided the transformation of renewable power sources (C3 AI, 2019)

Table 2. Challenges and Ethical Concerns in AI for Climate Change

Challenge/Ethical Concern	Description	Potential Risks	Solutions that may mitigate the risks	Stakeholders Involved
Energy Intensive	AI applications are highly computational and consume significant energy.	Increase in carbon footprint, potentially offsetting the benefits of AI.	Development and implementation of energy-efficient AI models and machinery.	Developers of AI, energy companies, and environmental regulatory bodies.
Biases in AI Models	AI models can inherit biases from the data used to develop them.	Unequal distribution of benefits and burdens of climate change.	Integrating fairness checks and utilizing diverse data sources within AI models.	AI scholars, ethicists, data scientists, big data professionals.
Opportunities and Threats	Lack of transparency in AI decision-making processes.	Difficulty in assessing impact and enforcing accountability.	Establishment of clear policies and guidelines for data transparency and access.	Ministries of health, NGOs, AI firms, international organizations.
Integration with Fossil Fuel Industry	AI is being used to enhance fossil fuel extraction, contrary to climate goals.	Continued reliance on fossil fuels, leads to environmental harm.	Implementation of policies and guidelines to limit AI's use in non-sustainable activities.	Environmental NGOs, policymakers, industry leaders, and business stakeholders.
Differentiation Geometry: Equity across Global and Intergenerational Dimensions	AI-based climate strategies may disproportionately affect certain regions and generations.	Extension and deepening of social inequalities and new social conflicts.	Internationalization and diversification of stakeholders in AI/ ICT policy formation.	International organizations, climate justice advocates, and nation-states.

Free's detractors have noted that big digital corporations are supportive of the fossil fuel industry. This is why The Guardian's Kirchgaessner (2019) documented the vast efforts these businesses dedicating towards fighting climate legislation and climate science denial. It has further been observed that while they have come out in the open supporting sustainability and climate change initiatives this has not stopped them from partnering with fossil fuel companies. There is very little transparency when it comes to the impact of these technologies on the environment. As you will see, there is very little data available to assist those researchers and politicians, who attempt to assess the impact of technologies on climate. Cook et al. (2017) have identified, based on the Greenpeace report, that a huge number of companies fail to disclose exact quantifications of their effects on the environment. This lack of openness complicates the possibility of naming and shaming the offending businesses and undertaking investigations necessary to determine the overall impact they are making on climate change. The coexistence of AI and the fossil fuel industry is a perfect illustration of this and other complex and often contradictory roles that technology serves in environmental sustainability. Despite the ability of AI to slice carbon emissions and utilize renewable energy, employing AI to support fossil fuel extraction cancels these benefits. To ensure that AI improves climate objectives or does not worsen the degradation of the environment, a need arises to increase accountability and transparency.

One problem that arises when assessing the impact of technology on climate change is issues with the non-transparency of information to researchers and policymakers. Cook et al. concerning Greenpeace claimed that hardly any companies report on their use of clean and renewable energy and dirty energy. While AWS controls more than half of the global cloud services market, it is rather tight-lipped about its energy consumption. This opaqueness contributes to the increased hardness of identifying the culprits when it comes to operational inefficiencies and hinders optimum energy usage across all the subfields of digital technology. There is a need to improve the consciousness level of users and people who work with big data to make AI more environmentally friendly. According to Wolff-Anthony, Kanding, and Selvan (2020), it is appropriate to present carbon/energy outcomes with regular manufacturing qualities. Privacy, security, responsibility, and potential for misuse are some of the ethical issues always associated with the use of artificial intelligence. IEEE and other organizations are currently heading global projects on ethical issues in AI and different ethical frameworks have been advanced by the scientists (Floridi et al., 2018; Dignum, 2019; Coeckelbergh, 2020a). These ethical issues must be solved irrespective of their promises and the capability of AI to solve climate change. As explained in a couple of articles, human stewardship seems to override self-improving technologies concerning undesirable outcomes (Guzman, 2021). That is, using AI in environmental environments can raise political issues.

AI can modify the ''architecture of choice'' to increase the likelihood of those choices that are more environmentally sustainable (Thaler and Sunstein, 2008). Even though climate nudging can increase beneficial environmental settings, it brings certain questions about autonomy and rationality. However, whether for environmental reasons, society is yet to accept such intercessions as (…). To ensure that the positive impacts of AI are optimized, while the negative impacts are mitigated, applications of artificial intelligence must be ethical and transparent. High-level strategies and the use of ethical practices are required to ensure that the growth of AI is done alongside sustainable development goals. It is best illustrated with the concept of creating a "green government" that will use Artificial Intelligence to govern people to achieve climate goals and targets. Some countries have managed to enact environmental laws that balance climate change and people's freedoms, though, it may be challenging to draw such demarcation especially if the country in question is a democratic one. The discrepancies in the views of the states on basic human rights further complicate the problem. Looking at the experience of central and local governments in developed countries and observing current trends, it can be concluded that states can no longer ignore the problem of climate change and need to understand how to combine governance and nudge to try to influence citizens. But this could mean that while some states undertake actions to solve this problem, other states may not – thus, generating challenges to international and intergenerational equity.

Justice – local global and intergenerational – is essential. Climate change impacts can be global, regional, and national and vary depending on the geographical area while people of the present generation can impact on people of future generations. Solutions for climate change, if worst addressed, act against development and increase inadequacies and injustices as the COMEST report highlights (COMEST, 2010). In fact, for political reasons, climate AI solutions must consider Justice concepts and address a range of communities, ages, nations, and cultures. Humans have the impulse to control everything as a result of which climate change is mainly a result of the Anthropocene epoch as argued by Crutzen (2006). As exciting and promising as it may sound to use AI to increase the number of beings in charge of the planet, it would be wiser to employ existing technology to reduce the excessive pressure that people exert on the planet. It is important to be rather cautious when it comes to specifying necessary technologies in climate policy. Dobbe and Whittaker (2019) focus on propagating open practices related to energy and carbon emission disclosures and their suggested technology-oriented and climate-sensitive technological policy prescriptions. As much as it is essential, the debate cannot be a fragment, but the whole full-stack supply chain. The Assets of an AI System map and the essay by Crawford and Joler (2018) show how much natural and human capital is needed to build and sustain a single Amazon Echo and where it will end up. According to the analyzed research, efficiency gains in computing may even increase dependency and thus intensify climate change (Coulombel et al., 2019). On the one hand, the comparison of relative efficiencies can to some extent be meaningful, while on the other hand, the absolute efficiency values are needed to ensure that energy metering is feasible and meaningful in terms of bringing real added values to the mitigation of climate change.

Understanding these interactions is crucial when designing autonomous climate policies that are both ethically correct and effective. Thus, to use AI and related technologies safely, policymakers need to find a balance between the tech industry development, fairness to all AI users or citizens, and protection of the environment. Following the observation that, given AI policy proposals developed in non-energy related industries, climate impacts are often dismissed, Selby, Cox, and Royston (2016) concluded that it is necessary to institutionalize the practice of 'non-energy policy' analysis. Therefore, these ambient effects of AI on the environment and ecology need to be institutionalized as normal policy levers in several policy sectors. As acknowledged, technology plays a very large role in climate change, and there is a need to pronounce new ecological transaction policies and technology regulations. To preserve the earth, it is important to avoid applying AI technology to increase the rate of fossil fuel extraction. McGlade and Ekins (2015) claim that to limit the global mean temperature rise to 2°C, a lot of reserves of oil, gas, and coal need to remain 'stranded'. There is a need for AI legislation to regulate or restrict the use of AI in the extraction of fossil fuel in a way that the gains outweigh the negative aspects that is on environmental harm. The Allen Institute agrees that it is possible to 'certify' AI techniques to differentiate carbon-neutral from non-carbon-neutral AI and has a role to play in shaping the design and deployment of AI systems through product certification, with the result of incentivizing best practices. The "data exchange" approach, which centralizes climate data for ease of access can enhance settings meant for AI learning as well as diminish costs (St. John, 2018).

Measures for reducing carbon emission remain scarce even with enhanced concern in the issue of climate change. We expect the contribution of AI to the formulation of climate plans that do not use the carbon budget to the fullest extent. But the fact is that AI is very material and energy-intensive, and companies do not report on their environmental impacts. We need to concerns that concern data privacy, attributions of responsibility, and the ethics of justice and norms about artificial intelligence. Consumers perpetuate the situation by consuming new gadgets and vehicles that are powered by oil resources. Global environment preservation goes hand in hand with innovation in climate-friendly AI and solving technical issues, which is the reason for their choice. This change thus has implications for society, the economy, and how people live their lives. AI technicians and users should be more aware of climate change, and AI energy and material consumption should be made visible. In response to these problems, scholars have endorsed what they call a climate-sensitive technology policy and a technology-oriented climate policy strategy. Nevertheless, this indicates that AI and the related analytic methods can be used despite this limitation to approach various climatic problems. Of those, some tech enterprises are using AI to address climate change, enhance efficiency in resource development, and research new fuels (Kates-Harbeck, Svyatkovskiy, and Tang, 2019). AI is also crucial in achieving the set goals of the European Green Deal, the SDGs, democratic processes, and social rights. In a now familiar refrain, AI outcomes have to be unambiguous, measurable, specific, and accurate and users and developers have to ensure this. Machine intelligence or AI is a new-age technology, any technology needs to go through the rigmarole of testing and DT inadvertent non-profit experiments These technologies must incur initial costs to become sustainable in the long term but also effective in preventing climate change.

Legal Regulation of AI Use in the EU and Ukraine

To achieve safe, 'moral' and accountable use of AI there must be a comprehensive set of legislation in place. These acts should regulate the conceptual and categorical proximities and state basic principles and standards of how to design, evaluate, deploy, and demarcate AI systems: any negative occurrences and remediation of harm rules and regulations should be part of such acts. During the 1970s and 1980s, what can be considered as the first attempts to apply artificial intelligence to law were initiated, one of which was the thesis entitled 'An Artificial Intelligence Approach to Legal Reasoning' by Anne Gardner. This was welcomed by the first International Conference on Artificial Intelligence and Law in 1987, and by the establishment of the International Association for Artificial Intelligence and Law in 1991. In the same year 1992, the publication of the journal entitled "Artificial Intelligence and Law" was started. Nevertheless, at present a solid legal foundation for AI has emerged only in technologically advanced countries of the EU, the USA, and East Asia. However, it is important to note that the European Union has come up with tremendous improvements in the same. Artificial intelligence is present in virtually every legal field, including intellectual property, privacy and data protection, product liability, antitrust rules, etc. The above legal fields should adapt to beat the new challenges resulting from the integration of AI.

As AI is a complete innovation, assessing legal risks requires both, a free creative approach and at the same time a good understanding of social needs. The formation of a legislative process for artificial intelligence which has been called for by leading organizations is an indication of the realistic prospect of AI. What can be offered at the moment are ideas and suggestions, and what is needed is a united legislation. It must be noted here that while approaches vary from one industry to the other and one country to the other, advancement is however being made (Mitchell et al., 2020). Thus, it is necessary to emphasize that despite a rather extensive list of AI applications, every industry should have a uniform legislative foundation that would make it possible to introduce individual developments. Human rights, privacy, and morality should all be protected under one law as can be seen from the above analysis. It should also need free and unhampered access to information as to how AI impacts people and the earth. For a firm ground and to ensure that there is an assurance of confidence in the various AI technologies, a comprehensive legal structure is needed. Set out below is the legislative act that has claimed attention: European Parliament Resolution on Civil Law Rules on Robotics adopted on February 16, 2017. It offers opinions on legal and non-legal measures concerning the emergence, deployment, and sustenance of robotics and artificial intelligence (AI) to the European Commission. The Resolution therefore urges the European Commission to present a legislative initiative that will lay down the legal framework of the obligations of robots and artificial intelligence. This includes the classification of intensive autonomous systems, smart autonomous robots as well as cyber-physical systems and their subcategories. Special EU agency for robotics and AI has set up a charter with procedures, model licenses for designers, the code of conduct for the research ethical committee, and the code of conduct for robotics engineers.

The Commission should also recognize electronic persons and, in particular, the most sophistically robots, grant them a specific legal status, and make them legally responsible for any damages they might inflict. This electronic personality would be used where robots act independently and where they have to interact with other persons on their own. The Resolution aims to free the EU and its related countries of obligations to agree on and adhere to standards set by other countries, allowing companies the 'right' to progress, invent, and create, and, at the same time, retain control over the law. The Resolution states, such rules, that should not hamper the procedures of investigation, creation, and development. It means that the new laws concerning AI and robot creation and utilization should not limit innovative thinking. The Resolution is broken down into several main sections, including social, economic, ethical, and legal challenges related to the development of Robotics & Artificial Intelligence, existing regulations regarding the development & usage of Robotics & AI, necessary Regulations & standards; Control Decision when using Robotics & AI technology, Institution of Control; Civil Liability Issues and Protection of Data & personal information using Robotics & AI. Although the Resolution is consultative, it is the first attempt to embark on a legislative harmonization of the standards in the development and deployment of AI. The European Parliament (2017) noted that even though the EU had to deal with jurisdictional legal, ethical, and social implications, it has been active in engaging in the issues involving AI and robots. It is envisaged to become a framework that is expected to aid future legislation bills.

Artificial Intelligence for Europe, often known as AI strategy, which presents the EU approach to the utilization of and regarding AI, was endorsed by the EC in 2018 (European Commission, 2018a). That is why the Horizon 2020 program in the framework of which the European Union invested €1. AI research and innovation funding of $1 billion during the years 2014, 2015, 2016, and 2017. This communication provides the current status of the legislation on safety and civil liability and emphasizes that the development and use of AI should conform to EU values and fundamental rights. In the first year, only the first Communication was published, while in 2018 the European Commission issued other communications (European Commission, 2018b, 2018c) and approved a broad strategy (European Commission,

2018a). "Ethics Guidelines for Trustworthy Artificial Intelligence" published by the European Commission in 2019 provides the postulates that can be followed when creating and using artificial intelligence which ought to be considered reliable (European Commission, 2019a). In the following suggestions, the conditions that AI must accomplish are described and they encompass superintendence and control by human beings, privacy and data protection, resilience and safety, inevitability, equality and no discrimination, accountability, explainability, and social and environmental welfare. In the case of the European rules, such requirements are the recognition of basic rights, meeting legal requirements, as well as the promotion of the elements that constitute the AI's purpose and reliability. Along with GDPR, these rules position the EU as a place where future rules for AI business will be set. To enhance a competitive robotics and artificial intelligence industry in Europe the European Commission has established the Robotics and Artificial Intelligence Unit. To understand and address the challenges of AI and generalize the European approach, in April 2018 the EU member states signed the Declaration of Cooperation on Artificial Intelligence (European Commission, 2018b).

A standard regulatory framework for artificial intelligence (AI) was outlined in the "White Paper on Artificial Intelligence: Recent advice from the European Commission, for instance, is a publication titled "A European Approach to Excellence and Trust," produced in February 2020 (European Commission, 2020b, 2020c). This document is seeking feedback and proposals for creating a brand-new field of legislation which is AI within the EU. The White Paper, albeit high-level in character, fails to address several important questions: Strikingly, the exact legal violations the AI White Paper seeks to counter are uncloaked. The problem of categorizing AI applications as safe and unsafe is not clear since the companies themselves often do not know which group belongs to their application until they face issues. It is a real possibility that new regulations will overlap with the existing ones in quite a similar manner to the GDPR, which already governs many AI solutions. These limitations raise awareness about the need to have better and more elaborate laws on the same to ensure that AI technology is developed and implemented in the right and safest way. On balance, the measures the European Commission has taken are useful; nonetheless, responding to all of these concerns is crucial for achieving a sound and integrated set of AI regulations. A few legal loopholes are highlighted in the European Commission's paper on the safety and liability of artificial intelligence and other smart technologies such as the Internet of Things and robotics (European Commission, 2020a). These gaps encompass the aspects of obtaining neural and accurate data for training of AI, the aspect of AI decision making, the security challenges occasioned by the interconnected and open characteristics of AI systems and products, as well as the aspects of products, systems, and value chains. In addition, operating system obscurity and legal nuances involved in product liability laws;

as well as general fault-based liability laws are ill-suited to AI systems capable of making decisions independently (Feindor-Schmidt, 2020).

The practical application of the White Paper may pose problems to various companies that the latter may encounter, though there might be a few benefits as well. The ignorance of the EU regulations is that the variations are different from one member state to another; the White Paper claims that it is easier to manage one set of regulations for the entire region. However, Mitchell et al. also state that AI can also be helpful to society (Mitchell et al., 2020). The EU plans to release a complete set of AI legal papers, along with new guidelines for creators and users of AI after they are deemed to have these gaps. Three of these resolutions may be part of this package; Relation of civil liability concerning AI, ethical aspects of AI, Robotics, and related technologies, and IP rights for development of AI technologies (European Parliament, 2020a, 2020b, 2020c). The above resolutions were passed on the twentieth of October, 2020. CSOs are of significant utility in AI technology development and employment since they are the backbone of this sector. With regards to AI governance in Europe, the White Paper on AI recommends that the framework should allow for maximum laster engagement including CSOs, and consult them on the creation of the structure as well as its implementation (European Commission, 2020b).

CSOs must be mindful of the fact that the emergence of AI is capable of precipitating the creation of additional social issues and finding a way of addressing such issues. In doing so, CSOs may be able to prepare for future problems, initiate and control the discourse on the generation of AI while minimizing the further negative impact on society, and participate in the development of the legal guidelines for AI. CSOs can assist those in authority in formulating the relevant laws and regulations by identifying algorithmic prejudice in the application of new algorithms by businesses and organizations. Their input is vital, such that the advancement and deployment of the technologies are not only ethically sound but also premised on understanding the extent of impact they are to have within the broader society. Today, the EU does not have a common set of AI regulations, but there are several legislations to govern AI creation and application. These are laws on human rights, computer crimes and unlawful acts, consumer protection or product liability, data protection, and infringement of trademarks and patents. Some of the other resolutions passed include; the AI White Paper that presents; regulatory priorities and key issues, and the framework for developing a complete AI legal regime. Taking into account how fast this field is progressing, the EU is likely to be among the first to have laws regulating the use of AI and hence set a pace for other countries, including Ukraine to follow.

The topic of legal regulation of Artificial Intelligence is rather novel in Ukraine. Digitalization has begun in several sectors, and it is the function of the authorities to pass legislation that will determine how the state will adopt and implement new sophisticated technology. The fundamental strategic plan for the development of the digital economic space for the years 2018 to 2020 and the work plan for its functioning were unveiled in Ukraine in 2018 by the relevant government decision (Parliament of Ukraine, 2018). The Concept for the Development of Artificial Intelligence in Ukraine was approved by the government in 2020 (Parliament of Ukraine, 2020). As already know, Ukraine has the highest concentration of AI companies in Eastern Europe, and therefore, it has all the prerequisites for the development of AI, noted the Minister of Digital Transformation. Outsourcing and merchandising enterprises from the USA, UK, Japan, and many other countries acquired businesses with Ukrainian origins such as Snap, Google, and Rakuten. The further desire of the government is to create AI-supportive conditions that will contribute to Ukrainian digitalization and economic growth. The potential for the use of AI in Ukraine to grow will enhance the competitiveness of the country in the global market (Fedorov, 2020).

The Concept for the Development of Artificial Intelligence in Ukraine was signed with the Ukrainian Cabinet of Ministers in December 2020 the 2030 implementation plan was included. According to this concept, AI is a set of organized information technologies with the option of applying algorithms and research-based methods to perform elaborate work. These technologies can independently generate and process data, and make use of knowledge representations, algorithms, and models for decision-making, as well as the means of how to go about their work. The situation concerning the AI development plan of Ukraine also speaks about the need to create appropriate legal conditions encouraging technical advances and reflecting moral norms and legal requirements. It will be imperative for the Ukrainian legislation on AI to fall in line with international tendencies to boost economic activity and foster competition in the global market as Ukraine continues with its ongoing process of the development of the digital economy. The Concept aims to provide a list and description of the key objectives for using artificial intelligence (AI) in Ukraine. The ones are the subjects of its activities: the development of public administration, the improvement of the economy, and the protection of the rights and interests of individuals and legal entities in the case of legal assistance, as stated in the constitution of Ukraine. Activities of socioeconomic, scientific, technical, legal, defense, and other national importance are in this category. In 2019 Ukraine, being a member of the Council of Europe's Special Committee for Artificial Intelligence, acceded to the OECD's Artificial Intelligence Recommendation (OECD, 2019).

The Concept establishes the basic principles for the utilization of AI and its application through the AI specifications of the OECD. These are embracing the principles of human and civil rights as well as fundamental freedoms, the rule of law, the importance of keeping information, and the usage of AI safe and responsible. These rules have to be adhered to by developers, implementers, and users of the AI systems for the latter to work as intended. Priority spheres of state policy in AI include education and vocational training, science and technology, economy, cybersecurity and information protection, defense and state security, public administration and law, regulation and ethical standards, and justice. It is quite striking that almost all aspects of life are affected indirectly by environmental matters, but they are not dealt with directly. This omission means that the legislative treatment of environmental aspects of artificial intelligence needs further supplements. Several activities are provided in the Concept to prevent violation of legal and ethical requirements. These are: harmonizing Ukrainian AI regulations with the EU practices; ensuring compliance with the standards set up in the Council of Europe – Recommendation CM/Rec (2020) 1; and, most importantly, introducing into the Ukrainian legal system the principles of the 2019 OECD, which refers to AI. The Concept also envisages forming technical standardization committees that correspond to international standards, cooperation with the subcommittees of the International Organization for Standardization and the International Electrotechnical Commission (ISO/IEC JTC 1/SC 42), and support of joint projects in the AI industry among major market participants.

Another emphasis is made to consider the AI Code of Ethics as something that should be created with input from a diverse group of people. This strategy aims to ensure that AI research that is conducted in Ukraine does not violate moral standards as well as promotes Ukraine's overarching socioeconomic and technological goals. By addressing these elements, the Concept wishes to create a basis for the development of AI that promotes freedom of creativity while respecting ethical principles and human rights. Ukrainian scholars are already concerned with the legal regulation issues of artificial intelligence (AI) in many legal fields, drawing on the EU standards, although the country adopted the first special normative act on AI only in 2020. O. There, criminal responsibility and AI legal entity is studied by the renowned academic E. Radutnyi. He went further to argue in the future that the criminal legislation of Ukraine should be updated to incorporate provisions that create criminal responsibility for 'electronic person' which would mean AI as a legal entity. As was recently outlined by Radutnyi (2018), all the discussions regarding AI liability will not make sense unless people remain in charge of AI technologies. In contrast, N. Martsenko is against the thought that AI should be considered as the subject of civil rights and states that this may lead to the emergence of vague legal situations. She argues that EU regulations set the definition of AI as an 'electronic person' too soon and that there is no full extensive legal framework that informalizes the status of the technology, civil liability, user protection, and data protection. According to the recommendations given by Martsenko to bring civil liability under the relationship with consumers, it is suggested to approach AI as an object of civil rights – the same as the product (Martsenko, 2019).

Scientists working on the possibility of the legal regulation of AI claim that the European Parliament resolutions are a shining light that points to the directions where legislation is needed and at the prospect of legal regulation at the EU and national levels. Besides guaranteeing non-endorsement of human rights in the formation of new patterns of social connections with autonomous instruments, it would help foster business in Europe to entail the European legislative standards for robotics and AI (Pozova, 2017). Some of the current uses of AI technologies being trialed in Ukraine are aimed at improving the environment. Nevertheless, there are not many scientific works concerning the legal regulation of AI in the framework of environmental protection], and there is a lack of legislation regulating the use of AI. Such differences underscore the need for immediate policy adoption and human rights, environmental safety, and the long-term and mutually beneficial transformation of the state. Presumably, EU standards are likely to become the framework about which Ukrainian legislation is being formed owing to Ukraine's tendency toward European integration. It is important that AI can be regulated by law so that it can benefit society in the long run. Definitions of 'artificial intelligence' even established by legislature bring into existence new niches and industries. However, there are basic standards that require legal harmonization such as the rights of persons within the use of AI, and ways of using AI in handling environmental issues stating efficiency and speed as core values. For any AI technology to be incorporated hence befitting the societally and environmentally sensitive setup, all-embracing and futuristic legal instruments have to be established.

CONCLUSION AND SUGGESTIONS

AI is one of the most revolutionary technologies that can have positive impacts on governance, business as well as in society. Besides, it also offers optimistic outcomes in overcoming important world issues such as democratic endangerment crime-fighting climate change, and ecological deterioration. Another principle is that human-centered AI has to ensure that human rights are respected during the creation of AI, the use of AI, and the management of AI systems. Human dignity has been accorded in EU Treaties and the Charter of Fundamental Rights in as much as the understanding that every individual is created with a unique and inalienable moral status must be respected alongside the need to preserve the environment for the future generation (Madiega, 2019). There are many uses of artificial intelligence (AI) in helping to fight climate change. AI can collect and analyze CO2 and temperature information, track environmental and natural calamities, and visualize the effects of climate change on the human environment. AI improves weather prediction, energy use, endangered animals' information, deforestation and industrial carbon details, oceanic conditions, periods of drought, precise farming, notions of intelligent recycling, sophisticated carbon dioxide capture and geoengineering, and encouraging green consumption.

However, several problems emerged with AI technology and its deployment that need to be solved. Some of the problems associated with AI are; it is a very material and energy-intensive technology, it is involved in the deployment of techniques that accelerate the extraction of fossil fuels, and it also relies heavily on some exotic minerals with adverse effects on the environment. An unfortunate aspect that most companies fail to put forward and clear is their impact on the ecological system around it; thus, it becomes harder to mitigate the negative impacts a business unit makes on the environment. Also, the social implications of AI are the issues of freedom in political terms – what AI may do for and to people, which regards direct enforcement of orders, and the threats to privacy and data security. Some of the ethical and political concerns of AI use are therefore global justice and intergenerational equity, and the larger question of the Anthropocene – the geological age that marks humanity's most intense geological intervention in the climate and ecosystem. To solve these issues, it is possible to employ a more or less effective approach that would sustain the positive effect of AI technology concerning environmental conservation but would also protect human rights and make a contribution to social justice. Therefore, developing comprehensive legal and regulatory instruments is obligatory for achieving optimistic AI results and minimizing the risks and other unintended consequences. Thus, political speech, public discussion, and innovative technology must still be connected from an institutional perspective. That is why this interaction is significant; people are interdependent and definitely influence climate change, and thus, they all must change their lifestyles and take responsibility for the Earth's future. To make sure that AI is used safely and effectively for environmental and other applications, a complex of laws is needed. They should set the legal measures for assuming liabilities towards negative outcomes and award damages. They should also control an unambiguous conceptual and categorical design and state basic principles and regulations for AI development, evaluation, deployment, utilization, and termination.

Emerging AI goods and services engage diverse fields of the legal system comprising intellectual property, individual privacy and data protection, and product liability as well as antitrust law. All these legal domains will have to evolve for the new conditions that arise as a result of AI systems integration. Since AI is a relatively new application of technology and, as mentioned earlier, legal risk analysis is still rare in most cases, the fight for compliance requires innovation and an understanding of the requirements of today's society. That is why the best companies demand the establishment of a working regulation of artificial intelligence as a recognition of its extraordinary opportunities. The call for regulation is people's recognition of the change that can be brought about by the existence of AI, as well as the fact that something needs to be done to mitigate the impact it brings. In the same argument, good regulation will ensure that the above risks are minimized, ensure that the development of AI is aligned with societal importance, and deal with other problems such as global warming. In the EU there are several resolutions and the AI White Paper to regulate emerging important topics but there is no single unified system of AI regulation yet. Taking into consideration the present pace of the promotion of artificial intelligence, the EU is likely to be one of the first to provide AI with the corresponding legal framework, which the states, such as Ukraine, will have to follow.

As for legislation, to this date, Ukraine has begun legal proceedings regarding AI, but there remain no laws prohibiting said technology. Additionally, to the call for contributions that are concerned with AI regulations and their effects, there is a lack of legal science studies on the effects of regulations on environmental protection, which is a critical component that defines suitable legislation. Therefore, the guarantee that machine learning, and artificial intelligence, their release and application in Ukraine increase compliance with the requirements of the protection of human rights, meet the demands of international legal environmental standards, and contribute to the advancement of the principles of sustainable development requires urgent and consistent actions. EU norms will form the basis of Ukraine's AI law because the country has been heading towards European integration. It will also contribute to the minimization of greenhouse gas emissions in Ukraine to fight climate change. When there is well-developed legislation, it will be easier to regulate the use of AI while at the same time realizing the advantages of AI and controlling for possible risks. By doing this, Ukraine will be in the right standing to harness AI for social and environmental benefits that align with the United Nations Sustainable Development Goals.

REFERENCES

Andrae, A. (2017). Total consumer power consumption forecast. Conference: Nordic Digital Business Summit. Available online: https://www.researchgate.net/publication/320225452_Total_Consumer_Power_Consumption_Forecast [Accessed on 21 May 2024].

Andrae, A., & Edler, T. (2015). On global electricity usage of communication technology: Trends to 2030. *Challenges*, 6(1), 117–157. DOI: 10.3390/challe6010117

Autodesk. (2016). Reimagining the future of air travel. Available online: https://www.autodesk.com/customer-stories/airbus [Accessed on 21 May 2024].

Belkhir, L., & Elmeligi, A. (2018). Assessing ICT global emissions footprint: Trends to 2040 & recommendations. *Journal of Cleaner Production*, 177, 448–463. DOI: 10.1016/j.jclepro.2017.12.239

Chen, Y., & Jin, S. (2023). Artificial intelligence and carbon emissions in manufacturing firms: The moderating role of green innovation. *Processes (Basel, Switzerland)*, 11(9), 2705. DOI: 10.3390/pr11092705

Cheong, S., Sankaran, K., & Bastani, H. (2022). Artificial intelligence for climate change adaptation. *Wiley Interdisciplinary Reviews. Data Mining and Knowledge Discovery*, 12(5), e1459. Advance online publication. DOI: 10.1002/widm.1459

Coeckelbergh, M. (2020a). *AI ethics*. Cambridge, MA: MIT Press Essential Knowledge series, 167-183.

Coeckelbergh, M. (2020b). AI for climate: Freedom, justice, and other ethical and political challenges. *AI and Ethics*, 1(1), 67–72. DOI: 10.1007/s43681-020-00007-2

COMEST. (2010). The ethical implications of global climate change. Available online: http://www.gci.org.uk/Documents/UNESCO_COMEST_.pdf [Accessed on 21 May 2024].

Cook, G., Lee, J., Tsai, T., Kong, A., Deans, J., Johnson, B., & Jardim, E. (2017). Clicking clean: Who is winning the race to build a green internet. Greenpeace Report. Available online: https://www.actu-environnement.com/media/pdf/news-28245-clicking-clean-2017.pdf [Accessed on 21 May 2024].

Cornell University. (2019). AI helps reduce Amazon hydropower dams' carbon footprint. *Science Daily,* September 19, 2019. Available online: www.sciencedaily.com/releases/2019/09/190919134703.htm [Accessed on 21 May 2024].

Coulombel, N., Boutueil, V., Liu, L., Viguié, V., & Yin, B. (2019). Substantial rebound effects in urban ridesharing: Simulating travel decisions in Paris, France. *Transportation Research Part D, Transport and Environment*, 71, 110–126. DOI: 10.1016/j.trd.2018.12.006

Crawford, K., & Joler, V. (2018). Anatomy of an AI system: The Amazon Echo as an anatomical map of human labor, data and planetary resources. Available online: https://anatomyof.ai [Accessed on 21 May 2024].

Crutzen, P. J. (2006). The "Anthropocene". In Ehlers, E., & Krafft, T. (Eds.), *Earth system science in the Anthropocene*. Springer., DOI: 10.1007/3-540-26590-2_3

December 2018 COM (2018) 795 final. Available online: https://eur-lex.europa.eu/legal-content/EN/TXT/?uri=CELEX%3A52018DC0795 [Accessed on 22 May 2024].

Dignum, V. (2019). Responsible artificial intelligence: How to develop and use AI in a responsible way. Artificial Intelligence: Foundations, Theory, and Algorithms. Cham.: Springer. https://doi.org/DOI: 10.1007/978-3-030-30371-6

Dobbe, R., & Whittake, M. (2019). AI and climate change: How they're connected, and what we can do about it. AI Now Institute, October 17, 2019. Available online: https://medium.com/@AINowInstitute/ai-and-climate-change-how-theyre-connected-and-what-we-can-do-about-it-6aa8d0f5b32c [Accessed on 21 May 2024].

Duff, A. S. (2015). *Information society. International Encyclopedia of the Social & Behavioral Sciences* (2nd ed.). Elsevier., DOI: 10.1016/B978-0-08-097086-8.95017-7

Efremova, N., Foley, C., Unagaev, A., Karimi, R., Averkin, A., & Gutiérrez-Romero, R. (2022). Monitoring large-scale regenerative grazing using artificial intelligence. https://doi.org/DOI: 10.21203/rs.3.rs-2136218/v1

European Commission. (2018a). EU declaration on cooperation on artificial intelligence. Available online: https://ec.europa.eu/jrc/communities/en/node/1286/document/eu-declaration-cooperation-artificial-intelligence [Accessed on 22 May 2024].

European Commission. (2018b). Communication from the commission to the European Parliament, the European Council, the Council, the European Economic and Social Committee and the Committee of the Regions. Artificial intelligence for Europe. Brussels, 25 April 2018. Available online: https://eur-lex.europa.eu/legal-content/EN/TXT/?uri=COM%3A2018%3A237%3AFIN [Accessed on 22 May 2024].

European Commission. (2018c). Communication from the commission to the European Parliament, the European Council, the Council, the European Economic and Social Committee and the Committee of the Regions. Coordinated plan on artificial intelligence, Brussels, 7

European Commission. (2019). A definition of artificial intelligence: Main capabilities and scientific disciplines. Available online: https://ec.europa.eu/digital-single-market/en/news/definition-artificial-intelligence-main-capabilities-and-scientific-disciplines [Accessed on 21 May 2024].

European Commission. (2019a). Ethics guidelines for trustworthy AI. Available online: https://digital-strategy.ec.europa.eu/en/library/ethics-guidelines-trustworthy-ai [Accessed 22 May 2024].

European Commission. (2020a). Robotics and artificial intelligence (Unit A.1). Available online: https://ec.europa.eu/digital-single-market/en/content/robotics-and-artificial-intelligence-innovation-and-excellence-unit-a1 [Accessed 22 May 2024].

European Commission. (2020b). White paper on artificial intelligence: A European approach to excellence and trust of 19 February 2020. Available online: https://ec.europa.eu/info/publications/white-paper-artificial-intelligence-european-approach-excellence-and-trust_en [Accessed on 21 May 2024].

European Commission. (2020c). Commission report on safety and liability implications of AI, the Internet of Things and robotics of 19 February 2020. Available online: https://ec.europa.eu/info/publications/commission-report-safety-and-liability-implications-ai-internet-things-and-robotics-0_en [Accessed on 22 May 2024].

European Parliament. (2017). Resolution of 16 February 2017 with recommendations to the commission on civil law rules on robotics. Available online: https://www.europarl.europa.eu/doceo/document/TA-8-2017-0051_EN.html [Accessed on 21 May 2024].

European Parliament. (2020a). Resolution of 20 October 2020 on intellectual property rights for the development of artificial intelligence technologies. Available online: https://www.europarl.europa.eu/doceo/document/TA-9-2020-0277_EN.html [Accessed on 22 May 2024].

European Parliament. (2020b). Resolution of 20 October 2020 with recommendations to the commission on a framework of ethical aspects of artificial intelligence, robotics and related technologies. Available online: https://www.europarl.europa.eu/doceo/document/TA-9-2020-0275_EN.html [Accessed on 22 May 2024].

European Parliament. (2020c). Resolution of 20 October 2020 with recommendations to the commission on a civil liability regime for artificial intelligence. Available online: https://www.europarl.europa.eu/doceo/document/TA-9-2020-0276_EN.html [Accessed on 22 May 2024].

Fedorov, M. (2020). By developing the sphere of artificial intelligence, we ensure Ukraine's competitiveness on the international market. December 2, 2020. Available online: https://www.kmu.gov.ua/news/mihajlo-fedorov-rozvivayuchi-sferu-shtuchnogo-intelektu-mi-zabezpechuyemo-konkurentospromozhnist-ukrayini-na-mizhnarodnomu-rinku [Accessed on 21 May 2024].

Feindor-Schmidt, U. (2020). Regulation of artificial intelligence in Europe - What's in the pipeline? Lexology, December 1, 2020. Available online: https://www.lexology.com/library/detail.aspx?g=d9f74ab9-139c-49e1-9d82-70de718af80f [Accessed on 22 May 2024].

Filippova, A. (2021). Current security issues in the information society. *SHS Web of Conferences, 109,* n. 01014. https://doi.org/DOI: 10.1051/shsconf/202110901014

Floridi, L., Cowls, J., Beltrametti, M., Chatila, R., Chazerand, P., Dignum, V., Luetge, C., Madelin, R., Pagallo, U., Rossi, F., Schafer, B., Valcke, P., & Vayena, E. (2018). AI4People—An ethical framework for a good AI society: Opportunities, risks, principles, and recommendations. *Minds and Machines*, 28(4), 689–707. DOI: 10.1007/s11023-018-9482-5 PMID: 30930541

Gardner, A. V. D. L. (1984). An artificial intelligence approach to legal reasoning. *Thesis (Ph.D.),* Stanford University. The MIT Press, p. 239.

Gates, B. (2020). COVID-19 is awful. Climate change could be worse. The blog of Bill Gates, August 04, 2020. Available online: https://www.gatesnotes.com/Energy/Climate-and-COVID-19 [Accessed on 22 May 2024].

Gent, E. (2020). This 'once-for-all' neural network could slash AI's carbon footprint. SingularityHub, May 4, 2020. Available online: https://singularityhub.com/2020/05/04/this-once-for-all-neural-network-could-slash-ais-carbon-footprint/ [Accessed on 21 May 2024].

Giles, M. (2019). Is AI the next big climate-change threat? We haven't a clue. MIT Technology Review, July 29, 2019. Available online: https://www.technologyreview.com/2019/07/29/663/ai-computing-cloud-computing-microchips/ [Accessed on 21 May 2024].

Guzman, A. (2021). Race after technology: Abolitionist tools for the new Jim code. *Information Communication and Society*, 24(13), 1989–1990. DOI: 10.1080/1369118X.2020.1844269

Hao, K. (2019). Here are 10 ways AI could help fight climate change. MIT Technology Review, June 20, 2019. Available online: https://www.technologyreview.com/2019/06/20/134864/ai-climate-change-machine-learning/ [Accessed 21 May 2024].

Hazas, M., Morley, J., Bates, O., & Friday, A. (2016). Are there limits to growth in data traffic?: On time use, data generation and speed. *Proceedings of the Second Workshop on Computing within Limits, 14,* 1–5. https://doi.org/DOI: 10.1145/2926676.2926690

Hern, A. (2018). Bitcoin's energy usage is huge – We can't afford to ignore it. *The Guardian,* January 17, 2018. Available online: https://perma.cc/2X2H-CF9V [Accessed on 21 May 2024].

Hughes, B. a GE company and C3.ai. (2019). Announce joint venture, June 24, 2019. Available online: https://c3.ai/baker-hughes-and-c3-ai-announce-joint-venture-to-deliver-ai-solutions [Accessed on 21 May 2024].

IPCC (Intergovernmental Panel on Climate Change). (2018). An IPCC special report on the impacts of global warming of 1.5°C. Available online: https://www.ipcc.ch/sr15/ [Accessed on 21 May 2024].

Ise, T., & Oba, Y. (2019). Forecasting climatic trends using neural networks: An experimental study using global historical data. *Frontiers in Robotics and AI*, 6, 32. DOI: 10.3389/frobt.2019.00032 PMID: 33501048

Kaack, L., Donti, P., Strubell, E., Kamiya, G., Creutzig, F., & Rolnick, D. (2022). Aligning artificial intelligence with climate change mitigation. *Nature Climate Change*, 12(6), 518–527. DOI: 10.1038/s41558-022-01377-7

Kates-Harbeck, J., Svyatkovskiy, A., & Tang, W. (2019). Predicting disruptive instabilities in controlled fusion plasmas through deep learning. *Nature*, 568(7753), 526–531. DOI: 10.1038/s41586-019-1116-4 PMID: 30996348

Kirchgaessner, S. (2019). Revealed: Google made large contributions to climate change deniers. *The Guardian,* October 11, 2019. Available online: https://amp.theguardian.com/environment/2019/oct/11/google-contributions-climate-change-deniers?twitter_impression=true [Accessed on 21 May 2024].

Madiega, T. (2019). EU guidelines on ethics in artificial intelligence: Context and implementation. *European Parliamentary Research Service,* 1-13. Available online: https://www.europarl.europa.eu/RegData/etudes/BRIE/2019/640163/EPRS_BRI(2019)640163_EN.pdf [Accessed on 22 May 2024].

Martins, N. R. B., Angelica, A., Chakravarthy, K., Svidinenko, Y., Boehm, F. J., Opris, I., Lebedev, M. A., Swan, M., Garan, S. A., Rosenfeld, J. V., Hogg, T., & Freitas, R. A.Jr. (2019). Human brain/cloud interface. *Frontiers in Neuroscience*, 13, 112. DOI: 10.3389/fnins.2019.00112 PMID: 30983948

Martsenko, N. (2019). Legal regime of artificial intelligence in civil law. *Aktual ni problemy pravoznavstva, 4,* 91-98. Available online: http://dspace.wunu.edu.ua/handle/316497/38382 [Accessed on 22 May 2024].

Matheson, R. (2020). Reducing the carbon footprint of artificial intelligence. MIT News, April 23, 2020. Available online: https://news.mit.edu/2020/artificial-intelligence-ai-carbon-footprint-0423 [Accessed on 21 May 2024].

McGlade, C., & Ekins, P. (2015). The geographical distribution of fossil fuels unused when limiting global warming to 2°C. *Nature*, 517(7533), 187–190. DOI: 10.1038/nature14016 PMID: 25567285

McGovern, A., Elmore, K., Gagne, D., Haupt, S., Karstens, C., Lagerquist, R., Smith, T., & Williams, J. (2017). Using artificial intelligence to improve real-time decision-making for high-impact weather. *Bulletin of the American Meteorological Society*, 98(10), 2073–2090. DOI: 10.1175/BAMS-D-16-0123.1

McQuade, S., & Monteleoni, C. (2012). Global climate model tracking using geospatial neighborhoods. *Proceedings of the AAAI Conference on Artificial Intelligence*, 26(1), 335–341. DOI: 10.1609/aaai.v26i1.8178

Microsoft News Center. (2018). Microsoft demonstrates the power of AI and Cloud to Oil and Gas players, at ADIPEC 2018. November 12, 2018. Available online: https://news.microsoft.com/en-xm/2018/11/12/microsoft-demonstrates-the-power-of-ai-and-cloud-to-oil-and-gas-players-at-adipec-2018/ [Accessed on 21 May 2024].

Milojevic-Dupont, N., & Creutzig, F. (2021). Machine learning for geographically differentiated climate change mitigation in urban areas. *Sustainable Cities and Society*, 64, 102526. DOI: 10.1016/j.scs.2020.102526

Mitchell, A., Dokei, T., Hickman, T., & Albagli, D. (2020). Regulation of artificial intelligence in Europe and Japan. White & Case LLP, August 24, 2020. Available online: https://www.whitecase.com/publications/insight/regulation-artificial-intelligence-europe-and-japan [Accessed on 22 May 2024].

Monteleoni, C., Schmidt, G., Saroha, S., & Asplund, E. (2011). Tracking climate models. *Statistical Analysis and Data Mining*, 4(4), 372–392. DOI: 10.1002/sam.10126

Mulvaney, K. (2019). Climate change report card: These countries are reaching targets. *National Geographic,* September 19, 2019. Available online: https://www.nationalgeographic.com/environment/article/climate-change-report-card-co2-emissions [Accessed on 21 May 2024].

Muraleedharan, S. (2021). Role of artificial intelligence in environmental sustainability. *EcoMENA,* January 30, 2021. Available online: https://www.ecomena.org/artificial-intelligence-environmental-sustainability/ [Accessed on 21 May 2024].

OECD. (2019). Recommendation of the council on artificial intelligence. OECD Legal Instruments. Available online: https://legalinstruments.oecd.org/api/print?ids=648&lang=en [Accessed on 22 May 2024].

Parliament of Ukraine. (2018). On approval of the concept of development of the digital economy and society of Ukraine for 2018-2020 and approval of the action plan for its implementation. Available online: https://www.kmu.gov.ua/npas/pro-shvalennya-koncepciyi-rozvitku-cifrovoyi-ekonomiki-ta-suspilstva-ukrayini-na-20182020-roki-ta-zatverdzhennya-planu-zahodiv-shodo-yiyi-realizaciyi [Accessed 22 May 2024].

Parliament of Ukraine. (2020). The concept of artificial intelligence development in Ukraine. Available online: https://zakon.rada.gov.ua/laws/show/1556-2020-%D1%80#Text [Accessed on 22 May 2024].

Pearce, F. (2018). Energy hogs: Can world's huge data centers be made more efficient? *Yale Environment 360,* April 3, 2018. Available online: https://perma.cc/J2H3-EL75 [Accessed on 21 May 2024].

Popkova, E., Sergi, B., & Bogoviz, A. (2023). Editorial: Evolution of environmental economics and management in the age of artificial intelligence for sustainable development. *Frontiers in Environmental Science*, 11, 1176612. Advance online publication. DOI: 10.3389/fenvs.2023.1176612

Pozova, D. (2017). Prospects of legal regulation of artificial intelligence under EU legislation. *Chasopys tsyvilistyky,* 27, 116-120.

Raban, D., Gordon, A., & Geifman, D. (2011). The information society. *Information Communication and Society*, 14(3), 375–399. DOI: 10.1080/1369118X.2010.542824

Radutnyi, A. (2018). Subjectivity of artificial intelligence in criminal law. *Pravo Ukrayiny*, 1, 123–136.

Rasp, S., Pritchard, M., & Gentine, P. (2018). Deep learning to represent subgrid processes in climate models. *Proceedings of the National Academy of Sciences of the United States of America*, 115(39), 9684–9689. DOI: 10.1073/pnas.1810286115 PMID: 30190437

Reichstein, M., Camps-Valls, G., Stevens, B., Jung, M., Denzler, J., Carvalhais, N., & Prabhat, . (2019). Deep learning and process understanding for data-driven earth system science. *Nature*, 566(7743), 195–204. DOI: 10.1038/s41586-019-0912-1 PMID: 30760912

Rissland, E., Ashley, K., & Loui, R. (2003). AI and law: A fruitful synergy. *Artificial Intelligence*, 150(1-2), 1–15. DOI: 10.1016/S0004-3702(03)00122-X

Ritchie, H. (2019). Number of people in the world without electricity falls below one billion. *Our World in Data,* January 18, 2019. Available online: https://ourworldindata.org/number-of-people-in-the-world-without-electricity-access-falls-below-one-billion [Accessed on 21 May 2024].

Roach, J. (2020). Microsoft finds underwater datacenters are reliable, practical and use energy sustainably. September 14, 2020. Available online: https://news.microsoft.com/innovation-stories/project-natick-underwater-datacenter/ [Accessed on 21 May 2024].

Rolnick, D., Donti, L., Kaack, L., Kochanski, K., Lacoste, A., Sankaran, K., Slavin Ross, A., Milojevic-Dupont, N., Jaques, N., Waldman-Brown, A., Luccioni, A., Maharaj, T., Sherwin, D. E., Mukkavilli, S. K., Kording, K. P., Gomes, C., Ng, A. Y., Hassabis, D., Platt, J. C., . . . Bengio, Y. (2019). Tackling climate change with machine learning. Available online: https://arxiv.org/abs/1906.05433 [Accessed on 21 May 2024].

Selby, J., Cox, E., & Royston, S. (2016). Impact of non-energy policies on energy systems. UK Energy Research Centre, London, November 2016. Available online: https://ukerc.ac.uk/publications/impact-of-non-energy-policies-on-energy-systems/ [Accessed on 21 May 2024].

Snow, J. (2019). How artificial intelligence can tackle climate change. *National Geographic,* July 18, 2019. Available online: https://www.nationalgeographic.com/environment/2019/07/artificial-intelligence-climate-change/ [Accessed on 21 May 2024].

St. John, J. (2018). Texas takes a big step in improving access to smart meter data. *Greentechmedia,* February 6, 2018. Available online: https://perma.cc/G4ZJ-L4LT [Accessed on 21 May 2024].

Stein, A. L. (2020). Artificial intelligence and climate. *Yale Journal on Regulation*, 37(3), 890–939.

Strubell, E., Ganesh, A., & McCallum, A. (2019). Energy and policy considerations for deep learning in NLP. Available online: https://arxiv.org/abs/1906.02243 [Accessed on 21 May 2024]. DOI: 10.18653/v1/P19-1355

Thaler, R., & Sunstein, C. (2008). *Nudge: Improving decisions about health, wealth and happiness*. Yale University Press.

UNESCO. (2019a). An integrated system for global real-time precipitation observation using PDIR. Available online: https://en.unesco.org/news/irain-newmobile-app-promote-citizen-science-and-support-water-management [Accessed on 21 May 2024].

UNESCO. (2019b). Artificial intelligence for sustainable development: Challenges and opportunities for UNESCO's science and engineering programmes. UNESCO Digital Library. Available online: https://unesdoc.unesco.org/ark:/48223/pf0000368028.locale=en [Accessed on 21 May 2024].

United Nations. (1992). Framework convention on climate change. Available online: https://www.un.org/ru/documents/decl_conv/conventions/climate_framework_conv.shtml [Accessed on 21 May 2024].

United Nations. (1997). Kyoto protocol to the United Nations framework convention on climate change. Available online: https://www.un.org/ru/documents/decl_conv/conventions/kyoto.shtml [Accessed on 21 May 2024].

United Nations. (2015). The Paris agreement. Available online: https://www.un.org/ru/climatechange/paris-agreement [Accessed on 21 May 2024].

United States Environmental Protection Agency. (2020). Sources of greenhouse gas emissions. Available online: https://www.epa.gov/ghgemissions/sources-greenhouse-gas-emissions [Accessed 21 May 2024].

Vinuesa, R., Azizpour, H., Leite, I., Balaam, M., Dignum, V., Domisch, S., & Nerini, F. (2020). The role of artificial intelligence in achieving the sustainable development goals. *Nature Communications*, 11(1), 233. Advance online publication. DOI: 10.1038/s41467-019-14108-y PMID: 31932590

Vos, J. (2009). Actions speak louder than words: Greenwashing in corporate America. *Notre Dame Journal of Law, Ethics & Public Policy*, 23(2), 673–697.

Wang, X., Ji, K., & Xie, T. (2023). AI carbon footprint management with multi-agent participation: A tripartite evolutionary game analysis based on a case in China. *Sustainability (Basel)*, 15(11), 9013. DOI: 10.3390/su15119013

Wang, Z., & Zhu, X. (2024). Enhancing nanocrystal synthesis: A comparative study of online artificial intelligence optimization and offline high-throughput experimentation in chemical material discovery. *ACS Applied Nano Materials*, 7(6), 6499–6505. DOI: 10.1021/acsanm.4c00255

WMO (World Meteorological Organization). (2019). The global climate in 2015–2019. Available online: https://library.wmo.int/index.php?lvl=notice_display&id=21522#.YNb_cmgzbIU [Accessed on 22 May 2024].

Wolff-Anthony, L., Kanding, B., & Selvan, R. (2020). Carbon Tracker: Tracking and predicting the carbon footprint of training deep learning models. Available online: https://arxiv.org/abs/2007.03051 [Accessed on 21 May 2024].

Zhang, P., Xiao, Y., Li, Y., Ge, X., Shi, G., & Yang, Y. (2024). Toward net-zero carbon emissions in network AI for 6G and beyond. *IEEE Communications Magazine*, 62(4), 58–64. DOI: 10.1109/MCOM.003.2300175

ADDITIONAL READING

Fusi, N., Lipton, Z. C., & Swan, D. (2021). *Artificial intelligence for environmental monitoring: Applications and challenges*. Springer. https://link.springer.com/article/10.1007/s00542-021-06487-8

Gailhofer, P., Tarin, C., & Vinuesa, R. (2021). *Artificial intelligence and climate change: Ethical issues*. Emerald Insight. https://www.emerald.com/insight/content/doi/10.1108/NEJE-05-2021-0032/full/html

Haussler, S., Schimel, D. S., & Turner, W. (2023). *Deploying artificial intelligence for climate change adaptation*. Science Direct. https://www.sciencedirect.com/science/article/pii/S095965262300788X

Rayhan, A., & Rayhan, S. (2023). *The role of artificial intelligence in climate change mitigation and adaptation*. ResearchGate. https://www.researchgate.net/publication/361987654_The_role_of_artificial_intelligence_in_climate_change_mitigation_and_adaptation

Sandalow, D. B., McCormick, C., Kucukelbir, A., Friedman, J., Nagrani, T., Fan, Z., Halff, A. M., d'Aspremont, A., Glatt, R., Méndez Leal, E., Karl, K., & Ruane, A. (2024). *Artificial Intelligence for Climate Change Mitigation Roadmap*. Columbia University. https://academiccommons.columbia.edu/doi/10.7916/cfn0-xj39

Vinuesa, R., Azizpour, H., Leite, I., Balaam, M., Dignum, V., Domisch, S., Fella, P., Langhans, S. D., Tegmark, M., & Nerini, F. F. (2020). The role of artificial intelligence in achieving the Sustainable Development Goals. *Nature Communications*, 11(1), 233. https://www.nature.com/articles/s41467-019-14108-y. DOI: 10.1038/s41467-019-14108-y PMID: 31932590

KEY TERMS AND DEFINITIONS

Artificial Intelligence: The attempt to emulate human intelligence in machines to make the machines solve problems that would otherwise call for the intelligence of man.

Climate Change Mitigation: Measures and plans that are designed towards minimizing or avoiding emissions of greenhouse gases to check the level of global warming and climate change.

Climate Change Adaptation: Maladaptive responses: changes made in natural or human systems about climate changes that have occurred or that have been anticipated to minimize the negative impacts of change, or to capitalize on its positive effects. Environmental Monitoring: The regular surfacing and recording of conditions of the environment that may include information concerning air, water, land quality, and species to identify changes for policy making.

Legal Frameworks: Systemized laws and rules that prescribe how people in organizations and societies should conduct themselves and offer directions where there are gaps in the kind of performance that is expected.

Sustainable Agriculture: Agriculture that sustains present consumption levels of food, at the same time enhancing the biological capacity of the Earth to feed the present and future population in economically, ecologically, and socially responsible manners.

AI Ethics: The convergence of Machine Learning, that is, the building of algorithms and models to recognize patterns from the data provided and the use of Ethical Principles in developing and implementing artificial intelligence to support actions that are often fair, transparent as well as effective to the people.

International Policy: International Relations refer to steering principles that are established and accepted by several countries to deal with world concerns and foster intercontinental nexus including treaties and trade.

Greenhouse Gas Emissions: Pollution by the emission of gases into the atmosphere like carbon dioxide and methane, which enhances the greenhouse effect and hence global warming and climate change.

Technological Innovation: The action that results in the creation of innovations that can be viewed as tools in the process of problem-solving, work optimization, and enhancement of the performance of various industries.

Chapter 7
Integrating Policy, Governance, and Law in Regulating AI for Sustainable Environmental Protection

Shashwata Sahu
https://orcid.org/0009-0003-6843-9554
KIIT University, Bhubaneswar, India

Navonita Mallick
https://orcid.org/0000-0003-3621-3764
KIIT University, Bhubaneswar, India

Sanghamitra Patnaik
https://orcid.org/0000-0002-4057-0572
KIIT University, Bhubaneswar, India

ABSTRACT

Artificial Intelligence (AI) transforms environmental conservation by enhancing sustainability and efficiency in addressing critical challenges. However, AI must be regulated within a robust policy, governance, and law framework to harness its full potential. This paper explores the complex interaction of these elements in regulating AI for sustainable environmental protection. Through a doctrinal methodology, it examines legal texts, policies, and governance structures to assess their adequacy in guiding AI applications in ecological contexts. The findings reveal significant gaps in current frameworks, underscoring the need for integrated, enforceable

DOI: 10.4018/979-8-3693-8684-2.ch007

guidelines that ensure ethical AI deployment. The research concludes by advocating for comprehensive regulatory tools to align AI with sustainability goals.

I. INTRODUCTION

> *"It has become appallingly obvious that our technology has exceeded our humanity."*
> *- Albert Einstein*

The development of Artificial Intelligence has opened up a new era of technological advancements. It represents a powerful tool applicable for various purposes, including environmental protection. AI's capabilities in large data set processing, predictive analytics, and decision-making render it an optimal solution for tackling environmental issues. (Xu et al., 2021) Indeed, various crucial environmental challenges, from climate change and biodiversity destruction to pollution and natural resource mismanagement, can be assisted by AI. (Chen et al., 2023). Therefore, driven by the need to battle these problems, AI has become critical for ensuring global sustainability.

Environmental issues have arisen as a result of anthropogenic processes taking place across the world, presenting a major crisis for ecosystems, human health, and global economic stability. Climate change, driven by the implementation of greenhouse gases, is at the forefront of negative environmental change. The anthropogenic process of climate change is already producing such negative effects as increased frequency and harshness of extreme weather phenomena, higher sea waters, and alterations in ecosystems (Shivanna, 2022). Furthermore, biodiversity is at its highest threat in the history of humankind, with the rate of extinction increasing due to pollution and climate change. Pollution, whether it be the air, water, or soil, remains among the major environmental challenges negatively affecting both nature and people. Additionally, humans are exploiting natural resources at an unprecedented and hopelessly unsustainable rate to the point of reserves eventually drying up, resulting in global disasters. AI tools show promise regarding these environmental challenges as well. First and foremost, it can be used to prevent greenhouse emissions by optimizing energy use in grids in a way that both decreases CO2 emissions and increases the efficiency of renewable sources. Secondly, it can assist with biodiversity preservation in terms of population and illegal activity monitoring and environmental change predictions. Finally, it is vital for pollution prediction and prevention as AI processes data from typically employed sensors and satellites and is capable of predicting pollution spreads.(Lima et al., 2020) However, the deployment of AI in environmental protection is not free from challenges. While simultaneously introducing new innovative solutions, AI generates new risks and introduces ethical concerns.

AI in decision-making regarding environmental management raises severe questions related to transparency, responsibility, and potential biases.(Osasona et al., 2024) Moreover, AI should be thoroughly evaluated for how it affects the environment, as data centers where AI systems are stored and applied are responsible for consuming huge amounts of energy, and other AI components yield a lot of carbon.

Given a combination of high potentials and significant hazards, a complex solution is required, which merges policy, governance, and law applicable to managing AI for environmental issues. The existing policy and law system regulating AI in environmental protection remains fragmented and incomplete. Policies and governance measures targeting AI are separated from the ones deployed for the environment, which results in a limited application of some measures. (Guidance, 2021) Some regulations lack upper oversight, and frequently environmental issues are neglected in applying measures of protection of AI applications. Aural, a complex solution is required to make sure AI is deployed efficiently, effectively, legally, ethically, and sustainably. The policy framework should highly encourage the development and application of AI technologies specifically to promote environmental causes and limit the application of those that might yield a negative impact.(Taghikhah et al., 2022) Governance measures should provide comprehensive control mechanisms for the transparent application of AI, reliable accountability measures, and the establishment of public trust. AI systems should be developed and operated in a way that respects the rights and needs of all parties involved, including vulnerable ecosystems and human communities.

II. OBJECTIVES OF RESEARCH

The research work is aimed to explore how are govern the applications of AI in the context of environmental protection policy, governance, and legal frameworks. The task is also to expose deficiencies and offer a recommendation on whether any of a spectrum or plurality or regulatory ways or others may prescribe a road or channel towards sustainability. The primary intention of the research is to conduct an inquiry into how the two fields interconnect to give insight and some food for thought on the ways to use AI in an efficacious effective, safe, and ethical way in the environmental strategies targeted to move us towards some objectives of sustainability.

III. RESEARCH METHODOLOGY

The research uses a doctrinal methodology, based on an analysis of the legal texts, policy, and governance of these legal texts. This legal interpretation will provide a proper basis for a later discussion about the nature of our law and its potential to regulate the deployment of AI for environmental conservation. Furthermore, some secondary sources will be used in the paper, such as academic literature, case studies, or policy analysis. In addition, some primary sources will be employed: statutes, regulations, or international agreements. With the methodology described earlier, a solid picture of the legislation and other regulations will arise regarding how much response it can give, what level of effectiveness it can have regarding AI and the challenges it brings about regarding environmental governance.

IV. AI APPLICATIONS IN ENVIRONMENTAL PROTECTION

The integration of AI into environmental and ecological protection has brought along several salient advantages, which allow the environmental sector to address the issues that have been highly complex and intricate. The goal of the application of AI on environmental protection is manifold, ranging from the improvement of efficiency to the more accurate prediction of climate change as part of models that can be developed using AI algorithms that, in turn, use machine learning to process big data. AI, as a subcategory of computer science, uses computer systems to learn and adapt to human nature. (Raman et al., 2024) Machine learning, as a subcategory of AI, is the technology that enables computer systems to learn and adapt based on new knowledge, such as the collection of large amounts of data, algorithms, and more. AIs have become an integral part of climate models, previously based on statistical methods and physical laws. (Sarker, 2021) They now target violent weather as part of the hostility-related catastrophes, which have become increasingly more salient and inadvertently frequent due to the characteristics of climate change. According to the "State of AI in the Enterprise" report by Deloitte, 58% of organizations use AI to fulfil their environmental, social, and governance initiatives within the spectrum of their mission, environmental efforts, and sustainability efforts. (Deloitte, n.d.) According to the "AI for the Planet" report by the "Boston Consulting Group," one of the results that have been achieved is that AI could reduce 5% to 10% of global greenhouse gases by 2030 (Feng & Liu, 2023).

AI's role in "Pollution Tracking" is another activity in which AI has been efficient. Pollution of air, water, and soil is done through human activities, such as burning tyres and releasing waste gases into the air, near or into a river. These are a threat not only to life but also to the fundamental basis for the existence of life.

AI applications help to monitor and predict the level of pollution. By tracking in real-time, communities can respond and take appropriate actions. According to the same source, sensors supported by AI can predict and monitor air pollution. Other sources state that AI algorithms identify shipping, mining, or industrial waste disposed of in river bodies, as well as other mining areas. AI technology can predict 25% of pollution in towns and cities (McKinsey & Company, 2023).

Furthermore, biodiversity conservation has significantly benefited from AI applications. With the increasing issues of habitat destruction, climate change, and human activity among others, global biodiversity is on the decline. As such, conservation approaches need to be innovative, and AI has played a crucial role in monitoring and detection. AI technologies commonly monitor wildlife populations, movement patterns, or migration, as well as the identification and analysis of illegal activities such as poaching and deforestation (Silvestro et al., 2022). AI technologies include drones and camera traps equipped with AI; the contribution of AI to delving is critical due to increased automation and precision. For example, "The International Union for Conservation of Nature (IUCN)" quantified that deployments of DBA-powered monitoring systems had improved species detectability by 40% in critical habitats and improved species detectability by 40%, which significantly supported biodiversity conservation. (*IUCN*, n.d.)

While progress has been made in this regard, the application of AI to environmental protection is not easy. This is largely because AI models have high sensitivity towards data, any poor quality or less occupied environmental dataset can lead an AI system astray. That said, using AI can also be power-hungry in and of itself deploying machine-learning services may require a lot of computational grunt work, which will have an environmental impact (Subramaniam et al., 2022). The data protection and privacy aspects, the dangers of algorithmic discrimination, or even simply fair access to AI benefits also need to be addressed. This shift is urgently needed: a UNEP report found that just 30% of countries surveyed have specific environmental policies related to AI, suggesting the existence of a considerable gap in the global governance landscape.(Environment, 2022)

The use of AI in environmental protection has already resulted in several successful implementations, and therefore, the effectiveness of AI in the area of environmental protection is apparent. Having taken into account some particular case examples, we will be able to better understand the nature of implications and effects that AI creates, and how it is possible to measure these social and environmental effects. One of the most prominent examples is the use of machine learning algorithms for the conservation of the environment.(Konya & Nematzadeh, 2024) In the Amboseli National Park in Kenya, AI-based drones are used for the monitoring of wildlife, elephants in particular. Lowland elephants in this park are considered to be the most threatened species because they are affected by poachers very actively. According to Suen and colleagues, drones help to protect elephants better, as in many other parks, they suffer from poaching, while in Amboseli, approximately 400 elephants live happily (Brickson et al., 2023). AI algorithms detect patterns and anomalies by analysing real-time data, and park rangers investigate these anomalies immediately if they suspect the activity of poaching. Therefore, AI effectively helps to save valuable wildlife, and it is one of the examples of how AI works in different contexts of social activities including nature conservation.

The use of AI for water and air pollution control is another prominent example of AI's positive effects on nature. When it comes to pollution control, the Chinese Ministry of Ecology and the Environment uses AI-based pollution monitoring. This AI system monitors air pollution in 338 of its cities, and it applies AI algorithms that analyse the data received from over 1000 monitoring stations. According to Molotch and Yang, moreover, these AI algorithms allow for predicting pollution levels and locating the greatest pollution sources (Neo et al., 2023). At the same time, the analysis of the data helps to understand the situation to which park rangers should respond quickly. AI algorithms in this system do not just automatically detect pollution, but they determine the relation of pollution to different sources which can be eliminated from and cleaned up with sustainable technology. The introduction of this information system led to cleaner air in approximately various Chinese cities (Hossain et al., 2023). In the Netherlands, meanwhile, the use of AI technology for the control of water and air pollution was also connected with smart grids. In particular, Amsterdam uses state-of-the-art grids with the help of which people can manage the consumption difference (Berry, 2021). In doing so, the smart grid can level the load and decrease loss because of the unpredictability of the difference between actual electricity consumption and its planning. In addition, the City of Amsterdam is becoming much more efficient with Fossil Free Supply, using AI, solar, and wind for more optimal distribution. At the same time, the use of AI for water and air pollution control also highlighted the challenges that were created by the use of AI. Therefore, it is very important to monitor the system we get carefully. Some major policy implications and considerations are required for further applications of AI in the prospective contexts.

Table 1. Overview of key opportunities of AI methods for sustainability

Opportunity	
Mitigate Climate Change Consequences	**AI can play a significant role in addressing climate change e.g. by optimizing energy consumption and supporting climate adaptation strategies.**
Governance	AI's versatility allows the adoption of insights from other fields and its integration into other domains to develop sustainable solutions across various disciplines.
Scientific Value	The responsible development and deployment of AI technologies can be assured by governance structures and policies that ensure ethical, transparent use in line with sustainability objectives.

V. POLICY ENVIRONMENT

The role of AI in environmental protection has spurred a range of policies in many nations that seek to either facilitate or regulate the application of AI tools in this area. However, these policies differ greatly between countries, reflecting the nature of the national priorities, regulatory systems, and levels of technology development present here.(Guidance, 2021)

In the US and China, two leading technological countries, AI policies have incorporated the role of AI in environmental protection. The US now has a general AI strategy developed by the "American AI Initiative". Environmental protection is mentioned as part of the technological advances that AI can provide; however, the emphasis made in the initiative is related to the questions of innovation and competitiveness here (Joshua P. Meltzer & Cameron F. Kerry, 2021). The application of AI tools in this area is only mentioned in terms of the Environmental Protection Agency, which is hardly an AI strategy to speak of, and means it is a rather fragmentary approach to the role of AI in the general environmental protection policy of the US. Such fragmentary strategies do not ensure that all AI risks are considered and minimized (Francisco & Linnér, 2023).

Meanwhile, China has been actively developing the implementation of AI in the context of environmental policies. The development outlined in the "Next Generation Artificial Intelligence Development Plan" addressed the issue of environmental protection, which is mentioned in the context of smart cities and a range of other applications, relevant to sustainable development. The Chinese government has applied AI in the area of environmental monitoring bordering on the massive, including the application of AI for air quality monitoring and pollution control. (Khunakornbodintr, 2024) However, due to the concentration of power in China and the generally reduced level of transparency and public participation permitted in this measure, there are reasons to believe that this technology rollout is at the risk of disregarding ethical concerns and the consideration of the rights of the population affected.

India increasingly recognizes AI as part of its digital transformation strategy with strong implications in environment conservation. The government's National Strategy on Artificial Intelligence, also known as #AIforAll, discusses the deployment of AI in multiple areas, such as agriculture, water management, and climate action. It seeks to facilitate India's industrial shift towards AI-facilitated resource savings, achieve better monitoring and better enforcement of environmental law reduce overall enforcement costs, employers and staff, and ensure the preservation of democratic freedom. (Aayog, n.d.) Moreover, the government demonstrates clear ambitions to increase the inclusion of AI into its Smart Cities Mission to reduce waste and electricity use, facilitate IoT for waste pickup, and perform analysis of urban development efficiency. Moreover, the Indian government sees the potential in deploying AI to reduce climate change impact, ultimately achieving a more effective governance model. (Bibri et al., 2024) However, the strategies do not specifically target the issues of ethics and public or administrative governance in AI applications for environmental causes. Nevertheless, the current model shows promise, laying the foundation for future policies that will support sustainable development of the environment.

European Union member states have the confidence to deploy AI in environmental condition management. The EU's Green Deal and AI Act clearly illustrate the nation's strong intentions to integrate AI in the management of its environmental preservation. The latter project aims specifically to create legislation that will create a contractual framework that will control AI management of environmental AI applications, especially those considered to be of high risk.(Eckert & Kovalevska, 2021) It is of note that the EU was the first to introduce legislation that costly emphasized governance buildings, transparency and accountability of AI, and use of the technology to enhance rather than replace human rights.(Díaz-Rodríguez et al., 2023) The model also emphasizes the importance of public involvement in environmental policy and that failures, in that matter, may result in violation of the principles of European Democracy and act on government activities.

The deployment of AI in environmental conservation is increasingly determined by a complex landscape of international policies and agreements. These frameworks are central to defining how AI technologies are developed, deployed, and governed on a worldwide scale, particularly regarding the most pressing environmental concerns, such as climate change, biodiversity loss, and pollution. Although several international agreements are influential in the area, two are recognized as the most significant. The first and the most important document that impacts AI deployment within the context is the Paris Agreement, adopted under the United Nations Framework Convention on Climate Change by its 196 participants. The central aim of the agreement is to keep the rise of global temperature this century well below 2 degrees Celsius above pre-industrial levels. Still, naturalists believe it should be limited to 1.5 degrees Celsius instead, and that objective cannot be achieved without the use of innovative technologies. By advancing climate modelling, optimizing energy usage, and monitoring emissions, AI can contribute to accomplishing this objective (Liu et al., 2020). However, the agreement also promotes sustainable development and responsible technology in general, calling for new frameworks that ensure AI application is conducted in ways that do not harm environmental integrity and social equality. The most meaningful in the present circumstances aspects of the technologies are remote applications and VR and AR development since they may facilitate energy expansion, and other valuable resources (Mondejar et al., 2021). Secondly, e-learning can be more widely introduced because it has already been positively practised in education. Another important agreement in the context of the relevant issue is the Convention on Biological Diversity. It is a legally binding international treaty covering all ecosystems, including terrestrial, marine, and other aquatic ecosystems. It is a complementary governmental process to protect ecosystems, species and sustainable use, pick, etc., genetic resources, not to forget the right of access to them and the distribution, security, fair and equitable part of benefits arising from those resources (William J. Snape III, 2010). The CBD, implicitly supports the use of AI in monitoring biodiversity, tracking species numbers, and population controls, and the monitoring of protected areas and local forests, while the CBDA implicitly supports it as the Convention is implicitly difficult if the appropriate products in each case are controlled wisely in close cooperation with indigenous knowledge and local communities.

One of the most applicable international schemes is the "Sustainable Development Goals," adopted by the United Nations in 2015. It is an extensive program aimed at sustainable development by 2030, and many of the goals include protecting the environment. For example, these are Goal 6, Clean Water and Sanitation; Goal 13, Climate Action; Goal 14, Life Below Water; and Goal 15, Life on Land. (*THE 17 GOALS | Sustainable Development*, 2024) AI actively contributes to their implementation thanks to the opportunities for environmental monitoring, resource management, and disaster prevention. At the same time, SDGs focus on the inclusion, transparency, and ethics to be observed in technologies, and these aspects are fundamental to environmental purposes (Regona et al., 2024). Therefore, AI should not be used to exacerbate existing environmental damages. The integration of AI into attempts to achieve these environmental goals has opened new opportunities for nature preservation, resource use, and climate change mitigation. Thus, the function of rapid data processing and analysis improves the monitoring of the environment. In particular, changes in the forests, decreased biodiversity, and pollution can now be observed more easily, thanks to artificial intelligence (Akter et al., 2024). In addition, new AI-powered models help in weather forecasting, and their increased accuracy contributes to more detailed knowledge of how the climate changes in different locations. It provides a better preparation for implementing changes in advance and preventing tragedies, as well as adapting to the changing environment.

Besides, AI contributes to the preservation of resources, optimization of their use, and increase in energy efficiency. As a result, waste generation decreases, and the effects of different industries become less of a problem on a large scale. Thus, in agriculture, precise farming, thanks to AI, increases the crop and reduces the amount of water used and the necessity for fertilizers and pesticides (Javaid et al., 2023). In cities, smart technologies define smart cities to affect water and energy use, transport load, and waste management. This aspect is particularly important in regions with rapid urbanization, where industrial waste and city pollution are severe (Herath & Mittal, 2022). It is in this context that the application of AI to environmental sustainability efforts must be evaluated. While existing applications of AI have already resulted in more efficient and effective methods of environmental conservation, particularly regarding combating deforestation and poaching, far more effective AI-driven solutions can yet be developed. In particular, new AI techniques, enabled by the development and increased availability of big data and computing resources, could be developed to include additional data sources and thus tackle a broad range of global challenges. The use of AI in solving such global challenges is of special relevance today, given the goals, targets, and recommendations concerning technological development established by the United Nations through the framework of their Sustainable Development Goals (Sarker, 2022). Thus, in analysing AI's potential impact on achieving these states, one must include its possible consequences with regard to environmental sustainability.

VI. GOVERNANCE FRAMEWORK

AI governance for environmental sustainability entails several stakeholders working in coalition with distinct but interconnected competencies. Those stakeholders include government, international organizations and the business sector; collectively they have a stake in establishing the national policies, frameworks or ethical standards that guide AI deployment. Governments are the core constructors of regulatory frameworks and public policies that control AI applications. Their mandate is basically to draft laws and regulations on how AI should be used in a way that corresponds with societal goals, including sustainability and eco-protection (Xue & Pang, 2022). To fulfil duties to operate AI responsibly, governments must establish the legal and regulatory framework in which AI operates e.g., data privacy/first-party claims if sensitive information is shared; intellectual property rights reserved for inventive algorithms or processes versus large-scale companies that merely develop tools; as well as liability concerns emanating from ethical considerations. Governments can also finance and provide research assistance, accelerate public-private partnerships or incorporate AI in multi-purpose installations such as intelligent grids and environment surveillance systems (Raab, 2020). This is in order to ensure AI will only be deployed for uses that serve the main interests of society and pose minimal hazards on both human beings and also the setting.

International organizations can help ensure that AI global governance is aligned across the world. Recognizing that many of AI technologies do not respect boundaries, a variety of international fora including the United Nations (UN), World Economic Forum and the Organisation for Economic Co-operation and Development (OECD) have sought settings to promote dialogues on common principles. They play a crucial role in building consensus around AI governance and exchanging knowledge on how to operationalise ethical guidelines that protect human rights, environmental sustainability, and global equity.(Feijóo et al., 2020) International organizations can also play a strong role in this sense as normative bodies, they are able to fill the gaps within different national jurisdictions and provide cohesion for regulatory frameworks on AI at an international level.

Figure 1. Flowchart of the integrated role of AI applied within corporate decision-making in promoting Sustainable Development Goals and CSR

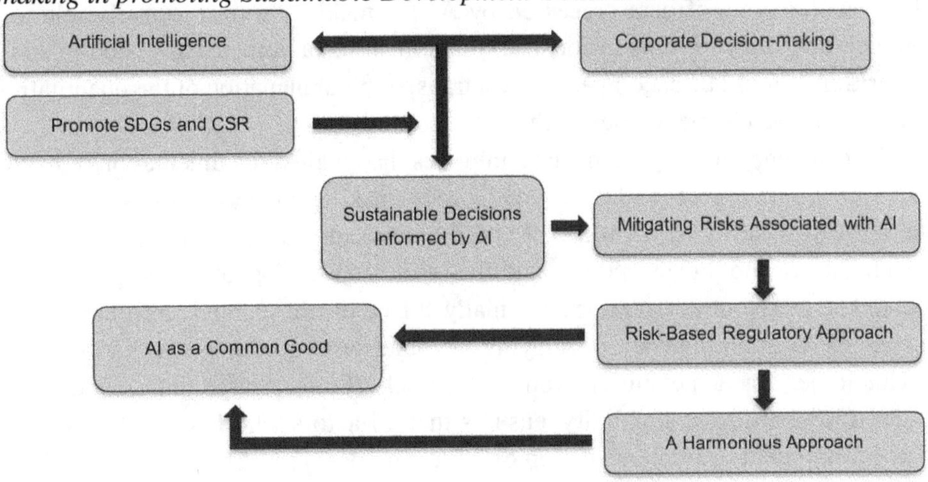

The private sector is one of the key stakeholders in the context of AI governance. Various actors are a part of the private sector, including technology companies, startups, and industry associations that are at the forefront of AI emergence. As they are responsible for the development, deployment, and commercialization of AI, the way the technology will be recognized in society and in the sphere of environmental protection will depend on their way of use (Füller et al., 2022). The private sector could take into consideration existing regulatory terms to act in compliance with them, suggest some industry standards or contribute to their development, and introduce corporate social responsibility practices considering the implementation of AI technology while prioritizing ethics as a part of AI development. In addition, they could invest in research and development and consider cooperation with governments and international organizations in order to support the need for development. (Kulkarni et al., 2024) However, their attempts should be sustained and controlled by the public sector, which will control the private sector's attempt to profit from AI while managing the considerations that could have a positive impact on human and environmental health.

Good governance of AI in environmental protection requires that the technology adheres to the foundational principles of good governance, particularly transparency, accountability, and public trust. These are central to the ethical development and deployment of AI technologies, ensuring the public confidence that AI is being utilized in ways satisfactory to ethical standards and that the outcomes are conducive to sustainable environmental protection.(Mylrea & Robinson, 2023) The first fundamental principle is transparency, which is particularly important in the context of AI because of the characteristic complexity of AI decision-making and these technologies expanded use in environmental protection.(Felzmann et al., 2020) The vast amounts of data processed by AI for making environmentally relevant predictions, drawing conclusions, or implementing automated decision-making algorithms should be accompanied by a transparent explanation of the quantitative and qualitative characteristics of the data used.

Transparency in AI governance comprises the detailed documentation of methodologies, data sources and the entire decision-making process. These steps should be open to scrutiny by a variety of stakeholders, such as policymakers, environmentalists, or the broader public, and the resolution of disputes over AI openness should be in favour of data release. Finally, a lack of transparency may tarnish the image of AI as an innovation, delaying the otherwise rapid implementation of AI technologies for achieving environmental goals (Larsson & Heintz, 2020). The second principle, accountability, ensures that all actors who have participated in the development or implementation of AI technologies must be exposed to the set of rewards or sanctions for the consequences of their actions or decisions, as appropriate. These actors can include algorithm developers, AI technology implementers, or regulators (Srinivasan & San Miguel González, 2022). Furthermore, it created mechanisms, such as regulation, for adverse consequences, including potential biases in AI decision-making, data privacy issues, or other unwanted environmental behaviours. In other situations, accountability may imply, for example, AI technologies auditing or ethical reviews. Finally, the third principle of good governance is public trust. AI technologies, even the most advanced ones, are unable to make an impact in the absence of public trust in such technologies, and, therefore, some other social technologies frequently subdue such technologies' potential effects (Robles & Mallinson, 2023). Thus, public trust should be based on the explicit ways AI technology is likely to be used, its limitations, decisions that involve stakeholders who may distrust AI, for their understanding, and the satisfaction of the outcomes' adherence to the public values.

Table 2. Key Governance Models for AI in Environmental Protection

Governance Model	Description	Strengths	Weaknesses
Public-Private Partnership	Collaboration between government and private sector	Leverages resources from both sectors	Risk of conflicting interests
Multi-Stakeholder Governance	Involvement of diverse stakeholders including civil society	Inclusive decision-making	Complex coordination
Decentralized Governance	Distribution of control across various entities	Flexible and adaptive	Potential for fragmentation
Centralized Governmental Control	Strong government regulation and oversight	Clear authority and enforcement	Risk of bureaucratic delays
Hybrid Governance	Combination of centralized and decentralized elements	Balanced approach	Potential for confusion in roles and responsibilities

VII. LEGAL FRAMEWORK

AI technologies' rapid development has made the creation of comprehensive legislative systems lagging behind their full deployment, particularly with respect to the environment. While the existing legislation already encompasses many aspects of AI use, it often fails to describe its application in environmental management (Sinha & Lee, 2024). In this section, different legal initiatives are analysed to indicate their limitations with respect to AI utilization in environmental protection.

On the global scale, most legislative initiatives aimed at controlling AI are still highly generalized and focus on personal data protection or cybersecurity, rather than climatology or biodiversity. Therefore, there is a range of initiatives aimed at data protection that can be highly relevant to the subject of study. For instance, the "European Union's General Data Protection Regulation is highly advantageous for respective research purposes as both its content and context are highly relevant. First, both AI and the environmental field must rely on a massive array of data, making appropriate protection necessary (Stahl et al., 2022). However, sunscreen regulation does not explicitly indicate the environmental context of big data use. The same observation can be made with respect to the United States while the country has already passed AI-related laws such as the "Algorithmic Accountability Act," these are still too generalized and do not have a highly relevant environmental context. As for environmental protection laws such as the U.S. Clean Air Act, the European Union's Environmental Liability Directive, or the Paris Agreement, their official purposes are limited to defining environmental standards and their violation penalties rather than controlling the use of AI for environmental protection.(Kaminski, 2022) Many laws, however, have been found inapplicable due to the nature of their regulations. The creation of AI technologies has introduced a range of new problems that are not addressed by the existing system. For example, while the collection of massive data from AI systems used for climate or biodiversity research is highly useful, it must be taught properly to avoid the big data issues of privacy or security (Nti et al., 2022). In addition, some AI systems may be not only forecasting but also advising actions based on the data collected, and the actors of these scenarios should not be fully informed about the AI operation specifics. The most relevant aspect of legislation influencing AI utilization for environmental protection I found and presented above after the analysis is the lack of specialized regulation for AI technologies' dual-use specifications.(Zuiderwijk et al., 2021) There are no laws controlling the use of systems designed for large critical system analysis data collection in other processes, such as identifying unreliable citizens or optimal oil fields.

The rapid development of artificial intelligence has already had a profound influence on the collection of data and its subsequent use, thus giving rise to a number of complicated issues, among which, probably, the most important ones are concerned with data privacy and intellectual property.(Xu et al., 2021) Owing to the scope of its potential application and the rate at which it is integrated into various sectors, these issues become quite a topical problem, as, on the one hand, there is a moral question of whether AI is being used ethically, and, on the other, the organizations that develop and integrate this technology into production have to abide by the laws governing these problems. One of the greatest problems in this sphere related to AI's use is the one of data privacy. Various AI technologies require large and diverse amounts of data to function in the first place; in the context of environmental protection, it could be, among other things, geospatial information or sensor data, and, in the extreme case, even data constituting personal information related to the impact of environmental problems on these persons.(Holzinger et al., 2023) The processing of this data and the way it is obtained raise numerous critical issues related to who helps it collect, how it is stored and processed, and under which conditions is the information provided, including such resonant issues as the precious little anonymity assured by the bulk of AI technologies available and the possibility of data misuse.

Clearly, this problem could be solved only on a legislative level and institutions' AI-related work should be governed by data protection laws ensuring the safety of data for each individual or community. Among the laws that are already in place and should be complied with, one can find the European Union-based General Data Protection Regulation, which establishes that the data should not be collected or processed unnecessarily and should be collected only with the data subject's explicit consent, and should be protected as effectively and reliably as possible, in addition to giving the said subject control over their data. Failure to comply with these regulations may result in notable legal issues for the developers and producers and diminish the public's trust in this technology (Macenaite & Kosta, 2017). There are several issues to consider in deploying AI, particularly in the context of environmental monitoring. Firstly, the issues that are intertwined with data collection are the implementation of AI and intellectual property. In terms of data privacy, institutions utilizing AI are coerced to consider the steps for collecting the data, which typically implies avoiding the disclosure of sensitive information, as other data may be aggregated. In addition, it may be unclear to what extent the data should be protected, as it is not clear what information should or should not be considered sensitive.(Ali et al., 2023) I would argue that the nature of AI, in which decisions made are often inexplicable, may raise additional concerns, as it is possible that the use of AI systems be biased towards the use of the training data. Moreover, the use of AI systems may imply that an institution is not implementing the principle of privacy by design and default and taking into account the extent to which data is being collected and whether data is being processed wherever possible. The second issue is the aspect of intellectual property, which heavily resides in the use of AI. Algorithms, software, and databases, as all these artefacts, are programmed or partially programmed by the accomplishers. The AI implementers must determine the manner in which the AI results will be disseminated, while using AI is relative to the aspect of the degree to which these results will be disseminated.(Foss-Solbrekk, 2021) However, these AI implementers must also consider that, in the case of intellectual property, the maintainer may extend the extent to which intellectual property rights may be bestowed or the extent to which these IPRs may be extended and the measures taken if these IPRs are not conferred. For example, AI implementers may be required to consider the extent to which new AI technologies developed might make it difficult for actors in the environmental preservation field to access such technologies. In conclusion, it is necessary to weigh the cost of open sourcing and, at the same time, the cost of unsubstantiated restrictions, for example, in attempting to wean away competitors, the benefit of protecting property during development as well as collaboration (Bibri et al., 2024).

VIII. FINDINGS AND RECOMMENDATIONS

The research tends to show substantive gaps in governmental policy, governance, and legal frameworks with respect to the use of AI for the protection of the environment. The most notable key findings include the fact that the policies are fractal and, furthermore, they are developed out of context of implementation, whereas the governance dimension lacks transparency, accountability, and stakeholder engagement.(Yigitcanlar et al., 2021) Legal frameworks, in turn, are ill-equipped for addressing such an application of AI, as the sphere poses a range of original issues such as privacy, bias, and the ecological footprint of AI technologies. On the whole, the distinction suggests that there is no comprehensive framework for oversight and the harmonization of artificial solutions and sustainability goals necessary.

Notably, the identified disagreements have severe implications for the sustainable usage of AI in environmental protection. Mainly, the lack of integrated policies and governance mechanisms may result in the deployment of AI that emphasizes technical efficiency over ethical and environmental considerations. Notably, this may aggravate existing inequalities and environmental issues. Moreover, there is a risk of deploying unchecked AI by any legal mechanisms if no legal measures are implemented to regulate its use (Uwaga Monica Adanma & Emmanuel Olurotimi Ogunbiyi, 2024). This may result in the loss of human oversight over sophisticated technologies and carbon footprints stemming from AI development and use itself. These findings suggest that government policies are unprepared to address the ethical controversies and environmental risks associated with implementing AI in addressing environmental problems.

Following these gaps, to ensure that the AI systems are ethically and effectively used for environmental protection, the paper will propose the following recommendations. Governments and international organizations must pave the way for designing integrated policy frameworks that directly involve environmental concerns in AI governance. These frameworks should be oriented towards promoting the alignment of AI with the sustainability objectives, as such inspiring the aspiration to preserve the environment as one of the central purposes of the AI technologies development and use (Gustafsson et al., 2019). Similarly, it is essential to promote the creation of a policy designed to encourage the use of AI for environmental justice, where the benefits of AI for the natural environment are equally distributed among the various communities and ecosystems. The improvement of the governance mechanism is integral to ensuring transparency, responsibility, and legitimacy of AI as the chief tool for addressing relevant environmental concerns. This can be achieved by creating multi-stakeholder committees each of which will involve as its members representatives of the local and national government, industry and civil society, faith-based organizations, and local population equally affected by the AI deployment (Alliance, 2024). The task of such committees would be to monitor the use of AI for environmental purposes and make sure it is transparent, inclusive, and directed towards the public interest. It is essential to provide that regular audits and impact assessments of AI's role in the environment will be a mandatory part of the governance mechanism. Specific legislative approaches have to be adopted to keep up with the emerging challenges of using AI in the field of environmental protection. Thus, new laws will have to be developed that will target such issues as the requirements for the data privacy standards observed in the course of implementing AI solutions, the transparency of the algorithms used to manage the use of the organization's resources, the accountability of the said data-based interventions, as well as the environmental friendliness of the developed application (Van Wynsberghe, 2021). To require environmental transparency of AI, e.g., a legal requirement will have to be established that users have to provide clear and elaborative reporting of the processes that it goes through and the results, to which it arrives as a tool in environmental protection (Laux et al., 2024). In addition, there could be a regulation limiting the negative environmental impact of AI, for instance, by requiring energy-efficient systems to be used or by introducing carbon offsetting requirements that Al systems might have to meet. In addition to the legislative and policy requirements, it is also necessary to develop ethical mandates for using AI to protect the environment. They should be developed based on interdisciplinary studies and consultations with the stakeholders to bear the consequences of raising AI. Such mandates should be adopted concerning the so-called bias problems of AI tools, fairness issues, and the overreliance on AI tool solutions. Professional associations and international organisations can target their efforts to establish and endorse the said guidelines (Marda, 2018). In such a way, it will be possible to create a regulatory environment that can maximize the innovative and technological potential of AI but which will require that this potential is in line with the principles of sustainability, ethics, and social justice.

CONCLUSION

This research has examined AI's critical intersection with environmental protection, where numerous opportunities and risks have been highlighted. This analysis focuses on the transformative role AI can play in addressing some of the most urgent environmental catastrophes today. Several points of interest are related to the gaps in current regulation. They are the most relevant since they focus on problems within approaches and no longer on opportunities. Intuitively, existing AI risks should be identified through inadequate policy documents, legislation, and governance systems.(Galaz et al., 2021) First, there are several important reasons why existing policies become a gap. Generally, they are fragmented and vary from one type of institution or jurisdiction to another. There is a moderate recognition of issues in low-level documents, strategies, and business policies. Besides, it shows that there are no sufficient governance solutions. Lastly, the lack of legal standards and formal research frameworks could pose threats (Cuntz et al., 2024). As a result, one of the most important legal maxims can be used to draw a relevant conclusion about all findings. The welfare of the people shall be the supreme law related to the critical reasoning that can be applied to the use of AI for environmental protection through surveillance and other means.

The necessity and importance of integrating policy, governance, and legal frameworks cannot be overestimated in the context of ethical and sustainable AI deployment to protect the environment. It is noteworthy that the analysis of the existing regulatory frameworks and approaches shows that they are developed independently, leading to inconsistencies and gaps. However, it is essential to incorporate all these aspects, i.e., policy, governance, and law, into a single framework (Mennella et al., 2024). On the one hand, such a framework would help create a favourable regulatory environment that would empower and enable innovation while minimising potential compromises in terms of ethical standards, accountability or transparency. On the other hand, the proposed integration would facilitate the establishment of a safer environment for environmental protection. In this way, AI technologies could be used to achieve socially determined goals, protect vulnerable communities or ecosystems, and mitigate various risks. At the same time, it would support the development of mechanisms addressing novel challenges, such as the environmental impact of AI technologies or the risks of algorithmic discrimination (Li et al., 2023). In addition to the propositions for policy development and future research outlined above, it is possible to suggest several other areas for further research. First of all, more comprehensive studies exploring the environmental impact of AI technologies and finding ways to mitigate it are needed. In particular, more attention should be paid to energy consumption and carbon emissions. The findings of these studies may be used to design energy-efficient AI solutions and systems or create new instruments for carbon offset calculation.(Olawade et al., 2024) In addition, it would be beneficial to continue the exploration of the ethical implications of the use of AI technologies to protect the environment. It is important to design the frameworks defining how AI solutions can be used for this purpose in ethical, fair, and transparent ways.(Gaur et al., 2023) Third, policy development should focus on creating more integrated and adaptive regulatory frameworks that can keep pace with the changing nature of AI technology. This refers to the development of policies that support AI governance while explicitly incorporating environmental considerations, as well as to the development of governance structures that are flexible enough to adapt to new challenges as they arise (De Almeida et al., 2021). Finally, there is a significant need for international cooperation to develop global standards and guidelines for the use of AI for environmental protection. As both AI technologies and the protection of the environment are global issues, loosely regulated transfer and use of AI may harm some countries more than others, creating additional potential risks that should be considered in developing a coordinated global approach in managing these technologies. In conclusion, the integration of policy, governance, and law is hardly just a necessity. In the context of AI technology, it could be called a crucial step in ensuring that AI will be used for the greater good of environmental protection. So far, the existing research evidence and gaps identified in that research suggest that this unity of efforts will need to address the most important concerns and follow the avenues for further action proposed in the current study.

REFERENCE

Aayog, N. (n.d.). National strategy for artificial intelligence, 2018. *URL:* Https://Niti. Gov. *in/Sites/Default/Files/2019-01/NationalStrategy-for-AI-Discussion-Paper. Pdf.*

AI and the governance of sustainable development. An idea analysis of the European Union, the United Nations, and the World Economic Forum—ScienceDirect. (n.d.). Retrieved August 15, 2024, from https://www.sciencedirect.com/science/article/pii/S1462901123002393

AI-Based Modeling: Techniques, Applications and Research Issues Towards Automation, Intelligent and Smart Systems | SN Computer Science. (n.d.). Retrieved August 15, 2024, from https://link.springer.com/article/10.1007/s42979-022-01043-x

Akter, S., Babu, M. M., Hani, U., Sultana, S., Bandara, R., & Grant, D. (2024). Unleashing the power of artificial intelligence for climate action in industrial markets. *Industrial Marketing Management*, 117, 92–113. DOI: 10.1016/j.indmarman.2023.12.011

Alhosani, K., & Alhashmi, S. M. (2024). Opportunities, challenges, and benefits of AI innovation in government services: A review. *Discover Artificial Intelligence*, 4(1), 18. DOI: 10.1007/s44163-024-00111-w

Ali, S., Abuhmed, T., El-Sappagh, S., Muhammad, K., Alonso-Moral, J. M., Confalonieri, R., Guidotti, R., Del Ser, J., Díaz-Rodríguez, N., & Herrera, F. (2023). Explainable Artificial Intelligence (XAI): What we know and what is left to attain Trustworthy Artificial Intelligence. *Information Fusion*, 99, 101805. DOI: 10.1016/j.inffus.2023.101805

Alliance, A. G. (2024). *Briefing Paper Series.* https://www3.weforum.org/docs/WEF_AI_Governance_Alliance_Briefing_Paper_Series_2024.pdf

Artificial intelligence technology, public trust, and effective governance—Penn State. (n.d.). Retrieved August 15, 2024, from https://pure.psu.edu/en/publications/artificial-intelligence-technology-public-trust-and-effective-gov

Berry, I. (2021, November 19). *10 ways AI can be used in Smart Cities.* https://aimagazine.com/top10/10-ways-ai-can-be-used-smart-cities

Bibri, S. E., Krogstie, J., Kaboli, A., & Alahi, A. (2024). Smarter eco-cities and their leading-edge artificial intelligence of things solutions for environmental sustainability: A comprehensive systematic review. *Environmental Science and Ecotechnology*, 19, 100330. DOI: 10.1016/j.ese.2023.100330 PMID: 38021367

Brickson, L., Zhang, L., Vollrath, F., Douglas-Hamilton, I., & Titus, A. J. (2023). Elephants and algorithms: A review of the current and future role of AI in elephant monitoring. *Journal of the Royal Society, Interface*, 20(208), 20230367. DOI: 10.1098/rsif.2023.0367 PMID: 37963556

Capraro, V., Lentsch, A., Acemoglu, D., Akgun, S., Akhmedova, A., Bilancini, E., Bonnefon, J.-F., Brañas-Garza, P., Butera, L., Douglas, K. M., Everett, J. A. C., Gigerenzer, G., Greenhow, C., Hashimoto, D. A., Holt-Lunstad, J., Jetten, J., Johnson, S., Kunz, W. H., Longoni, C., & Viale, R. (2024). The impact of generative artificial intelligence on socioeconomic inequalities and policy making. *PNAS Nexus*, 3(6), pgae191. Advance online publication. DOI: 10.1093/pnasnexus/pgae191 PMID: 38864006

Chen, L., Chen, Z., Zhang, Y., Liu, Y., Osman, A. I., Farghali, M., Hua, J., Al-Fatesh, A., Ihara, I., Rooney, D. W., & Yap, P.-S. (2023). Artificial intelligence-based solutions for climate change: A review. *Environmental Chemistry Letters*, 21(5), 2525–2557. DOI: 10.1007/s10311-023-01617-y

Cuntz, A., Fink, C., & Stamm, H. (2024). Artificial Intelligence and Intellectual Property: An Economic Perspective. *World Intellectual Property Organization (WIPO) Economic Research Working Paper Series, 77.* https://papers.ssrn.com/sol3/papers.cfm?abstract_id=4757971

De Almeida, P. G. R., Dos Santos, C. D., & Farias, J. S. (2021). Artificial Intelligence Regulation: A framework for governance. *Ethics and Information Technology*, 23(3), 505–525. DOI: 10.1007/s10676-021-09593-z

Deloitte. (n.d.). *Deloitte Survey: Analytics and Data-driven Culture Help Companies Outperform Business Goals in the "Age of With."* Retrieved August 15, 2024, from https://www.prnewswire.com/news-releases/deloitte-survey-analytics-and-data-driven-culture-help-companies-outperform-business-goals-in-the-age-of-with-300889703.html

Díaz-Rodríguez, N., Del Ser, J., Coeckelbergh, M., López de Prado, M., Herrera-Viedma, E., & Herrera, F. (2023). Connecting the dots in trustworthy Artificial Intelligence: From AI principles, ethics, and key requirements to responsible AI systems and regulation. *Information Fusion*, 99, 101896. DOI: 10.1016/j.inffus.2023.101896

Eckert, E., & Kovalevska, O. (2021). Sustainability in the European Union: Analyzing the Discourse of the European Green Deal. *Journal of Risk and Financial Management*, 14(2), 2. Advance online publication. DOI: 10.3390/jrfm14020080

Environment, U. N. (2022, February 10). *Frontiers 2022: Noise, Blazes and Mismatches*. https://www.unep.org/resources/frontiers-2022-noise-blazes-and-mismatches

Feijóo, C., Kwon, Y., Bauer, J. M., Bohlin, E., Howell, B., Jain, R., Potgieter, P., Vu, K., Whalley, J., & Xia, J. (2020). Harnessing artificial intelligence (AI) to increase wellbeing for all: The case for a new technology diplomacy. *Telecommunications Policy*, 44(6), 101988. DOI: 10.1016/j.telpol.2020.101988 PMID: 32377031

Felzmann, H., Fosch-Villaronga, E., Lutz, C., & Tamò-Larrieux, A. (2020). Towards Transparency by Design for Artificial Intelligence. *Science and Engineering Ethics*, 26(6), 3333–3361. DOI: 10.1007/s11948-020-00276-4 PMID: 33196975

Feng, S., & Liu, S. (2023). Does AI Application Matter in Promoting Carbon Productivity? Fresh Evidence from 30 Provinces in China. *Sustainability (Basel)*, 15(23), 23. Advance online publication. DOI: 10.3390/su152316261

Foss-Solbrekk, K. (2021). Three routes to protecting AI systems and their algorithms under IP law: The good, the bad and the ugly. *Journal of Intellectual Property Law & Practice*, 16(3), 247–258. DOI: 10.1093/jiplp/jpab033

Francisco, M., & Linnér, B.-O. (2023). AI and the governance of sustainable development. An idea analysis of the European Union, the United Nations, and the World Economic Forum. *Environmental Science & Policy*, 150, 103590. DOI: 10.1016/j.envsci.2023.103590

Full article: Governance of artificial intelligence. (n.d.). Retrieved August 15, 2024, from https://www.tandfonline.com/doi/full/10.1080/14494035.2021.1928377

Füller, J., Hutter, K., Wahl, J., Bilgram, V., & Tekic, Z. (2022). How AI revolutionizes innovation management – Perceptions and implementation preferences of AI-based innovators. *Technological Forecasting and Social Change*, 178, 121598. DOI: 10.1016/j.techfore.2022.121598

Galaz, V., Centeno, M. A., Callahan, P. W., Causevic, A., Patterson, T., Brass, I., Baum, S., Farber, D., Fischer, J., Garcia, D., McPhearson, T., Jimenez, D., King, B., Larcey, P., & Levy, K. (2021). Artificial intelligence, systemic risks, and sustainability. *Technology in Society*, 67, 101741. DOI: 10.1016/j.techsoc.2021.101741

Gaur, L., Afaq, A., Arora, G. K., & Khan, N. (2023). Artificial intelligence for carbon emissions using system of systems theory. *Ecological Informatics*, 76, 102165. DOI: 10.1016/j.ecoinf.2023.102165

Guidance, W. H. O. (2021). Ethics and governance of artificial intelligence for health. *World Health Organization*. https://iris.who.int/rest/bitstreams/1352854/retrieve?ref=thinkml.ai

Gustafsson, S., Hermelin, B., & Smas, L. (2019). Integrating environmental sustainability into strategic spatial planning: The importance of management. *Journal of Environmental Planning and Management*, 62(8), 1321–1338. DOI: 10.1080/09640568.2018.1495620

Herath, H. M. K. K. M. B., & Mittal, M. (2022). Adoption of artificial intelligence in smart cities: A comprehensive review. *International Journal of Information Management Data Insights*, 2(1), 100076. DOI: 10.1016/j.jjimei.2022.100076

Holzinger, A., Keiblinger, K., Holub, P., Zatloukal, K., & Müller, H. (2023). AI for life: Trends in artificial intelligence for biotechnology. *New Biotechnology*, 74, 16–24. DOI: 10.1016/j.nbt.2023.02.001 PMID: 36754147

Hossain, M., Nijhum, I. R., Shad, M. T. M., Sadi, A. A., Peyal, M. M. K., & Rahman, R. M. (2023). An end-to-end pollution analysis and detection system using artificial intelligence and object detection algorithms. *Decision Analytics Journal*, 8, 100283. DOI: 10.1016/j.dajour.2023.100283

How AI Can Help Save Endangered Species | Scientific American. (n.d.). Retrieved August 15, 2024, from https://www.scientificamerican.com/article/how-ai-can-help-save-endangered-species/

Impact of artificial intelligence on human loss in decision making, laziness and safety in education—PMC. (n.d.). Retrieved August 15, 2024, from https://www.ncbi.nlm.nih.gov/pmc/articles/PMC10251321/

Information privacy, impact assessment, and the place of ethics—ScienceDirect. (n.d.). Retrieved August 15, 2024, from https://www.sciencedirect.com/science/article/pii/S0267364920300091

Javaid, M., Haleem, A., Khan, I. H., & Suman, R. (2023). Understanding the potential applications of Artificial Intelligence in Agriculture Sector. *Advanced Agrochem*, 2(1), 15–30. DOI: 10.1016/j.aac.2022.10.001

Joshua, P. Meltzer, & Cameron F. Kerry. (2021). *Strengthening international cooperation on artificial intelligence*. Brookings. https://www.brookings.edu/articles/strengthening-international-cooperation-on-artificial-intelligence/

Kaminski, M. E. (2022). Regulating the Risks of AI. SSRN *Electronic Journal*. DOI: 10.2139/ssrn.4195066

Khunakornbodintr, S. (2024). Examining the impact of green technological specialization and the integration of AI technologies on green innovation performance: Evidence from China. *Frontiers in Artificial Intelligence*, 6, 1237285. Advance online publication. DOI: 10.3389/frai.2023.1237285 PMID: 38725486

Ki-moon, B. (2011). *Address to the 66th General Assembly*. We the Peoples.

Konya, A., & Nematzadeh, P. (2024). Recent applications of AI to environmental disciplines: A review. *The Science of the Total Environment*, 906, 167705. DOI: 10.1016/j.scitotenv.2023.167705 PMID: 37820816

Kulkarni, A. V., Joseph, S., & Patil, K. P. (2024). Artificial intelligence technology readiness for social sustainability and business ethics: Evidence from MSMEs in developing nations. *International Journal of Information Management Data Insights*, 4(2), 100250. DOI: 10.1016/j.jjimei.2024.100250

Larsson, S., & Heintz, F. (2020). Transparency in artificial intelligence. *Internet Policy Review*, 9(2). Advance online publication. https://policyreview.info/concepts/transparency-artificial-intelligence. DOI: 10.14763/2020.2.1469

Laux, J., Wachter, S., & Mittelstadt, B. (2024). Three pathways for standardisation and ethical disclosure by default under the European union artificial intelligence act. *Computer Law & Security Report*, 53, 105957. DOI: 10.1016/j.clsr.2024.105957

Li, W., Yigitcanlar, T., Browne, W., & Nili, A. (2023). The Making of Responsible Innovation and Technology: An Overview and Framework. *Smart Cities*, 6(4), 1996–2034. DOI: 10.3390/smartcities6040093

Lima, M. A., Mendes, L. F. R., Mothé, G. A., Linhares, F. G., de Castro, M. P. P., da Silva, M. G., & Sthel, M. S. (2020). Renewable energy in reducing greenhouse gas emissions: Reaching the goals of the Paris agreement in Brazil. *Environmental Development*, 33, 100504. DOI: 10.1016/j.envdev.2020.100504

Liu, W., McKibbin, W. J., Morris, A. C., & Wilcoxen, P. J. (2020). Global economic and environmental outcomes of the Paris Agreement. [i]. *Energy Economics*, 90, 104838. DOI: 10.1016/j.eneco.2020.104838

Macenaite, M., & Kosta, E. (2017). Consent for processing children's personal data in the EU: Following in US footsteps? *Information & Communications Technology Law*, 26(2), 146–197. DOI: 10.1080/13600834.2017.1321096

Marda, V. (2018). Artificial intelligence policy in India: A framework for engaging the limits of data-driven decision-making. *Philosophical Transactions. Series A, Mathematical, Physical, and Engineering Sciences*, 376(2133), 20180087. DOI: 10.1098/rsta.2018.0087 PMID: 30323001

McKinsey & Company. (2023). *The state of AI in 2023: Generative AI's breakout year | McKinsey*. https://www.mckinsey.com/capabilities/quantumblack/our-insights/the-state-of-ai-in-2023-generative-ais-breakout-year

Mennella, C., Maniscalco, U., De Pietro, G., & Esposito, M. (2024). Ethical and regulatory challenges of AI technologies in healthcare: A narrative review. *Heliyon*, 10(4), e26297. DOI: 10.1016/j.heliyon.2024.e26297 PMID: 38384518

Mondejar, M. E., Avtar, R., Diaz, H. L. B., Dubey, R. K., Esteban, J., Gómez-Morales, A., Hallam, B., Mbungu, N. T., Okolo, C. C., Prasad, K. A., She, Q., & Garcia-Segura, S. (2021). Digitalization to achieve sustainable development goals: Steps towards a Smart Green Planet. *The Science of the Total Environment*, 794, 148539. DOI: 10.1016/j.scitotenv.2021.148539 PMID: 34323742

Mylrea, M., & Robinson, N. (2023). Artificial Intelligence (AI) Trust Framework and Maturity Model: Applying an Entropy Lens to Improve Security, Privacy, and Ethical AI. *Entropy (Basel, Switzerland)*, 25(10), 1429. DOI: 10.3390/e25101429 PMID: 37895550

Nahar, S. (2024). Modeling the effects of artificial intelligence (AI)-based innovation on sustainable development goals (SDGs): Applying a system dynamics perspective in a cross-country setting. *Technological Forecasting and Social Change*, 201, 123203. DOI: 10.1016/j.techfore.2023.123203

Neo, E. X., Hasikin, K., Lai, K. W., Mokhtar, M. I., Azizan, M. M., Hizaddin, H. F., Razak, S. A., & Yanto, . (2023). Artificial intelligence-assisted air quality monitoring for smart city management. *PeerJ. Computer Science*, 9, e1306. DOI: 10.7717/peerj-cs.1306 PMID: 37346549

New AI-Powered Drone Technology Aids Elephant Conservation | Tufts Now. (2023, March 8). https://now.tufts.edu/2023/03/08/new-ai-powered-drone-technology-aids-elephant-conservation

Nti, E. K., Cobbina, S. J., Attafuah, E. E., Opoku, E., & Gyan, M. A. (2022). Environmental sustainability technologies in biodiversity, energy, transportation and water management using artificial intelligence: A systematic review. *Sustainable Futures : An Applied Journal of Technology, Environment and Society*, 4, 100068. DOI: 10.1016/j.sftr.2022.100068

Olawade, D. B., Wada, O. Z., David-Olawade, A. C., Fapohunda, O., Ige, A. O., & Ling, J. (2024). Artificial intelligence potential for net zero sustainability: Current evidence and prospects. *Next Sustainability*, 4, 100041. DOI: 10.1016/j.nxsust.2024.100041

Osasona, F., Amoo, O., Atadoga, A., Abrahams, T., Farayola, O., & Ayinla, B. (2024). REVIEWING THE ETHICAL IMPLICATIONS OF AI IN DECISION MAKING PROCESSES. *International Journal of Management & Entrepreneurship Research*, 6(2), 322–335. DOI: 10.51594/ijmer.v6i2.773

Paris Agreement—An overview | ScienceDirect Topics. (n.d.). Retrieved August 15, 2024, from https://www.sciencedirect.com/topics/earth-and-planetary-sciences/paris-agreement

(PDF) Artificial Intelligence Regulation: A framework for governance. (n.d.). Retrieved August 15, 2024, from https://www.researchgate.net/publication/351039094_Artificial_Intelligence_Regulation_a_framework_for_governance

Raab, C. D. (2020). Information privacy, impact assessment, and the place of ethics. *Computer Law & Security Report*, 37, 105404. DOI: 10.1016/j.clsr.2020.105404

Raman, R., Pattnaik, D., Lathabai, H. H., Kumar, C., Govindan, K., & Nedungadi, P. (2024). Green and sustainable AI research: An integrated thematic and topic modeling analysis. *Journal of Big Data*, 11(1), 55. DOI: 10.1186/s40537-024-00920-x

Regona, M., Yigitcanlar, T., Hon, C., & Teo, M. (2024). Artificial intelligence and sustainable development goals: Systematic literature review of the construction industry. *Sustainable Cities and Society*, 108, 105499. DOI: 10.1016/j.scs.2024.105499

Research Journal, 5(5), 1178–1209. DOI: 10.51594/csitrj.v5i5.1156

Robles, P., & Mallinson, D. J. (2023). Artificial intelligence technology, public trust, and effective governance. *Review of Policy Research*, ropr.12555. DOI: 10.1111/ropr.12555

Sarker, I. H. (2021). Machine Learning: Algorithms, Real-World Applications and Research Directions. *SN Computer Science*, 2(3), 160. DOI: 10.1007/s42979-021-00592-x PMID: 33778771

Sarker, I. H. (2022). AI-Based Modeling: Techniques, Applications and Research Issues Towards Automation, Intelligent and Smart Systems. *SN Computer Science*, 3(2), 158. DOI: 10.1007/s42979-022-01043-x PMID: 35194580

s*Balancing the scale: Navigating ethical and practical challenges of artificial intelligence (AI) integration in legal practices | Discover Artificial Intelligence*. (n.d.). Retrieved August 15, 2024, from https://link.springer.com/article/10.1007/s44163-024-00121-8

Shivanna, K. R. (2022). Climate change and its impact on biodiversity and human welfare. *Proceedings of the Indian National Science Academy. Part A, Physical Sciences*, 88(2), 160–171. DOI: 10.1007/s43538-022-00073-6

Silvestro, D., Goria, S., Sterner, T., & Antonelli, A. (2022). Improving biodiversity protection through artificial intelligence. *Nature Sustainability*, 5(5), 415–424. DOI: 10.1038/s41893-022-00851-6 PMID: 35614933

Sinha, S., & Lee, Y. M. (2024). Challenges with developing and deploying AI models and applications in industrial systems. *Discover Artificial Intelligence*, 4(1), 55. DOI: 10.1007/s44163-024-00151-2

Smart Cities | Free Full-Text | The Making of Responsible Innovation and Technology: An Overview and Framework. (n.d.). Retrieved August 15, 2024, from https://www.mdpi.com/2624-6511/6/4/93

Srinivasan, R., & San Miguel González, B. (2022). The role of empathy for artificial intelligence accountability. *Journal of Responsible Technology*, 9, 100021. DOI: 10.1016/j.jrt.2021.100021

Stahl, B. C., Rodrigues, R., Santiago, N., & Macnish, K. (2022). A European Agency for Artificial Intelligence: Protecting fundamental rights and ethical values. *Computer Law & Security Report*, 45, 105661. DOI: 10.1016/j.clsr.2022.105661

Strengthening international cooperation on artificial intelligence. (n.d.). Brookings. Retrieved August 15, 2024, from https://www.brookings.edu/articles/strengthening-international-cooperation-on-artificial-intelligence/

Subramaniam, S., Raju, N., Ganesan, A., Rajavel, N., Chenniappan, M., Prakash, C., Pramanik, A., Basak, A. K., & Dixit, S. (2022). Artificial Intelligence Technologies for Forecasting Air Pollution and Human Health: A Narrative Review. *Sustainability (Basel)*, 14(16), 9951. DOI: 10.3390/su14169951

Sustainable AI: AI for sustainability and the sustainability of AI | AI and Ethics. (n.d.). Retrieved August 15, 2024, from https://link.springer.com/article/10.1007/s43681-021-00043-6

Taghikhah, F., Erfani, E., Bakhshayeshi, I., Tayari, S., Karatopouzis, A., & Hanna, B. (2022). Artificial intelligence and sustainability: solutions to social and environmental challenges. In *Artificial intelligence and data science in environmental sensing* (pp. 93–108). Academic Press.

THE 17 GOALS | Sustainable Development. (n.d.). Retrieved August 15, 2024, from robin

Van Wynsberghe, A. (2021). Sustainable AI: AI for sustainability and the sustainability of AI. *AI and Ethics*, 1(3), 213–218. DOI: 10.1007/s43681-021-00043-6

vinx*IUCN*. (n.d.). Retrieved August 15, 2024, from https://iucn.org/

William J. Snape III. (2010). *Joining The Convention On Biological Diversity: A Legal and Scientific Overview of Why the United States Must Wake Up*.

Xu, Y., Liu, X., Cao, X., Huang, C., Liu, E., Qian, S., Liu, X., Wu, Y., Dong, F., Qiu, C.-W., Qiu, J., Hua, K., Su, W., Wu, J., Xu, H., Han, Y., Fu, C., Yin, Z., Liu, M., & Zhang, J. (2021). Artificial intelligence: A powerful paradigm for scientific research. *Innovation (Cambridge (Mass.))*, 2(4), 100179. DOI: 10.1016/j.xinn.2021.100179 PMID: 34877560

Xue, L., & Pang, Z. (2022). Ethical governance of artificial intelligence: An integrated analytical framework. *Journal of Digital Economy*, 1(1), 44–52. DOI: 10.1016/j.jdec.2022.08.003

Yigitcanlar, T., Mehmood, R., & Corchado, J. M. (2021). *Green Artificial Intelligence: Towards an Efficient*. Sustainable and Equitable Technology for Smart Cities and Futures.

Zuiderwijk, A., Chen, Y.-C., & Salem, F. (2021). Implications of the use of artificial intelligence in public governance: A systematic literature review and a research agenda. *Government Information Quarterly*, 38(3), 101577. DOI: 10.1016/j.giq.2021.101577

Chapter 8
Machine Leaning (ML) on Digital Consumer Transforming Choice Recommendation Model: Projecting Smart Cities Fostering Sustainability

Bhupinder Singh
https://orcid.org/0009-0006-4779-2553
Sharda University, India

Anjali Raghav
https://orcid.org/0009-0004-0248-7956
Sharda University, India

Kittisak Jermsittiparsert
Shinawatra University, Thailand

ABSTRACT

The rapid evolution of Machine leaning (ML) has ushered in a transformative era in the digital consumer landscape. This research paper delves into the intricate interplay between AI and digital consumers, focusing specifically on the profound influence of AI in reshaping consumer recommendation models and, consequently, molding consumer shopping decisions. This chapter unveils a comprehensive understanding of the influence of AI on digital consumers and the subsequent transformation of consumer recommendation models. The impact on consumer shopping decisions is not merely technological but extends into the realms of psychology, trust, and

DOI: 10.4018/979-8-3693-8684-2.ch008

personalization. As businesses grapple with the imperative to adapt, the insights from this research provide a roadmap for navigating the evolving landscape of AI-driven consumer interactions.

1. INTRODUCTION

The symbiotic relationship between technology and consumer behavior has undergone a seismic shift, with Machine leaning (ML) emerging as a transformative force in shaping the digital consumer experience (Ullah et al., 2024) (Singh, 2024) (Singh & Kaunert, 2024) (Singh & Kaunert, 2024). At this paradigm shift falls in the intricate interplay between AI and consumer recommendation models, which have evolved from rudimentary algorithms to sophisticated systems capable of personalization and tailored engagement (Singh & Kaunert, 2024) (Singh & Kaunert, 2024). This chapter endeavors to delve into the multifaceted dimensions of AI's influence on the digital consumer, with a specific focus on its role in redefining consumer recommendation models (Singh et al., 2024) (Singh & Kaunert, 2024) (Hasan et al, 2021). The concept of "stimulus AI," characterized by its advanced algorithms and personalized learning capabilities, becomes the focal point of this investigation. As consumers increasingly navigate a vast digital marketplace, characterized by an abundance of choices and information, understanding how AI stimuli influence their shopping decisions becomes paramount for businesses and researchers alike (Wong et al., 2022). By exploring the nuanced dynamics of this interaction, the research aims to uncover the mechanisms through which stimulus AI not only informs but fundamentally transforms the consumer decision-making process, thereby revolutionizing traditional paradigms of commerce (Huynh-The et al., 2023).

AI-powered revolution that is permeating every facet of contemporary life from virtual assistants anticipating user needs to recommendation algorithms tailoring content and products to individual preferences, ML has become an omnipresent force in the digital ecosystem. Traditional consumer recommendation models, which relied on generalized preferences and historic data, are undergoing a metamorphosis as stimulus ML introduces a level of personalization and responsiveness that was once deemed unattainable. As the consumer landscape becomes increasingly dynamic and intricate, understanding the implications of this transformation on businesses and consumers is imperative for navigating the evolving terrain of digital commerce (Nashold Jr, 2020).

1.1 Objectives of the Paper

The objectives of this paper are two-fold: firstly, to investigate the nuanced influence of ML on consumer shopping decisions, and secondly, to dissect the role of stimulus ML in transforming existing consumer recommendation models. The study acknowledges that as consumers become more discerning and technologically savvy, the one-size-fits-all approach of yesteryears is no longer tenable. In its place, personalized, adaptive, and context-aware recommendation models, underpinned by stimulus ML, are emerging as the vanguard of a new era in consumer engagement.

This paper aims to contribute to the growing body of knowledge surrounding AI's impact on the digital consumer, offering insights that are not only theoretically rich but also practically relevant for businesses seeking to thrive in an era defined by unprecedented technological convergence. The exploration of stimulus ML influence on consumer shopping decisions promises to illuminate not just the present state of the digital consumer landscape but also chart a course for businesses to navigate the future of consumer engagement in an ML-driven world.

1.2 Background of the Chapter

The rapid evolution of the digital landscape has witnessed an unprecedented integration of Machine leaning (ML) into every facet of the consumer experience, fundamentally reshaping how individuals interact with digital platforms and make purchasing decisions. The background of this research is rooted in the recognition that AI has transcended its role as a mere technological tool and has become a dynamic force redefining the consumer journey (Singh, 2023). Traditional consumer recommendation models, characterized by static algorithms and generalized preferences, have given way to a more intricate and personalized paradigm. This term encapsulates the advanced algorithms and machine learning capabilities that drive a new era of consumer engagement, where AI not only anticipates but actively stimulates consumer choices (Smith et al., 2021). The background of this study is contextualized in the recognition that as consumers navigate an increasingly complex digital marketplace, characterized by an abundance of choices and information, the integration of stimulus ML into consumer recommendation models becomes not just a technological advancement but a transformative shift in how commerce is conducted. Understanding this backdrop is pivotal for comprehending the nuanced interplay between AI and the digital consumer, as well as for anticipating the profound implications of such interactions on the future landscape of consumer behavior and commerce (Clancy, 2021).

1.3 Rise of ML in Digital Consumer Landscape: Environment Preservation

The rise of Machine leaning (ML) in the digital consumer landscape marks a transformative epoch, where technology and consumer interactions have become intricately entwined. Over the past decade, AI has transitioned from a theoretical concept to a pervasive and dynamic force that shapes the way individuals engage with digital platforms. Its ascent is underpinned by advancements in machine learning algorithms, data analytics, and computational power, enabling AI systems to comprehend, learn, and adapt to user behaviors (Behera & Dadra, 2023). In the digital consumer arena, AI manifests in various forms, from chatbots and virtual assistants providing instant customer support to sophisticated recommendation engines tailoring content and product suggestions based on individual preferences (Wirtz & Lovelock, 2021). The deployment of AI in consumer-centric applications has not only expedited processes but has also ushered in a new era of personalization and predictive analytics (Mori, 2021). This overview underscores the omnipresence of AI in the digital consumer landscape, where its influence permeates daily interactions, offering unprecedented convenience, customization, and efficiency. As AI continues to advance, its role in shaping the digital consumer experience is poised to evolve further, presenting both opportunities and challenges for businesses and consumers alike (Butcher et al., 2021) (Diamandis & Kotler, 2020).

1.4 Identification of Challenges in Traditional Recommendation Systems

Traditional recommendation systems, while foundational, grapple with several challenges that have spurred the evolution toward more advanced models, including those driven by Artificial Intelligence. The key challenge lies in the limited ability of traditional systems to provide truly personalized recommendations. Conventional approaches often rely on basic collaborative filtering or content-based methods, which may struggle to capture the intricate nuances of individual user preferences in diverse and dynamic environments. Scalability is another hurdle, as the exponential growth of data poses constraints on traditional algorithms, impacting their efficiency and response times. Cold-start problems, wherein recommendations for new or less active users are inherently challenging, present a significant hurdle. The static nature of traditional models also impedes adaptability to evolving user behaviors and preferences over time (Rodgers & Nguyen, 2022). The challenge of the 'bubble effect' arises, where users are confined to a narrow set of recommendations based on their past interactions, limiting serendipity and the discovery of new items. As the digital landscape continues to expand and user expectations evolve, addressing these challenges becomes crucial for recommendation systems to remain relevant and effective in meeting the dynamic demands of the modern digital consumer. The inadequacies of traditional recommendation systems underscore the need for more sophisticated, adaptive, and personalized approaches, prompting the exploration of innovative models such as those infused with Artificial Intelligence.

1.5 Need for AI-driven Solutions to Enhance Consumer Experiences

The escalating complexity of the digital consumer landscape, coupled with the limitations of traditional recommendation systems, underscores an imperative shift towards AI-driven solutions to elevate consumer experiences. Artificial Intelligence, particularly machine learning algorithms, has demonstrated unparalleled capabilities in deciphering intricate patterns within vast datasets. Unlike conventional systems, AI-driven solutions excel in discerning subtle nuances in user behavior, allowing for highly personalized and context-aware recommendations. The dynamic nature of consumer preferences necessitates adaptive systems that evolve with user interactions, a quality inherent in AI models (Beyari, & Garamoun, 2022). The scalability of AI facilitates the processing of massive datasets in real-time, ensuring swift and accurate recommendations even in the face of growing digital ecosystems. AI-driven solutions also address the long-standing challenge of the cold-start problem by swiftly adapting to the preferences of new users. The transformative potential lies not only in enhancing the relevance and accuracy of recommendations but also in cultivating a more engaging, intuitive, and satisfying consumer journey. By leveraging the capabilities of AI, businesses can unlock a realm of possibilities for elevating user satisfaction, fostering loyalty, and ultimately thriving in the competitive digital marketplace (Dimitrieska et al., 2018). The adoption of AI-driven solutions represents not just a technological evolution but a strategic imperative for businesses aspiring to stay ahead in delivering unparalleled and personalized consumer experiences.

2. MACHINE LEARNING APPLICATIONS IN CONSUMER CHOICES: SMART CITIES FOSTERING SUSTAINABILITY

Machine leaning (ML) applications in consumer recommendations have emerged as a game-changer in the digital landscape, revolutionizing how businesses engage with their customers. At the forefront of this transformation are advanced machine learning algorithms that empower recommendation systems to go beyond conventional approaches. The prominent applications lies in the domain of personalized recommendations. ML, particularly through techniques like collaborative filtering and content-based filtering, can analyze vast amounts of user data, including purchase history, browsing behavior, and demographic information. This allows for the creation of highly personalized recommendations that reflect individual preferences with remarkable accuracy. Consequently, users are presented with products or content that align closely with their tastes, fostering a sense of personalization that enhances the overall consumer experience (Dellaert et al., 2020).

AI-driven recommendation systems excel in addressing the cold-start problem, a persistent challenge in traditional models. By leveraging techniques like hybrid models and deep learning, these systems can make meaningful predictions for new or less active users by drawing insights from similar user profiles. The ability of AI to adapt and learn from user interactions over time ensures that recommendations remain relevant and reflective of evolving preferences. This adaptability is particularly crucial in dynamic markets where consumer behaviors and trends are subject to frequent changes. Another noteworthy application of AI in consumer recommendations is the exploration of contextual awareness. Advanced algorithms can analyze real-time contextual information, such as location, device type, and current user activity, to tailor recommendations accordingly. This ensures that recommendations are not only personalized but also delivered at the right time and in the right context, enhancing their effectiveness. For instance, a consumer browsing on a mobile device in a specific location might receive recommendations that align with their immediate needs or interests (Gochhait et al., 2020). The applications of AI in consumer recommendations transcend the limitations of traditional models, offering businesses the means to deliver highly personalized, adaptive, and contextually aware experiences to their users (Busman & Ananda, 2022). As these AI-driven systems continue to evolve, the synergy between advanced algorithms and consumer data promises to redefine the dynamics of digital commerce, providing users with a seamless and enriching journey through personalized recommendations.

2.1 Examination of AI-driven Recommendation Algorithms and Their Effectiveness

The examination of AI-driven recommendation algorithms and their effectiveness reveals a profound shift in the landscape of consumer engagement and decision-making processes. At the core of this transformation are sophisticated machine learning algorithms that leverage vast datasets to deliver highly tailored recommendations. Collaborative filtering, a prevalent approach, analyzes user behavior and preferences to identify patterns and similarities among users, enabling the system to suggest items based on the preferences of like-minded individuals. Content-based filtering, on the other hand, considers the attributes of items and users to generate recommendations that align with a user's historical choices (Sharma & Singh, 2022).

The effectiveness of AI-driven recommendation algorithms is further underscored by their capacity to tackle the cold-start problem. As these algorithms integrate hybrid models and deep learning techniques, they can make informed predictions for new users or those with limited historical data by drawing insights from broader datasets. This versatility ensures that even users with sparse interaction histories receive meaningful and contextually relevant recommendations (Luo et al., 2019) (Allal-Cherif et al., 2021).

2.2 Impact on Consumer Shopping Decisions

The impact of AI-driven recommendation algorithms on consumer shopping decisions is substantial, ushering in a new era of personalized and informed decision-making. By analyzing vast datasets, these algorithms have the capability to understand individual preferences, past behaviors, and even contextual factors, thereby presenting users with highly relevant and tailored suggestions. This personalized approach not only streamlines the shopping experience but also influences consumers to explore products or services that align closely with their tastes. The adaptability of these algorithms, continuously learning and evolving with each user interaction, ensures that recommendations remain finely tuned to changing preferences over time. the efficiency and scalability of AI-driven recommendations enhance the overall user experience, providing timely suggestions even in the face of expanding product catalogs and growing datasets. As a result, consumers find themselves making more informed choices, discovering new items, and ultimately deriving greater satisfaction from their shopping journeys. The impact extends beyond individual choices; it also contributes to increased user engagement, brand loyalty, and the overall competitiveness of businesses in the dynamic landscape of e-commerce (Sharma et al., 2022) (Yang & Tan, 2022).

3. AI IN CONSUMER RECOMMENDATION MODELS

Machine leaning (ML) has emerged as a revolutionary force in the realm of consumer recommendation models, fundamentally altering the dynamics of how individuals discover, engage with, and make decisions about products and services in the digital landscape. At the heart of this transformation are advanced machine learning algorithms that leverage vast datasets to provide highly personalized and contextually relevant recommendations. Traditional recommendation models, limited by simplistic rule-based systems or collaborative and content-based filtering, have given way to more sophisticated AI-driven models capable of comprehensively understanding individual preferences.

The strengths of AI in consumer recommendation models lies in its ability to analyze and interpret complex patterns within enormous datasets. Through techniques like collaborative filtering, where the behavior and preferences of users with similar tastes are considered, and content-based filtering, which considers the attributes of items and users, AI models can generate remarkably accurate recommendations (Bag et al., 2022). The integration of deep learning and neural networks empowers these systems to discern intricate nuances in user behavior, enabling them to make predictions that are not just personalized but also adaptive to evolving preferences over time.

A critical enhancement offered by AI in consumer recommendations is its capacity to address the long-standing challenges faced by traditional models. The cold-start problem, wherein new or less active users receive suboptimal recommendations, is effectively mitigated by AI through the utilization of hybrid models and sophisticated algorithms that draw insights from broader datasets (Mogaji et al., 2020). Also, the scalability of AI-driven solutions ensures that even in the face of an exponentially growing volume of data, these models can efficiently process and analyze information in real-time, maintaining the accuracy and relevance of recommendations. Context-aware recommendations represent another frontier where AI shines. With incorporating real-time contextual information such as user location, device type, and current activity, these algorithms deliver suggestions that are not only personalized but also timed and situated for maximum impact (Ivanov & Webster, 2017). The transformative power of AI in consumer recommendations is not without its challenges. Ethical considerations, privacy concerns, and the need for algorithmic transparency are increasingly recognized as crucial aspects. The explainability of AI models, ensuring that users can understand how recommendations are generated, becomes imperative for building trust and addressing concerns related to the responsible use of personal data (Ameen, 2021) (Pereira et al., 2022).

3.1 Overview of Collaborative Filtering, Content-Based Filtering and Hybrid Recommendation Models

Collaborative filtering, content-based filtering, and hybrid recommendation models represent key strategies in personalized recommendation systems. Collaborative filtering relies on user behavior patterns, recommending items based on the preferences of similar users. Content-based filtering, in contrast, considers intrinsic item characteristics to make suggestions aligned with a user's past choices. Hybrid recommendation models integrate these approaches, combining collaborative and content-based methods to harness their respective strengths. Collaborative filtering excels in capturing user preferences, while content-based filtering accommodates scenarios with rich item attributes. By merging these techniques, hybrid models aim to provide more accurate and diverse recommendations, offering a comprehensive solution to the challenges faced by individual recommendation strategies (Bhagat et al., 2023).

3.2 Discussion on How AI Enhances Capabilities of These Models

Machine leaning (ML) plays a transformative role in enhancing the capabilities of collaborative filtering, content-based filtering, and hybrid recommendation models. In collaborative filtering, AI algorithms bring a heightened level of accuracy by employing advanced machine learning techniques. Deep learning architectures, such as neural networks, allow collaborative filtering models to discern intricate patterns within user behavior data, providing more nuanced and precise recommendations. This enables the system to adapt dynamically to changing user preferences, ensuring a personalized experience.

For content-based filtering, AI contributes by automating the extraction and analysis of item features. Natural Language Processing (NLP) and computer vision techniques, powered by AI, enable the model to comprehend textual descriptions, images, or other attributes of items. This results in a more sophisticated understanding of item characteristics, improving the relevance and accuracy of suggestions to users.

Hybrid recommendation models, which integrate collaborative and content-based approaches, benefit immensely from AI. Machine learning algorithms optimize the weighting and combination of the two recommendation strategies based on real-time user interactions. These algorithms adapt to evolving user preferences and dynamically adjust the balance between collaborative and content-based recommendations. Additionally, AI facilitates the incorporation of contextual factors, such as user location or temporal considerations, making hybrid models more adaptive and responsive to diverse user scenarios (Nalini et al., 2021).

AI also addresses inherent challenges in these recommendation models. It aids in mitigating the cold-start problem, ensuring that new users or items receive meaningful recommendations. AI-driven models can effectively handle sparse datasets, drawing insights from broader data landscapes to provide relevant suggestions. Furthermore, ethical considerations and user privacy concerns are being addressed through explainable AI (XAI) techniques, allowing users to understand how recommendations are generated and fostering trust in the recommendation system. The infusion of AI into collaborative filtering, content-based filtering, and hybrid recommendation models elevates their capabilities to unprecedented levels. From intricate pattern recognition to dynamic adaptation based on user interactions, AI-driven enhancements empower recommendation systems to provide more accurate, personalized, and ethically sound suggestions, thereby enriching the overall user experience in the digital landscape.

3.3 Comparative Analysis of Traditional Recommendation Systems and AI-enhanced Models

A comparative analysis of traditional recommendation systems and AI-enhanced models reveals a profound evolution in the landscape of personalized user suggestions. The traditional recommendation systems often rule-based or relying on basic collaborative, content-based filtering, exhibit simplicity and transparency. However, they are constrained by challenges such as the cold-start problem, scalability issues, and a lack of adaptability to dynamic user preferences. In contrast, AI-enhanced models harness the power of advanced machine learning algorithms, offering a quantum leap in recommendation capabilities (Yau, 2021).

Traditional systems, while straightforward, struggle to capture the intricate patterns in user behavior. They often provide generic recommendations and fall short in adapting to individual preferences over time. AI-enhanced models, driven by sophisticated algorithms like deep learning and neural networks, excel in deciphering complex user interactions, enabling a more accurate understanding of preferences and ensuring adaptive recommendations.

Scalability is a critical factor in the digital era where data volumes are massive. Traditional models may falter in processing and analyzing extensive datasets in real-time, limiting their effectiveness. AI-enhanced models, leveraging the computational prowess of machine learning, efficiently handle large datasets, ensuring timely and relevant recommendations for users in dynamic and expanding digital ecosystems. The cold-start problem, particularly prevalent in traditional systems when dealing with new users or items, is a significant limitation. AI-enhanced models employ hybrid approaches, matrix factorization, and other advanced techniques to address this challenge, ensuring meaningful recommendations even for users with sparse interaction histories.

The interpretability and transparency of traditional models often fall short when compared to AI-enhanced systems. Explainable AI (XAI) techniques embedded in advanced models allow users to understand how recommendations are generated, addressing concerns related to the opacity of algorithms and fostering user trust. While traditional recommendation systems have their merits in simplicity and transparency, the era of AI-enhanced models brings a paradigm shift. AI models, with their ability to handle complexity, scale, and adaptability, redefine the standards for personalized recommendations, offering users a more sophisticated, dynamic, and context-aware experience in the ever-evolving digital landscape (Rabby et al., 2021).

4. IMPACT ON CONSUMER SHOPPING DECISIONS

The impact of technology, particularly Machine leaning (ML), on consumer shopping decisions is profound and multifaceted. With the advent of AI-driven recommendation systems, consumers are now exposed to a personalized and tailored shopping experience. These systems analyze vast datasets, including past purchase history, browsing behaviors, and demographic information, to generate recommendations that align closely with individual preferences. This level of personalization has a direct influence on consumer decision-making, as it streamlines the shopping process and introduces users to products or services that are more likely to resonate with their tastes. The adaptability of AI-driven systems, continuously learning from user interactions, ensures that recommendations evolve in tandem with shifting preferences, contributing to an ongoing, dynamic consumer journey.

AI impacts consumer decisions by addressing common challenges in the shopping experience. The overwhelming abundance of choices in the digital marketplace can lead to decision fatigue for consumers. AI steps in by curating options, presenting users with a more manageable selection based on their preferences, thereby simplifying the decision-making process. This not only saves time for consumers but also enhances their overall satisfaction with the shopping experience (Cheng & Jiang, 2022). ML influences shopping decisions through its ability to provide real-time, context-aware recommendations. With considering factors such as location, time of day, and device type, ML systems offer suggestions that are not only personalized but also relevant to the immediate circumstances of the consumer. For instance, a user browsing a mobile shopping app in the evening might receive recommendations suited for leisure or evening activities, showcasing the contextual awareness embedded in ML-driven recommendation engines.

The impact of ML on consumer shopping decisions extends to the realm of trust and brand loyalty. When consumers perceive that a platform understands their preferences and consistently delivers relevant recommendations, it fosters a sense of trust. This trust, in turn, contributes to increased brand loyalty as users are more likely to return to platforms that consistently enhance their shopping experiences. The impact of ML on consumer shopping decisions is transformative, redefining the traditional retail landscape. The integration of personalized recommendations, adaptability to evolving preferences, and context-aware suggestions not only streamline the decision-making process but also contribute to a more satisfying and engaging shopping experience for consumers in the digital age (Stone et al., 2020).

4.1 Personalization and Customization

Personalization and customization are paramount in the contemporary consumer experience, shaping preferences and fostering deeper engagement. Personalization tailors products, services, or content based on individual user behaviors, preferences, and demographics. By leveraging data analytics and machine learning algorithms, businesses can curate unique experiences, providing users with content or recommendations that align with their tastes. On the other hand, customization allows users to actively modify or personalize products according to their preferences, reflecting a more interactive and user-centric approach. Both personalization and customization not only enhance user satisfaction by catering to individual needs but also contribute to brand loyalty and long-term customer relationships in an era where tailored experiences are increasingly valued (Nagy & Hajdú, 2021).

4.2 AI Enables Personalized and Customized Recommendations: Exploration

Machine leaning (ML) has revolutionized the landscape of personalized and customized recommendations, offering businesses unprecedented capabilities to understand and respond to individual user preferences. AI leverages advanced algorithms and data analytics to explore vast datasets containing user behaviors, purchase histories, and preferences. Personalization, in this context, involves tailoring recommendations based on an individual's unique characteristics and past interactions. AI-driven recommendation systems can discern intricate patterns within user data, allowing for the delivery of suggestions that are highly relevant to a user's tastes and preferences. For instance, e-commerce platforms utilize AI to analyze a user's browsing history and purchase patterns, offering product recommendations that align with their preferences (Mahmoud et al., 2020).

AI enhances personalization and customization through the dynamic adaptation of recommendation models. Machine learning algorithms embedded in AI systems continuously learn from user interactions, adapting and refining recommendations over time. This adaptability ensures that as user preferences evolve, the recommendations remain accurate and aligned with current tastes. The contextual awareness of AI contributes significantly to personalized and customized experiences. By considering factors such as location, time of day, and device type, AI systems provide recommendations that are not only personalized but also relevant to the immediate circumstances of the user. For example, a music streaming service may suggest upbeat songs in the morning and relaxing tunes in the evening, demonstrating the contextual understanding enabled by AI.

4.3 Analysis of Consumer Responses to personalized recommendations

The analysis of consumer responses to personalized recommendations underscores the significant impact such tailored experiences have on user engagement and satisfaction. Personalized recommendations, driven by AI algorithms that discern user preferences, often result in positive consumer responses. Users appreciate the convenience of discovering products, services, or content that aligns closely with their tastes, streamlining their decision-making process. The sense of relevance and the feeling of being understood by the platform foster a more favorable perception, contributing to increased user satisfaction. Additionally, personalized recommendations often lead to a more immersive and enjoyable user experience, enhancing brand loyalty as consumers are more likely to return to platforms that consistently provide relevant suggestions. But, it's essential for businesses to strike a balance, ensuring that the personalization efforts are transparent, respect user privacy, and offer users the ability to control and customize their recommendations to avoid any potential backlash or concerns (de Marcellis-Warin et al., 2022).

5. TRUST AND TRANSPARENCY

Trust and transparency are foundational pillars in the relationship between businesses and consumers, especially in an era dominated by digital interactions. Trust is the bedrock upon which customer loyalty, satisfaction, and long-term relationships are built. When consumers trust a brand, they are more likely to engage, make purchases, and advocate for the brand. Transparency plays a crucial role in fostering this trust. It involves open communication, honesty, and clarity in business practices. When businesses are transparent about their processes, policies, and data usage, consumers feel more confident and secure in their interactions (Leyer et al., 2020).

In the digital domain, where personal data is often exchanged for personalized services, trust and transparency become paramount. Consumers expect businesses to handle their data responsibly, ensuring privacy and security. Clearly communicating how data is collected, used, and protected builds trust and helps users make informed decisions. Transparency also extends to business operations, ethical practices, and customer service. When businesses are open about their values, corporate responsibility, and commitment to customer satisfaction, it creates a positive perception among consumers. The trust and transparency are mutually reinforcing elements in the relationship between businesses and consumers. Establishing trust requires transparency in communication, actions, and the responsible handling of data. By prioritizing these principles, businesses can cultivate a loyal customer base and navigate the evolving landscape of consumer expectations in the digital age.

5.1 Discussion on Role of Trust and Transparency in AI-driven Recommendation Systems

The role of trust and transparency is pivotal in the deployment and acceptance of AI-driven recommendation systems. As these systems become increasingly integral to the digital consumer experience, establishing and maintaining trust is fundamental. Trust begins with transparency in how these systems operate, how user data is utilized, and how recommendations are generated. Firstly, transparency in the functioning of AI algorithms is crucial. Users should have a clear understanding of how recommendations are determined, what factors influence them, and the criteria used by the system. This transparency mitigates concerns related to algorithmic opacity and ensures that users can comprehend the decision-making process behind the suggestions they receive.

The responsible handling of user data is another key element. AI-driven recommendation systems often rely on vast amounts of personal data to provide personalized suggestions. Ensuring transparency in how this data is collected, stored, and utilized is paramount. Clear communication about data privacy measures, security protocols, and user control over their data instills confidence in users, fostering a sense of security and trust. Also, transparency is essential in addressing potential biases in AI models. Users should be made aware of how these systems account for diversity, ensuring that recommendations are fair and inclusive. This transparency not only builds trust but also aligns with ethical considerations surrounding AI deployment.

To enhance trust further, businesses can incorporate explainable AI (XAI) techniques into their recommendation systems. XAI enables users to understand the rationale behind AI-generated recommendations, providing insights into how the system arrived at a particular suggestion. This transparency aids in demystifying complex algorithms and reinforces the credibility of the recommendation engine. The trust and transparency are integral components in the successful adoption and continued use of AI-driven recommendation systems. As users entrust these systems with their data and decision-making processes, businesses that prioritize transparency build a foundation of trust, fostering positive user experiences and long-term relationships. The interplay between trust, transparency, and ethical considerations is crucial in navigating the evolving landscape of AI technologies in the consumer realm (Kopalle, 2022).

5.2 Addressing Concerns relating to Privacy and Data Security

Addressing concerns related to privacy and data security is paramount in the deployment of AI-driven recommendation systems. Users entrust these systems with personal information to receive personalized suggestions, and safeguarding this data is essential for building and maintaining trust. Robust privacy measures, including encryption protocols, secure data storage, and adherence to regulatory frameworks such as GDPR, demonstrate a commitment to protecting user information. Transparent communication about how user data is collected, used, and stored helps to allay privacy concerns, fostering a sense of control for users. Implementing stringent data security practices not only safeguards sensitive information but also contributes to the overall credibility and acceptance of AI-driven recommendation systems among users (Rani & Sundaram, 2022).

6. IMPLICATIONS FOR BUSINESSES AND CONSUMERS

The widespread adoption of AI-driven recommendation systems has profound implications for both businesses and consumers, reshaping the dynamics of the digital marketplace. For businesses, the integration of these systems translates into enhanced customer engagement, increased conversion rates, and improved brand loyalty. By leveraging AI to provide personalized and context-aware recommendations, businesses can create a more tailored and satisfying experience for consumers, leading to higher retention rates and repeat business. Moreover, the data generated by these recommendation systems offers invaluable insights into consumer preferences and behaviors, enabling businesses to refine their marketing strategies, optimize product offerings, and stay agile in response to changing market trends.

Along with these benefits come responsibilities the businesses must prioritize transparency and ethical use of data, addressing concerns related to privacy and security. The implications of mishandling user data can be severe, eroding trust and damaging brand reputation. Therefore, a commitment to ethical AI practices and clear communication about data usage are essential for sustaining positive relationships with consumers (Lingam, 2018). For consumers, the implications of AI-driven recommendation systems are reflected in an increasingly personalized and convenient digital experience. These systems streamline decision-making processes, introducing users to products and services that align with their preferences. However, there are considerations related to privacy and the potential for algorithmic biases. Consumers must be aware of how their data is being used and have the ability to control and customize their recommendations. The implications also extend to the empowerment of consumers, who have more agency in shaping their digital experiences through the customization features offered by AI-driven systems.

The implications of AI-driven recommendation systems are transformative for both businesses and consumers. For businesses, it's an opportunity to enhance customer relationships and drive growth, but it requires a commitment to responsible AI practices. For consumers, the implications involve a more personalized, convenient, and interactive digital experience, but there is a growing need for awareness and control over the use of personal data. As AI continues to evolve, navigating these implications will be essential for creating a balanced and mutually beneficial digital ecosystem (Ho, 2023).

6.1 Business Strategies

The business strategies encompass the comprehensive plans and actions a company undertakes to achieve its goals and objectives. These strategies serve as roadmaps, guiding decision-making processes to attain sustainable growth and competitive advantage. Business strategies often involve a combination of market positioning, innovation, cost management, and customer-centric approaches. In today's dynamic landscape, adapting to technological advancements, including the integration of AI-driven technologies, has become a strategic imperative. Successful businesses align their strategies with changing consumer expectations, industry trends, and emerging technologies, leveraging these factors to stay agile and relevant. Whether through diversification, differentiation, or operational excellence, effective business strategies are crucial for navigating complexities, fostering resilience, and securing long-term success in the ever-evolving business environment (Nazir et al., 2023).

6.2 Examination Businesses Leverage: AI Improving Consumer Engagement and Loyalty

Businesses can strategically leverage Machine leaning (ML) to elevate consumer engagement and foster brand loyalty. AI-driven technologies, such as recommendation systems and chatbots, provide personalized and seamless interactions, enhancing the overall customer experience. By analyzing consumer behavior, preferences, and historical data, AI enables businesses to deliver tailor-made recommendations, anticipating and meeting individual needs. Additionally, chatbots powered by AI offer real-time assistance, resolving queries promptly and efficiently (Rustholkarhu et al., 2022). Such proactive and personalized engagement builds a stronger connection between the business and the consumer. Furthermore, AI facilitates predictive analytics, enabling businesses to anticipate consumer trends and preferences, leading to more informed decision-making in product development and marketing strategies.

In essence, businesses that strategically deploy AI not only enhance consumer engagement but also cultivate lasting loyalty by consistently providing value and personalized interactions (van Esch & Cui, 2021).

6.3 Strategic Implications for Marketing, Sales and Customer Service

The integration of Machine leaning (ML) holds strategic implications across key business functions, transforming marketing, sales, and customer service. In marketing, AI enables data-driven insights, facilitating targeted and personalized campaigns. Machine learning algorithms analyze consumer behavior, optimizing content delivery and enhancing the effectiveness of marketing strategies (Singh, 2024). In sales, AI-driven tools streamline lead generation, automate routine tasks, and provide predictive analytics, empowering sales teams to focus on high-value activities. AI-powered chatbots and virtual assistants revolutionize customer service by offering real-time support, resolving queries, and enhancing the overall customer experience. The strategic adoption of AI in these functions not only improves operational efficiency but also unlocks new levels of customer engagement, contributing to sustained business growth and competitiveness in the dynamic market landscape (Singh, 2024).

7. CONSUMER PERSPECTIVES

Consumer perspectives on Machine leaning (ML) are shaped by a dynamic interplay of expectations, experiences, and concerns. On one hand, consumers appreciate the convenience and personalization that AI-driven technologies, such as recommendation systems and virtual assistants, bring to their digital experiences (Singh & Kaunert, 2024). The ability of AI to understand preferences, provide relevant suggestions, and streamline tasks resonates positively with many users. On the other hand, there are apprehensions related to privacy, data security, and the potential for algorithmic biases. Consumers increasingly seek transparency from businesses regarding how their data is used and are mindful of the ethical considerations surrounding AI. There's a growing awareness of the impact of AI on the job market and concerns about the potential loss of human-centric roles. While consumers appreciate the efficiency and innovation that AI introduces, there's a desire for a balanced integration that preserves human involvement and ethical considerations (Singh, 2024).

The evolving landscape of AI applications influences consumer perspectives. As businesses continue to refine AI technologies, incorporating user feedback, addressing ethical concerns, and prioritizing transparency are pivotal for shaping positive consumer attitudes towards AI. Ultimately, understanding and respecting consumer perspectives are essential for fostering trust and ensuring the responsible deployment of AI technologies in the digital age (Singh & Kaunert, 2024).

7.1 Consumers Perceive and Respond to AI-driven Recommendations: Gathering Insights

Gathering insights into how consumers perceive and respond to AI-driven recommendations is crucial for businesses aiming to enhance user experiences. With leveraging surveys, user feedback, and analytics, businesses can gain valuable information on consumer attitudes towards personalized suggestions. Understanding whether users find recommendations helpful, relevant, or intrusive allows businesses to fine-tune their AI algorithms (Singh & Kaunert, 2024). Also, exploring how consumers respond to the level of personalization, the transparency of recommendation processes, and concerns about privacy provides a nuanced understanding. This feedback loop empowers businesses to align their AI-driven recommendation strategies with user expectations, fostering a positive reception and greater satisfaction among their customer bas (Singh & Kaunert, 2024)e.

7.2 Addressing Potential Ethical Considerations from Consumer Standpoint

Addressing potential ethical considerations from the consumer standpoint is an imperative for businesses deploying AI-driven technologies. As consumers increasingly interact with AI in various facets of their digital experiences, concerns related to privacy, transparency, and fairness become prominent. To navigate these ethical considerations, businesses must prioritize transparency in their AI systems, clearly communicating how consumer data is collected, processed, and utilized. Implementing robust security measures to safeguard personal information is equally essential to build consumer trust (Singh & Kaunert, 2024).

Businesses should address issues related to algorithmic biases that may inadvertently favor certain groups over others. Striving for fairness and inclusivity in AI algorithms ensures that recommendations and interactions are equitable for all consumers. Providing users with control over their data and the ability to customize their AI-driven experiences enhances the sense of autonomy and mitigates privacy concerns (Singh & Kaunert, 2024). The businesses that proactively address ethical considerations from the consumer standpoint not only build trust but also contribute to the responsible and sustainable deployment of AI technologies, fostering positive and long-term relationships with their user base (Singh & Kaunert, 2024).

8. CONCLUSION

The integration of Machine leaning (ML) in various facets of business operations has ushered in a transformative era, reshaping how businesses interact with consumers and how consumers navigate the digital landscape. The strategic use of AI in recommendation systems, marketing, sales, and customer service offers unprecedented opportunities for personalization, efficiency, and data-driven decision-making. However, this technological advancement is not without its challenges. Businesses must navigate issues of privacy, transparency, and algorithmic biases to ensure the ethical deployment of AI technologies.

From the consumer standpoint, there is a delicate balance between appreciating the convenience and personalization offered by AI and harboring concerns about privacy and the potential loss of human-centric roles. Businesses that actively seek and respond to consumer insights, address ethical considerations, and prioritize transparency are better positioned to build trust and foster positive consumer perceptions.

As AI continues to evolve, the strategic implications for businesses are vast, requiring adaptability and a commitment to responsible AI practices. The journey forward involves not only embracing the technological advancements that AI brings but also understanding and addressing the evolving needs, expectations, and ethical considerations of consumers. In this dynamic landscape, the synergy between technological innovation and consumer-centric values will shape the future trajectory of AI in business.

9. FUTURE SCOPE

The future scope of Machine leaning (ML) in business holds immense potential for transformative growth and innovation. As technology advances, AI is poised to play an increasingly central role across various industries. In business operations, AI is likely to enhance efficiency through automation, optimizing workflows, and providing real-time data-driven insights for more informed decision-making.

The field of AI-driven recommendation systems is expected to evolve, becoming even more personalized and context-aware. The integration of natural language processing and sentiment analysis may lead to more sophisticated systems that understand and respond to user preferences and emotions. AI's role in marketing is anticipated to expand, with predictive analytics and machine learning algorithms enabling businesses to anticipate consumer trends, personalize marketing campaigns, and optimize customer engagement strategies. In sales, AI-driven tools may continue to streamline lead generation, improve sales forecasting, and enhance customer relationship management.

The convergence of AI with other technologies, such as the Internet of Things (IoT) and augmented reality, opens new frontiers. AI-powered systems may play a pivotal role in analyzing and interpreting the vast amounts of data generated by IoT devices, leading to more intelligent and responsive ecosystems. The ethical considerations and responsible AI practices are expected to gain prominence, with businesses increasingly focusing on transparency, fairness, and user privacy. Explainable AI (XAI) techniques may become more prevalent to address concerns related to algorithmic opacity.

The future also holds potential for AI to contribute significantly to sustainability efforts. Smart systems may optimize resource utilization, reduce energy consumption, and facilitate eco-friendly practices in various industries. AI in business is characterized by continual innovation and integration across diverse sectors. As businesses harness the power of AI responsibly, aligning technological advancements with ethical considerations, the trajectory points towards a future where AI becomes an integral and beneficial part of our evolving digital landscape.

REFERENCES

Allal-Chérif, O., Simón-Moya, V., & Ballester, A. C. C. (2021). Intelligent purchasing: How artificial intelligence can redefine the purchasing function. *Journal of Business Research*, 124, 69–76. DOI: 10.1016/j.jbusres.2020.11.050

Ameen, N., Tarhini, A., Reppel, A., & Anand, A. (2021). Customer experiences in the age of artificial intelligence. *Computers in Human Behavior*, 114, 106548. DOI: 10.1016/j.chb.2020.106548 PMID: 32905175

Bag, S., Srivastava, G., Bashir, M. M. A., Kumari, S., Giannakis, M., & Chowdhury, A. H. (2022). Journey of customers in this digital era: Understanding the role of artificial intelligence technologies in user engagement and conversion. *Benchmarking*, 29(7), 2074–2098. DOI: 10.1108/BIJ-07-2021-0415

Behera, C. K., & Dadra, R. (2023). Understanding young consumers' attitude formation for new-age fintech credit products: An SOR framework perspective. *Journal of Financial Services Marketing*, •••, 1–15.

Beyari, H., & Garamoun, H. (2022). The Effect of Artificial Intelligence on End-User Online Purchasing Decisions: Toward an Integrated Conceptual Framework. *Sustainability (Basel)*, 14(15), 9637. DOI: 10.3390/su14159637

. Bhagat, R., Chauhan, V., & Bhagat, P. (2023). Investigating the impact of artificial intelligence on consumer's purchase intention in e-retailing. *foresight, 25*(2), 249-263.

Busman, S. A., & Ananda, N. A. (2022). Artificial Intelligence and Digital Marketing Role in Increasing Consumer Purchase Intention. [AIJBM]. *American International Journal of Business Management*, 5(01), 63–68.

. Butcher, N., Wilson-Strydom, M., & Baijnath, M. (2021). Artificial intelligence capacity in sub-Saharan Africa: Compendium report.

Cheng, Y., & Jiang, H. (2022). Customer–brand relationship in the era of artificial intelligence: Understanding the role of chatbot marketing efforts. *Journal of Product and Brand Management*, 31(2), 252–264. DOI: 10.1108/JPBM-05-2020-2907

Clancy, M. (2021). *Reflections on the financial and ethical implications of music generated by artificial intelligence* (Doctoral dissertation, PhD Thesis. Trinity College, Dublin).

de Marcellis-Warin, N., Marty, F., Thelisson, E., & Warin, T. (2022). Artificial intelligence and consumer manipulations: From consumer's counter algorithms to firm's self-regulation tools. *AI and Ethics*, 2(2), 259–268. DOI: 10.1007/s43681-022-00149-5

Dellaert, B. G., Shu, S. B., Arentze, T. A., Baker, T., Diehl, K., Donkers, B., & Steffel, M. (2020). Consumer decisions with artificially intelligent voice assistants. *Marketing Letters*, 31(4), 335–347. DOI: 10.1007/s11002-020-09537-5

Diamandis, P. H., & Kotler, S. (2020). *The future is faster than you think: How converging technologies are transforming business, industries, and our lives*. Simon & Schuster.

Dimitrieska, S., Stankovska, A., & Efremova, T. (2018). Artificial intelligence and marketing. *Entrepreneurship*, 6(2), 298–304.

Gochhait, S., Mazumdar, O., Chahal, S., Kanwat, P., Gupta, S., Sharma, R., & Sachan, R. (2020, May). Role of Machine leaning (ML) in understanding the behavior pattern: a study on e-commerce. In *ICDSMLA 2019: Proceedings of the 1st International Conference on Data Science, Machine Learning and Applications* (pp. 1600-1606). Singapore: Springer Singapore. DOI: 10.1007/978-981-15-1420-3_166

Hasan, R., Weaven, S., & Thaichon, P. (2021). Blurring the line between physical and digital environment: The impact of artificial intelligence on customers' relationship and customer experience. In *Developing digital marketing* (pp. 135–153). Emerald Publishing Limited. DOI: 10.1108/978-1-80071-348-220211008

Ho, S. P. S., & Chow, M. Y. C. (2023). The role of artificial intelligence in consumers' brand preference for retail banks in Hong Kong. *Journal of Financial Services Marketing*, •••, 1–14.

Huynh-The, T., Pham, Q. V., Pham, X. Q., Nguyen, T. T., Han, Z., & Kim, D. S. (2023). Artificial intelligence for the metaverse: A survey. *Engineering Applications of Artificial Intelligence*, 117, 105581. DOI: 10.1016/j.engappai.2022.105581

Ivanov, S. H., & Webster, C. (2017, June). The robot as a consumer: a research agenda. In *Marketing: experience and perspectives* (pp. 29–30). Conference.

Jain, S., & Gandhi, A. V. (2021). Impact of artificial intelligence on impulse buying behaviour of Indian shoppers in fashion retail outlets. *International Journal of Innovation Science*, 13(2), 193–204. DOI: 10.1108/IJIS-10-2020-0181

Kopalle, P. K., Gangwar, M., Kaplan, A., Ramachandran, D., Reinartz, W., & Rindfleisch, A. (2022). Examining Machine leaning (ML) technologies in marketing via a global lens: Current trends and future research opportunities. *International Journal of Research in Marketing*, 39(2), 522–540. DOI: 10.1016/j.ijresmar.2021.11.002

Leyer, M., Oberlaender, A., Dootson, P., & Kowalkiewicz, M. (2020, June). Decision-making with artificial intelligence: Towards a novel conceptualization of patterns. In *PACIS* (p. 224).

Lingam, Y. K. (2018). The role of Machine leaning (ML) in making accurate stock decisions in E-commerce industry. *Int. J. Adv. Res. Ideas Innov. Technol*, 4(3), 2281–2286.

Luo, X., Tong, S., Fang, Z., & Qu, Z. (2019). Frontiers: Machines vs. humans: The impact of artificial intelligence chatbot disclosure on customer purchases. *Marketing Science*, 38(6), 937–947. DOI: 10.1287/mksc.2019.1192

Mahmoud, A. B., Tehseen, S., & Fuxman, L. (2020). The dark side of artificial intelligence in retail innovation. In *Retail Futures* (pp. 165–180). Emerald Publishing Limited. DOI: 10.1108/978-1-83867-663-620201019

. Mogaji, E., Soetan, T. O., & Kieu, T. A. (2020). The implications of artificial intelligence on the digital marketing of financial services to vulnerable customers. *Australasian Marketing Journal*, j-ausmj.

Mori, M. (2021). AI-powered virtual assistants in the realms of banking and financial services. In *Virtual Assistant*. IntechOpen. DOI: 10.5772/intechopen.95813

Nagy, S., & Hajdú, N. (2021). Consumer acceptance of the use of artificial intelligence in online shopping: Evidence from Hungary. *Amfiteatru Economic*, 23(56), 155–173. DOI: 10.24818/EA/2021/56/155

Nalini, M., Radhakrishnan, D. P., Yogi, G., Santhiya, S., & Harivardhini, V. (2021). Impact of Machine leaning (ML) on marketing. *International Journal of Aquatic Science*, 12(2), 3159–3167.

Nashold, D. B., Jr. (2020). *Trust in Consumer Adoption of Artificial Intelligence-Driven Virtual Finance Assistants: A Technology Acceptance Model Perspective* (Doctoral dissertation, The University of North Carolina at Charlotte).

Nazir, S., Khadim, S., Asadullah, M. A., & Syed, N. (2023). Exploring the influence of artificial intelligence technology on consumer repurchase intention: The mediation and moderation approach. *Technology in Society*, 72, 102190. DOI: 10.1016/j.techsoc.2022.102190

Pereira, A. M., Moura, J. A. B., Costa, E. D. B., Vieira, T., Landim, A. R., Bazaki, E., & Wanick, V. (2022). Customer models for artificial intelligence-based decision support in fashion online retail supply chains. *Decision Support Systems*, 158, 113795. DOI: 10.1016/j.dss.2022.113795

Rabby, F., Chimhundu, R., & Hassan, R. (2021). Artificial intelligence in digital marketing influences consumer behaviour: A review and theoretical foundation for future research. *Academy of Marketing Studies Journal*, 25(5), 1–7.

Rani, V. S., & Sundaram, N. (2022). Collaborative social media marketing in small scale business using artificial intelligence. *ECS Transactions*, 107(1), 5175–5182. DOI: 10.1149/10701.5175ecst

Rodgers, W., & Nguyen, T. (2022). Advertising benefits from ethical artificial intelligence algorithmic purchase decision pathways. *Journal of Business Ethics*, 178(4), 1043–1061. DOI: 10.1007/s10551-022-05048-7

Rusthollkarhu, S., Toukola, S., Aarikka-Stenroos, L., & Mahlamäki, T. (2022). Managing B2B customer journeys in digital era: Four management activities with artificial intelligence-empowered tools. *Industrial Marketing Management*, 104, 241–257. DOI: 10.1016/j.indmarman.2022.04.014

Sharma, A., & Singh, B. (2022). Measuring Impact of E-commerce on Small Scale Business: A Systematic Review. *Journal of Corporate Governance and International Business Law*, 5(1).

Sharma, S., Islam, N., Singh, G., & Dhir, A. (2022). Why do retail customers adopt Machine leaning (ML) based autonomous decision-making systems? *IEEE Transactions on Engineering Management*.

Singh, B. (2019). Profiling Public Healthcare: A Comparative Analysis Based on the Multidimensional Healthcare Management and Legal Approach. *Indian Journal of Health and Medical Law*, 2(2), 1–5.

Singh, B. (2022). Relevance of Agriculture-Nutrition Linkage for Human Healthcare: A Conceptual Legal Framework of Implication and Pathways. *Justice and Law Bulletin*, 1(1), 44–49.

Singh, B. (2022). COVID-19 Pandemic and Public Healthcare: Endless Downward Spiral or Solution via Rapid Legal and Health Services Implementation with Patient Monitoring Program. *Justice and Law Bulletin*, 1(1), 1–7.

Singh, B. (2023). Blockchain Technology in Renovating Healthcare: Legal and Future Perspectives. In *Revolutionizing Healthcare Through Artificial Intelligence and Internet of Things Applications* (pp. 177-186). IGI Global.

Singh, B. (2023). Federated Learning for Envision Future Trajectory Smart Transport System for Climate Preservation and Smart Green Planet: Insights into Global Governance and SDG-9 (Industry, Innovation and Infrastructure). *National Journal of Environmental Law*, 6(2), 6–17.

Singh, B. (2023). Blockchain Technology in Renovating Healthcare: Legal and Future Perspectives. In Kaushik, K., Dahiya, S., Aggarwal, S., & Dwivedi, A. (Eds.), *Revolutionizing Healthcare Through Artificial Intelligence and Internet of Things Applications* (pp. 177–186). IGI Global., DOI: 10.4018/978-1-6684-5422-0.ch012

Singh, B. (2024). Cherish Data Privacy and Human Rights in the Digital Age: Harmonizing Innovation and Individual Autonomy. In Pucelj, M., & Bohinc, R. (Eds.), *Balancing Human Rights, Social Responsibility, and Digital Ethics* (pp. 199–226). IGI Global., DOI: 10.4018/979-8-3693-3334-1.ch007

Singh, B. (2024). Legal Dynamics Lensing Metaverse Crafted for Videogame Industry and E-Sports: Phenomenological Exploration Catalyst Complexity and Future. *Journal of Intellectual Property Rights Law*, 7(1), 8–14.

Singh, B. (2024). Lensing Legal Dynamics for Examining Responsibility and Deliberation of Generative AI-Tethered Technological Privacy Concerns: Infringements and Use of Personal Data by Nefarious Actors. In Ara, A., & Ara, A. (Eds.), *Exploring the Ethical Implications of Generative AI* (pp. 146–167). IGI Global., DOI: 10.4018/979-8-3693-1565-1.ch009

Singh, B. (2024). Social Cognition of Incarcerated Women and Children: Addressing Exposure to Infectious Diseases and Legal Outcomes. In Reddy, K. (Ed.), *Principles and Clinical Interventions in Social Cognition* (pp. 236–251). IGI Global., DOI: 10.4018/979-8-3693-1265-0.ch014

Singh, B. (2024). Evolutionary Global Neuroscience for Cognition and Brain Health: Strengthening Innovation in Brain Science. In Prabhakar, P. (Ed.), *Biomedical Research Developments for Improved Healthcare* (pp. 246–272). IGI Global., DOI: 10.4018/979-8-3693-1922-2.ch012

Singh, B., Dutta, P. K., Gautam, R., & Kaunert, C. (2024). Uncapping the Potential of Quantum Computing Towards Manufacturing Optimization: Routing Supply Chain Projecting Sustainability. In Hassan, A., Bhattacharya, P., Dutta, P., Verma, J., & Kundu, N. (Eds.), *Quantum Computing and Supply Chain Management: A New Era of Optimization* (pp. 395–419). IGI Global., DOI: 10.4018/979-8-3693-4107-0.ch024

Singh, B., Dutta, P. K., & Kaunert, C. (2024). Boosting Quantum Computing to Redefine Supply Chain in the Hospital Industry: Pitch the Vision for Medical and Healthcare Needs. In Hassan, A., Bhattacharya, P., Dutta, P., Verma, J., & Kundu, N. (Eds.), *Quantum Computing and Supply Chain Management: A New Era of Optimization* (pp. 196–219). IGI Global., DOI: 10.4018/979-8-3693-4107-0.ch014

Singh, B., Dutta, P. K., & Kaunert, C. (2024). Replenish Artificial Intelligence in Renewable Energy for Sustainable Development: Lensing SDG 7 Affordable and Clean Energy and SDG 13 Climate Actions With Legal-Financial Advisory. In Derbali, A. (Ed.), *Social and Ethical Implications of AI in Finance for Sustainability* (pp. 198–227). IGI Global., DOI: 10.4018/979-8-3693-2881-1.ch009

Singh, B., Jain, V., Kaunert, C., Dutta, P. K., & Singh, G. (2024). Privacy Matters: Espousing Blockchain and Artificial Intelligence (AI) for Consumer Data Protection on E-Commerce Platforms in Ethical Marketing. In Saluja, S., Nayyar, V., Rojhe, K., & Sharma, S. (Eds.), *Ethical Marketing Through Data Governance Standards and Effective Technology* (pp. 167–184). IGI Global., DOI: 10.4018/979-8-3693-2215-4.ch015

Singh, B., & Kaunert, C. (2024). Human-Machine Nexus for Digital Rebound Fostering Futuristic Energy-Efficiency: Ecological Footprint of Technology for Smart-Sustainable Urban Mobility. In González-Lezcano, R., & Sansaniwal, S. (Eds.), *Design Strategies for Efficient and Sustainable Building Facilities* (pp. 222–244). IGI Global., DOI: 10.4018/979-8-3693-3200-9.ch010

Singh, B., & Kaunert, C. (2024). Hidden Gems Breakthrough Dynamic Landscape of Adaptive Learning Technologies for Higher Education: Bridging the Gap Between Theoretical and Practical Knowledge Projecting Student Learning Outcomes. In Minh Tung, T. (Ed.), *Adaptive Learning Technologies for Higher Education* (pp. 222–247). IGI Global., DOI: 10.4018/979-8-3693-3641-0.ch010

Singh, B., & Kaunert, C. (2024). Embryonic Machine-Deep Learning, Smart Healthcare and Privacy Deliberations in Hospital Industry: Lensing Confidentiality of Patient's Information and Personal Data in Legal-Ethical Landscapes Projecting Futuristic Dimensions. In Kumar, P., Singh, P., Diwakar, M., & Garg, D. (Eds.), *Healthcare Industry Assessment: Analyzing Risks, Security, and Reliability. Engineering Cyber-Physical Systems and Critical Infrastructures* (Vol. 11). Springer., DOI: 10.1007/978-3-031-65434-3_7

Singh, B., & Kaunert, C. (2024). AI-Driven Strategies for Customer Engagement, Market Segmentation, and Resource Optimization: Projecting End User Satisfaction and Futuristic Growth of Business. In Nadda, V., Tyagi, P., Singh, A., & Singh, V. (Eds.), *AI Innovations in Service and Tourism Marketing* (pp. 104–128). IGI Global., DOI: 10.4018/979-8-3693-7909-7.ch006

Singh, B., & Kaunert, C. (2024). Vertical Assimilation of Artificial Intelligence and Machine Learning in Safeguarding Financial Data. In Naz, F., & Karim, S. (Eds.), *Safeguarding Financial Data in the Digital Age* (pp. 169–197). IGI Global., DOI: 10.4018/979-8-3693-3633-5.ch010

Singh, B., & Kaunert, C. (2024). Computational Thinking for Innovative Solutions and Problem-Solving Techniques: Transforming Conventional Education to Futuristic Interdisciplinary Higher Education. In Fonkam, M., & Vajjhala, N. (Eds.), *Revolutionizing Curricula Through Computational Thinking, Logic, and Problem Solving* (pp. 60–82). IGI Global., DOI: 10.4018/979-8-3693-1974-1.ch004

Singh, B., & Kaunert, C. (2024). Wind and Solar Energy as Renewable Energy for Sustainable Global Future: Projecting Future Multi-Sector Sustainable Policies and Innovation. In Ara, A., & Thakore, R. (Eds.), *Promoting Multi-Sector Sustainability With Policy and Innovation* (pp. 210–245). IGI Global., DOI: 10.4018/979-8-3693-2113-3.ch009

Singh, B., & Kaunert, C. (2024). Aroma of Highly Smart Internet of Medical Things (IoMT) and Lightweight EdgeTrust Expansion Medical Care Facilities for Electronic Healthcare Systems: Fortified-Chain Architecture for Remote Patient Monitoring and Privacy Protection Beyond Imagination. In Hassan, A., Bhattacharya, P., Tikadar, S., Dutta, P., & Sagayam, M. (Eds.), *Lightweight Digital Trust Architectures in the Internet of Medical Things (IoMT)* (pp. 196–212). IGI Global., DOI: 10.4018/979-8-3693-2109-6.ch011

Singh, B., & Kaunert, C. (2024). Augmented Reality and Virtual Reality Modules for Mindfulness: Boosting Emotional Intelligence and Mental Wellness. In Hiran, K., Doshi, R., & Patel, M. (Eds.), *Applications of Virtual and Augmented Reality for Health and Wellbeing* (pp. 111–128). IGI Global., DOI: 10.4018/979-8-3693-1123-3.ch007

Singh, B., & Kaunert, C. (2024). Salvaging Responsible Consumption and Production of Food in the Hospitality Industry: Harnessing Machine Learning and Deep Learning for Zero Food Waste. In Singh, A., Tyagi, P., & Garg, A. (Eds.), *Sustainable Disposal Methods of Food Wastes in Hospitality Operations* (pp. 176–192). IGI Global., DOI: 10.4018/979-8-3693-2181-2.ch012

Singh, B., & Kaunert, C. (2024). Revealing Green Finance Mobilization: Harnessing FinTech and Blockchain Innovations to Surmount Barriers and Foster New Investment Avenues. In Jafar, S., Rodriguez, R., Kannan, H., Akhtar, S., & Plugmann, P. (Eds.), *Harnessing Blockchain-Digital Twin Fusion for Sustainable Investments* (pp. 265–286). IGI Global., DOI: 10.4018/979-8-3693-1878-2.ch011

Singh, B., & Kaunert, C. (2024). Harnessing Sustainable Agriculture Through Climate-Smart Technologies: Artificial Intelligence for Climate Preservation and Futuristic Trends. In Kannan, H., Rodriguez, R., Paprika, Z., & Ade-Ibijola, A. (Eds.), *Exploring Ethical Dimensions of Environmental Sustainability and Use of AI* (pp. 214–239). IGI Global., DOI: 10.4018/979-8-3693-0892-9.ch011

. Singh, B., & Kaunert, C. (2024). Unmanned aerial vehicle. *Cognitive Machine Intelligence: Applications, Challenges, and Related Technologies*, 108.

Singh, B., & Kaunert, C. Reinventing Artificial Intelligence and Blockchain for Preserving Medical Data. In *Ethical Artificial Intelligence in Power Electronics* (pp. 77–91). CRC Press. DOI: 10.1201/9781032648323-6

Singh, B., Kaunert, C., & Malviya, R. (2024). Coupling Human-Computer Interface Lensing Artificial Emotional Intelligence: Transforming for Futuristic Cognitive Recognition for Decision-Marking in Industry 5.0. In Kumar, N., Pal, S., Agarwal, P., Rosak-Szyrocka, J., & Jain, V. (Eds.), *Human-Machine Collaboration and Emotional Intelligence in Industry 5.0* (pp. 233–250). IGI Global., DOI: 10.4018/979-8-3693-6806-0.ch013

Smith, A., Harvey, J., Goulding, J., Smith, G., & Sparks, L. (2021). Exogenous cognition and cognitive state theory: The plexus of consumer analytics and decision-making. *Marketing Theory*, 21(1), 53–74. DOI: 10.1177/1470593120964947

Stone, M., Aravopoulou, E., Ekinci, Y., Evans, G., Hobbs, M., Labib, A., & Machtynger, L. (2020). Machine leaning (ML) in strategic marketing decision-making: A research agenda. *The Bottom Line (New York, N.Y.)*, 33(2), 183–200. DOI: 10.1108/BL-03-2020-0022

Ullah, A., Qi, G., Hussain, S., Ullah, I., & Ali, Z. (2024). The role of llms in sustainable smart cities: Applications, challenges, and future directions. *arXiv preprint arXiv:2402.14596.*

van Esch, P., & Cui, Y. (2021). Does consumer promiscuity influence purchase intent? The role of Machine leaning (ML), change seeking, and pride. *Journal of the Association for Consumer Research*, 6(3), 394–401. DOI: 10.1086/714503

Wirtz, J., & Lovelock, C. (2021). *Services marketing: People, technology, strategy*. World Scientific. DOI: 10.1142/y0024

Wong, L. W., Tan, G. W. H., Ooi, K. B., Lin, B., & Dwivedi, Y. K. (2022). Artificial intelligence-driven risk management for enhancing supply chain agility: A deep-learning-based dual-stage PLS-SEM-ANN analysis. *International Journal of Production Research*, •••, 1–21.

Yang, G., Ji, G., & Tan, K. H. (2022). Impact of artificial intelligence adoption on online returns policies. *Annals of Operations Research*, 308(1-2), 1–24. DOI: 10.1007/s10479-020-03602-y

Yau, K. L. A., Saad, N. M., & Chong, Y. W. (2021). Artificial intelligence marketing (AIM) for enhancing customer relationships. *Applied Sciences (Basel, Switzerland)*, 11(18), 8562. DOI: 10.3390/app11188562

Chapter 9
Role of Artificial Intelligence in the Study of the Blue Economy and Marine Ecosystem:
Use of AI in preservation of Marine EcoSystem

Sonia Mukherjee
https://orcid.org/0000-0002-1293-1342
MIT World Peace University, India

Gopal Wamane
Dr. Vishwanath Karad MIT World Peace University, India

Vishwajeet Pundkar
University College Dublin, Ireland

ABSTRACT

There is no universal definition of the term Blue Economy. However, the term is being widely used in the recent times in the context of climate change and environmental degradation. The rapidly increase in the plastic pollution and damage caused to the aquatic lives including coral reefs have now been a prominent cause of concern. Worldwide all the notable economies derive a good amount of revenue from the marine ecosystem. However, the increase in the environmental damage have impacted on the livelihoods of many individuals relying on the marine ecosystem for their survival. All the economies are looking into all possible forms of remedies to save the degradation caused to the blue economy. Artificial Intelligence (AI) is

DOI: 10.4018/979-8-3693-8684-2.ch009

being gradually adopted by all the economies in the world for generation of clean energy, monitoring the data used for examining the marine ecosystem and minimizing the harm done through destructive fishing practices and detecting the damage from plastic pollution. The use of AI will help to minimize such activities by alerting the authorities in a timely manner.

1. INTRODUCTION

The term "Blue economy" refers to the supportable usage of the marine resources in order to enhance the economic growth, healthier living, and good employment prospects. This includes the conserving the overall health and well-being associated with the marine ecosystems. It promotes the sustainable use of the existing resources. The idea of blue economy (BE) was first inserted in the year 2012. However, the definition of blue economy varies as per the place and category of the establishment or companies. The World Bank defined the blue economy as "*the use of the marine and ocean resources in a sustainable way to benefit the economies, source of revenue and marine ecology health*". The European Union further redefines BE as "*The different forms of economic happenings connected and associated with the oceans, seas and shorelines*". It covers a wide range of interconnected deep-rooted and emergent sectors". The Center for the Blue Economy defines the term Blue economy as a term widely used *around the world with three associated but separate meanings-the overall role of the oceans to the economies, the requirement to report the ecological and biological sustainability of the marine and the water economy as a mounting prospect for both the manufacturing and emerging country*". The United Nations defines the Blue Economy which encompasses an ambit of economic segments and accompanying strategies that collected regulate whether the usage of the marine resources are sustainable in nature or not". Sometimes the term BE is referred to as the "*ocean economy*". The blue economy is estimated to be worth more than US$1.5 trillion per year globally. Furthermore, a Sustainable Development Goal (SDG) devoted to the oceans and the aim is to preserve the marine resources and aquatic life. As per reports, the progress towards reaching these SDG has not progressed much.

The idea of Blue Economy had originated in the year 2012 in Rio Janeiro. The United Nations Conference on Sustainable Development had coined the concept of BE (UNCTAD, 2014). The United Nations presented the overall definition of the term Blue Economy as representative of an ocean economy aiming at "the upgrading and enhancing of the welfare of the human and helping in social equity. In addition to this, it reduces the environmental risks and ecological shortages (UN, 2014, p2). The World Bank termed the concept of "Blue Economy" in the form of supportable usage of aquatic lives for boosting the economic development, better chances of livings, and occupations while conserving the healthiness and well-being of the marine ecosystem (World Bank 2017, p.6). The connotation is a complete notion and acceptance of the marine sustainability extending from maintainable fisheries to environment well-being and thwarting contamination.

2. LITERATURE REVIEW

In today's times, the blue economy is facing numerous threats from events such as environmental change and tragedies, mainly demonstrated through the disturbance of marine-based business and the damages which are associated with the marine-related environments and the associated biodiversity (Karanja and Saito, 2018). Climate change have spread to modifications in the ocean weather, rise in the sea level, the increase in the acidity, water movement, and circulation pattern of ice, nonstop affecting seaside ecosystems and the animal life indispensable for the generation of seaside financial profits (Gaines et al., 2019). Furthermore, seaside threats include many more climate related events such as Tsunamis, downpours, and tremors do have a deep impact on the substantial encounters and on the resilience and sustainability issues. The expected rise in the occurrence and strength of hydrological, and climatic disasters including torrents, tropical tempests and scarcities in additional aggravates the helplessness in the context of the marine economy (IPCC, 2019). Moreover, with the onset of the Covid-19 pandemic worldwide have further deepened the problems associated with the blue economy and been harsh on the squeezed coastal tourism, fisheries, sea food manufacture and maritime conveyance (UNCTAD, 2020). Kathijotes (2013) stresses that the goal of the Blue Economy models is alteration of resources from shortage to plenty and to start undertaking urgent issues that is causing the ecological related problems. Mulazzani et al. (2016) had put onward the organization tool based on ecology service outline to resolve the coastal blue growth. Soma et al. (2018) study portrays the attainment of the long-term supportable blue growth in the form of teamwork, enclosure, and faith in the maritime segment. Sarker et al. (2018) also established an organization to highlight the growth in the blue economy and how to endorse further blue growth for the attainment of sustainable development goals (SDGs). Howard (2018) had an in-depth discussion on the part of shareholders in maintainable and supportable development. He found that the junction of the blue economy and maritime ecosystem, ecosystem bookkeeping is meticulously connected to blue growth (Häyhä and Franzese, 2014; Lillebø et al., 2017).

The Blue economy plays a vital part in society and focuses on improving human life and decreasing environmental dangers. Fisheries have created more than 350 million jobs worldwide, 80% of global trading is done by marine transport and aquaculture supplies 50% of the fish people consume. The concept has grown a lot in recent years and *Artificial Intelligence (AI)* can help achieve the goals the blue economy has set out. It can aid in pr*eserving marine life, create sustainable economic growth and enhance resource management.* With the help of AI, humans can locate the most suitable install locations. AI can look at factors such as weather-related incidents.

In recent years, there has been a rise of AI for dealing with severe environmental issues such as climate change. Hence, there is a need for strategies for dealing with the complex issues pertaining to climate change and loss of biodiversity. AI has been developed over the times as an effective tool which facilitate the formation of novel methods to adjusting to environment change. AI is known to have the potential to significantly help in numerous fields together with the recognition of high-risk places prone to environmental hazards and calamities and further formation of plans for societies (Rutenberg et al., 2021). The different AI-enabled remedies offer numerous advantages for environment change including locating all the vulnerable areas and inventing newer measures to protect organization and inhabitants from the effects of weather change (Bag et al., 2023). One application case is monitoring of the effects of climate events using devices and drones powered by AI technology including use of the predictive models developed using machine learning and algorithms. Additionally, AI has the capacity to maximize the use of resources efficiently and minimize the emissions and their harmful impact on climate (Zhang et al., 2020; Mihiretu et al., 2023. The primary advantage of AI mainly lies in precisely evaluating the vast quantities of data. It becomes a crucial tool and helps in minimizing the effects of climate changes in numerous forms such as flooding, landslides, or drought (Nost and Colven, 2022). It is also being increasingly used to assess data such as climate models satellite images, and weather patterns and permits government and societies to produce particular approaches dealing with the precise dangers (Chen et al., 2021). The use of Artificial Intelligence plans can estimate the possible impacts of increasing sea levels in numerous regions by following the patterns and connections. Such notable effort is FloodNet. The program examines the satellite photos and predicts real time flooding using the AI technique (Gangisetty and Rai, 2022). The river and waterbody pictures are examined using the deep learning algorithms. The system can estimate flooding situation depending on the water level, and landscape. For example, CoastalDEM project (Kulp and Strauss, 2018) which is used for AI to generate seaside elevation charts. AI can predict the upcoming temperature and precipitation shifts. AI models can estimate the possible impacts of weather change in numerous regions by examining huge databases of weather model output by means of machine learning algorithms (Jain et al., 2023). The recent projections available from the OECD point to the fact that by the year 2030, the marine economy has the potential to contribute over USD 3.0 trillion to worldwide GDP, including noteworthy growth predicted in maritime aquaculture, seaward wind energy, fish processing and ship building and restoration (OECD, 2016; Barbier, 2023; Betin and Titova, 2023; Doerr et al., 2023).

The usage of AI and generation of renewable energy offers incredible potential. However, it has numerous challenges. An important obstacle is the know-how gap for any effective AI projects. It requires the need for AI proficiency and a deeper understanding of the energy systems. However, in some situations the availability of faulty data can lead AI to give wrong solutions. The notable challenges should be taken into account for implementing AI technique in situations dealing with maritime conservation and biodiversity loss. The synthesis of artificial intelligence and renewable energy offers alternative sources for power generation. There is an urgent need for using clean energy for promoting sustainability in the future span of time. The rapid adoption of AI can make this change happen in an efficient manner.

Keeping this backdrop in mind, the study examines the role and contribution of blue economy in achieving the Sustainable Development Goals. Furthermore, the study would further analyse the role of AI for boosting the growth of blue economy and restoring the marine eco-system and list down notable challenges coming in the way of adoption of AI for the preservation of the marine eco-system. The study uses numerous case study to show the adoption, applications and challenges of using AI for restoring the marine ecosystem. The findings conclude that AI is significantly helping the planet in terms of restoration of the marine eco-system.

3. BLUE ECONOMY AND SUSTAINABLE DEVELOPMENT GOALS

Blue Economy has the potential to unlock manifold ecological, societal, and monetary benefits and helps in promoting the sustainable development. With the increase in the growth of the human population, there has been an increase in the demand for resources. The present resources in the form of seas, coastal areas, and inland waters are facing a crisis in the form of environmental change, experiencing higher rate of pollution and other forms of biodiversity harm putting the well-being of the present and forthcoming generations in jeopardy. Nations are progressively looking at the possible solutions to look into their marine economies. Simultaneously, the countries are addressing the tragedy and trying to achieve the goals set by 2030 in the form of Sustainable Development Goals.

The blue economy is equal to the world's 7th main economy and it has a yearly economic value projected at around US$2.5 trillion. The Blue Economy will be sustainable in nature provided it has the possibility to correct for all the pressing issues related to the disengaged way our marine, shorelines, and fresh water ecologies are presently handled as per the earth's economic, social and environmental needs. Furthermore, it delivers important assistances for present and upcoming generations; and it helps in restoring, protecting and maintaining the diverse, creative and strong environments. It will have the idea of promoting clean technologies, and focus on renewable energy. A hale and hearty marine sea is crucial for a sustainable upcoming future both for the individuals and the earth. There is an urgent need to harness the resources of our sea, and seaside ecosystems in meeting the future demand, and further ensure sustainability in the long run.

The fast unsustainable growth of marine related sectors is posing a greater environmental risk and is quickly finishing the sea's resource base. Developing countries are facing numerous challenges as they depend deeply on marine-based resources and are facing the consequences of rapid coastal and environmental change. A sustainable blue economy means more decent jobs, availability of cleaner energy, enhanced food safety and higher safety of the world's marine coastlines and local waters. The issue of climate resilience is a first focus of a sustainable blue economy. Supporting policy-coherence, and planning are significant ways to shape this resilience.

The term blue economy is essentially related to the Sustainable Development Goals (SDGs) 14. It is characterized as "*Life below Water*" which mainly focuses on the conservation and use of the marine resources in a sustainable way for the future sustainable development. To realise the SDG goals, The United Nations has acknowledged the decade "2021-2030" as the era of "*Ocean Science for Sustainable Development*". The rationale behind this is to support the consistent efforts in order to improve the marine health and gather marine stakeholders on a global scale to develop a mutual agenda. In relation to marine resources, the World Bank have emphasized on the key notion of Blue Economy (World Bank, 2017, p.4). The health and condition of the marine resources especially the oceans have radically deteriorated due to increase in the human activities in the forms of illegal industrial activities leading to higher pollution in addition to unmanageable ecological and biotic ruin (United Nations, 2016).

In one of the agenda of World Economic Forum Davos 2022, it was acknowledged that the G20 countries mainly have 45 percent of worldwide coastline and account for over 21% of the World's Exclusive Economic Zone. Furthermore, recent estimates suggest that marine sectors and events pay around USD 2.5 trillion to the worldwide economy. The traditional maritime sectors beginning to explore newer avenues for growth, and at the same time, the delicate state of the marine environment stresses on embracing the concept of a Blue Economy and further preserving the marine resources and building a resilient future. The paybacks of implementation of Blue Economy spread far outside the ecological sustainability as the approach is with line with the United Nation's Sustainable Development Goals (SDGs). As an example, the concept of supportable and manageable fisheries management supports SDG 2 (Zero Hunger). Coastal tourism executed sensibly will help in attaining the SDG 8 (Decent Work and Economic Growth) by generating employment prospects and improving home-grown economies. Marine preservation, ecologies and renovation efforts will help in achieving the SDG 14 (Life below Water), and protecting the maritime bionetworks. However, as per latest statistics, the SDG 14 faces a fund crunch. According to reports by the World Economic Forum, approximately $175 billion per year is required to attain the SDG 14 by the year 2030. Lastly, for achieving the SDG goals in the context of the Blue Economy, AI tool can be used to assist in reducing the environmental hazards and environmental stress growing in the present decade.

4. USE OF AI AND ITS ROLE IN BLUE ECONOMY

Artificial Intelligence (AI) has multiple uses in today's world. It can assist in making innovative marine accuracy models to predict the effects of environment change on oceans and improve the delivery and transport routes. These models are expected to support business group for achieving their own sustainability goals. AI can also help improve their efficiency in terms of maximum energy output and look into factors such as weather and ocean data and the latest AI technologies can reduce the dangerous aspects of the renewable energy systems posing for the marine life. The use of computer vision cameras, a form of AI technology, birds hovering close to the wind turbine can be noticed. Destructive fishing practise and illegal fishing activities continue to persist in today's world. The use of AI, will minimize such activities by alerting the authorities in a timely manner.

The use of Artificial Intelligence, can help marine biologists find the accurate number of plankton and fish stocks, using the latest technology to find out the exact number of fish, their size and further differentiate the different forms of fish from one another. This will be helping the scientists to study and recognize the lifecycle and lifespan of the marine species. Deep learning models can provide assistance to ecologists and scientists to count seals, based on archival photos, and monitor their populations and helping the scientists to recognize the evolution of the seal population more accurately.

AI helps in studying of the migration of marine species with the help of satellite tags. The satellite tags can collect latest data on animal location, depth of the water and temperature. AI furthermore helps in processing of data on relocation routes, their patterns and sends them to environmentalists. With the help of Machine Learning algorithms all kinds of future movements can be tracked providing information about the state of the ocean and how to protect the maritime environment and other aquatic species.

The use of sensors on animal tags, in the forms of accelerometers and cameras, AI can detect and infer the feeding behavior of species in the wild. The well-being of any marine populations in a right way depends on their feeding behavior. For example, throughout the study of penguin feeding behavior, accelerometers noticed shocks which happened when the penguins grasped their kill. Artificial Intelligence analyzed the data composed from the devices and provided a summary of the timetable of mealtimes, eating, extent of meals, etc.

Any normal or human caused devastation is capable of affecting the marine fauna. Therefore, it is important to identify these diverse forms of variations in a stretch and implement approaches to alleviate the adverse penalties. For example, Artificial Intelligence can be used to find interdependences in the coral reef ecosystems. This in turn will decide the level of priority conservation and the different sustainability measures. The Allen's coral atlas, are using Machine Learning approaches, charts and are monitoring the present situation of coral reefs through satellite images and making a comparison with the field images. Coral reefs are roughly one of the earth's most varied ecologies. As per the National Ocean Service, they are more than 4000 species and 800 hard corals and other forms of sea life. The coral reefs provide shelter for fishes in addition to protection of coastal areas. Regrettably, this ecology poses a threat from the rise in plastic pollution, negative fishing practises and environmental change. AI robots can then help in reconstruction of these crucial ecosystems and monitor their health.

An essential aspect of the blue economy is generating renewable energy and reducing environmental impact. One way to do this is through offshore wind farms that generate clean power. Additional methods of generating electricity through the help of ocean resources are tidal power plants and wave energy converters. With the help of AI, humans can locate the most suitable install locations. AI can look at factors such as weather and ocean data to determine where these clean-energy generating systems will provide the most benefit. AI technologies can also reduce the dangerous aspects these renewable energy systems can present for sea and other animal life. For example, computer vision cameras a form of AI technology can detect if birds are flying too close and stop the future turbine to keep the animal safe. In 2022, the German government implemented these AI cameras to protect endangered eagles. In addition to finding the ideal location to install clean generating systems,

AI can also help improve their efficiency in terms of maximum energy output. AI encompasses the modification and innovation of the systems and has the power to perform tasks predictably requiring through the application of humanoid intelligence.

Monitoring ocean ecosystems and collecting data is a complex, time-consuming and expensive task. However, autonomous underwater vehicles can quickly gather valuable insights humans can use in marine conservation. Destructive fishing practises have a significant impact on ocean biodiversity. If these illegal fishing activities persist, they can cause damage to marine ecosystems and to the people who depend on the ocean for their livelihood. With AI analysing different data, it can detect suspicious activity and alert authorities of the situation. AI can also assist the marine biotechnology industry as well as other blue economy sectors. For example, AI tool can be used to develop biofuel and pharmaceuticals. While most drugs currently come from land-based organisms, marine ecosystems could provide an excellent source for new ones. AI can enhance the screening process of components and materials gathered from these diverse ecosystems to aid in developing new medicine.

The application of AI from time to time can provide immense support in making newer and latest smart grids gradually over the time. Furthermore, it has the tendency to actively regulate the fluctuations and needs in the changes in the energy supply and demand. The usage of the past data and other actual factual statistics helps in an extensive variation of foundations, AI based grid management systems is capable of forecasting the energy consumption and vigorously adjusting the power generation and distribution. As a result, in the future there will be lesser reliance on fossil fuels as a backup option.

From optimizing the fish farming practises to reducing overfishing and improving overall fish health, AI can be used to predict the amount, worth and detailed dietary needs of fish inhabitants. Such analysis can reduce the environmental footprint of fishing happenings and further promote the sustainable fishing practices in the longer run. Maritime operators can use AI to improve ship routes, lessen fuel consumption and lessen the incidents of environmental emissions. This will further contribute to the decarbonization of the seas, water bodies and oceans, bringing into line with worldwide sustainability goals and plummeting the environmental effect of maritime transport. The key to mitigating water pollution is to identify the incidents related to pollution;

The maintenance of wind turbines and solar panels are one of the biggest for generation of renewable energy. The wind turbines and solar panels are frequently situated in isolated and inaccessible areas making labour-intensive inspections costly in nature. AI-driven predictive maintenance systems can analyse data from sensors embedded in these assets to forecast in a timely manner and further guaranteeing dependable energy delivery to customers.

Oceanographers and maritime environmentalists repeatedly face numerous difficulties and most of the problems are connected to the hazards to the human body. The profound oceans are unreachable, lacking access to light and are subject to life-threatening pressure. This is unswervingly disturbing research and leading to higher risk and resource intensive in nature. The responsibilities faced by outdated methods need enormous capitals and time, and are unsafe for individuals. Additionally, numerous vagaries and spectacle are indistinguishable to the human eye. Old-style approaches in the form of diver surveys and boat sampling need substantial expenses, time, and capitals but are still restricted to minor local ranges. Hence, a complete considerate of maritime ecologies necessitates big scale models and can be provided by AI.

4.1 Case Study: Use of AI in Coral Reef Restoration and Plastic Pollution

Water temperature in and around the Australia's Great Barrier Reef have risen to their warmest in the past 400 years and making the largest coral reef the most vulnerable species under threat as per a new study. The coral reef stretches for around 2400 km off the coast of the northern state of Queensland. Ocean temperatures in and around Australia's Great Barrier reef have hit the warmest level in over 400 years and pose an essential threat to the iconic reef. In a new study, an Australian team of scientists reconstructed sea surface temperature data from 1618 to 1995 by examining the coral skeleton samples in and around the reef and coupled the data set with recorded observations from 1990 to 2024. They found that in 2024, 2020, and 2017, the Coral Sea which lies off Australia's northeast coast and contains several reefs including the Great Barrier Reef experienced the warmest temperatures recorded in 400 years with 2024 the warmest year on record.

The coral reefs are denoted as the "*rainforest of the sea*" and is found to exist on the planet for approximately 450 million years. Coral reefs perform a vital role in the marine ecosystem. Nearly, thousands of the aquatic kind are found to be existing in the coral reef. For example, "*the Great Barrier Reef*" encompasses a number of around 400 coral types, 1500 fish kind, 4000 mollus kind and around six of the biosphere's seven turtle types. As per the statement by the Natural History Museum, around millions of unexplored types of creatures and kinds are existing in and everywhere on the coral reefs.

The coral reefs are known to deliver economic goods and services of value of around $375 billion every year. Approximately 500 million individuals around the world rest on on the coral reefs for their diet, revenue, and coastal defense from squalls and downpours. Coral reefs can engage up to 97% of the liveliness from surfs, squalls, and floods and further thwarts loss of life, possessions damage and soil corrosion.

Coral reefs remain one of the most valuable natural possessions and are perilous in nature. All these varied ecologies are homebased to approximately one fourth of the global marine life. They help us in protecting the coastlines from numerous storms coming from time to time, and offer sustenance and revenue for almost 100 million individuals. The annual economic value is around $36 billion. However, over the time they have been endangered and vulnerable by damaging fishing procedures to heating temperatures and further unmanageable or unsupportable coastal growth. Restoring the health of the reefs depend first upon the accuracy of predicting their health. The traditional monitoring efforts are unreliable and often yield erroneous results.

The prominent company Accenture had made partnership in terms of technology to a leading coral conservation foundation in *El Nido, Phillippines.* The objective is to develop a ground-breaking key for bringing back the health and well-being of the coral reefs. The project has been names as CORail. This is an AI-powered system for intensive care of the fish population and was first installed in a Coral reef in the Philippines in the year 2019. The tactically located photographic camera, fitted underwater and armed with the *Accenture Video Analytics Services Platform* and *Intel Hardware.* This will further allow for updated observations and help in detecting and photographing the marine life further classify and count the fish numbers. Till date, the project has composed more than 71000 images. This will help the scientists and researchers to gauge the reef health in real time. Furthermore, it could be used in studying the migration rate of the fish to colder climates and observing the disturbance in the threatened subaquatic zones.

In another similar incident (after the catastrophic coral bleaching incident in Australia), the prominent coral environmentalist Dr. Taryn Foster launched a firm termed as "Coral Maker". To restore the health of the coral reefs lost including the marine ecosystem and biodiversity, the firm, Coral Maker, has partnered with the expert firm Autodesk. The purpose is to create a ground-breaking answer for reef restoration using the latest AI technology ranging from vision systems and cobots. AI is being rapidly used to quicken the coral proliferation and bring back delicate maritime ecologies. The process of coral reef restoration is done through transplantation of tiny corals, cultured in nurseries and inserted on the injured reef. Autodesk have started using their software to give training to robots to choose up the minute corals and hence restore the coral restoration process.

The World's Oceans Day was an occasion to recap us around the accountability of defending our marine ecosystem at a worldwide measure. The sea's health is energetic and crucial in supporting us all. The United Nations have been calling for worldwide partnership to work in an organized way to generate ground-breaking and actual resolutions. Presently, researchers, scientists, and trendsetters are utilizing AI in a number of applications and methodologies to recognize and examine the conditions of our marine ecosystem. For example, underwater robots have been launched to understand and examine the climate change modelling with the help of the application of AI leading to creating solutions for our oceans.

The increase in the plastic pollution is indicating to the buildup of artificial or semi-synthetic plastic ingredients and its contamination in the atmosphere that is posing a threat to the creatures and towards the well-being of the human beings. The plastic elements may be of any form and size. They are also well connected to the other plastic associated materials. The extensive usage of plastics has led to substantial buildup of trash that is a danger to the bionetwork. Aquatic plastic contamination is a foremost cause of concern as even the utmost distant oceanic zones are precious. Presently, approximately 150 million heaps of materials made of plastic and their leftover are moving in the aquatic ecosystem. The increasing presence of the plastic polymers in the atmosphere in the long run will be dependent on the type of plastic used in the production process.

Plastics has the feature that it is durable in nature and play an important character in creating aquatic mess. In attendance, there is a rising apprehension about the effect of plastic left-over in seas, oceans and on the other forms of aquatic life and the consequence it bears on the individuals and other alive organisms that depend on the seafood (Abott and Sumaila 2019; Jambeck et al., 2015; Williams-Wynn and Naidoo, 2020). Aquatic ecosystems and marine networks are having important amounts of microplastics and other forms of plastics. In some situations, microplastics are seen in the different types of marine species as well as seafood (Akhbarizadeh et al., 2020; Zakeri et al., 2020).

In the coming years, the marine ecosystem will be affected through the increase in the cases of plastic pollution and the cases related to the ecological problem will be growing in numbers from the large proportion of scientific sections of the populations. The plastic leftover is accounting to nearly 41% of the marine waste. The latest statistics show that a projected 12 million heaps of plastic is inward bound towards the oceans annually and 142 million lots of trash are presently deposited in the maritime reserves (Horton, 2022). Plastic contamination has serious consequences for the aquatic and oceanic ecosystems. The absorption of plastics by flora and fauna and turtles has been extensively premeditated and at least 44% of the type of seabirds consuming plastics have been recognized (Xuan Son, 2021). Different forms of \fish and other marine creatures eat as much as 24,000 heaps of plastic every year and 25% of the fish are being available for human eating (Schmaltz et al., 2020; Selvaranjan, 2021).

The origin of the diverse plastic pollution in the marine ecosystem is heterogenous in nature. It was estimated that an approximate number lying between 307 and 925 million loads of waste has been discarded from Europe into the sea yearly with the plastic component representative of around 82% of the detected substrate, in numerous forms such as bottles, wrapping and bags. (Gonzalez-Fernandez et al., 2021). Approximately 60% of the marine mess off the coastline of the European nation is in the form of plastic waste. It has an estimated price to the European Union economy of around euro 259-695 million annually (Roberts, 2021). There is an urgent requirement in the form of multi-faceted approach in addition to policy intervention and using latest innovation technology for reducing the pollution.

AI has the potential and is helping in different forms to conserve the marine ecosystem in addition to fighting against the increase in plastic pollution with the help of machine learning. Machine learning is capable of tackling this problem of plastic pollution and its root causes. The ecosphere is generating more than 430 million tons of plastic and nearly two-third of them become left-over after the first use. As per the Ocean Conservatory, around eleven million metric loads of plastic are entering the sea and nearly 200 million metric tons are flowing into the sea. There are different technologies in the form of AI and ML have considerable potential and can tackle the problems associated with the ocean plastic pollution. Researchers from Wageningen University and EPFL in their study developed an AI based detector capable of tracking the marine debris through the satellite pictures. Furthermore, with the help of this technology plastic waste can be removed through the assistance of ships. The European Space Agency's Sentinel-2 satellite are recording the accumulation of garbage in the marine ecosystem. With the help of AI, these images can be examined and measures taken accordingly.

Razer created an automated robot that works based on the technologies ML and AI in the year 2021. The technology is capable of detection the marine plastic within a distance of two meters in choppy waters. The robot can collect around 250 kg of plastic in just one cycle. The AI models were competent to spot waste. Another non-profit organization *Ocean Cleanup*, has removed the plastic debris from the ocean with the help of advanced AI system. The main purpose is to preserve the marine life especially endangered sea turtles. In recent years, the ocean is facing many issues such as increase in the level of pollution, fishing of unsustainable nature and many issues related to the climate change. The AI technologies have been deployed to tackle all these major issues and leading to enhanced wellbeing of the ocean's health. *Overfishing* poses a crucial threat to the aquatic ecosystem. AI can support sustainable fisheries administration by examining the data on the majority of the fisheries population, migration patterns and other fishing activities. One of the models termed "Predictive models" is capable of forecasting the fish stock levels and further helping in reducing the issues related to overfishing.

AI improves in the prediction of the climate models and improved models help in more accurate predictions and further helps the policy makers to grow operative plans to alleviate environment change influences on timberland, seas, and oceans. FishialAI is a revolutionary attempt to use the power of AI to transform and further redevelop the fish preservation. The initiative has been backed by the Wye Foundation. The goal of the project is to produce a extremely precise, and open-ended AI models in order to identify the different variety of fish types on global scale. Over 851,678 fish images have been collected and approximately 100,954 of them have been curated. Around 289 species have been identified using the AI model. The FishialAI is an initiative to create a global community. Furthermore, it empowers everybody to partake in research related to marine and aquatic life, ecological preservation and the larger thoughtful of the marine ecosystems. The usage of the latest technology is an important step for the future goal towards conservation of marine species especially fishes.

AI is a powerful tool in terms of preservation of the marine ecosystem. It's role in terms of environmental preservation is indispensable. In the future, Ai will be offering new and innovative ways to preserve the natural resources and will ensure their health and vitality for the future.

AI technology is transforming the way we monitor underwater protected areas. With the help of AI driven automated systems, marine conservationists can collect and analyse data more efficiently and accurately than ever before. These advanced tools allow us to monitor vast ocean spaces with high resolution processing data from remote sources such as sensors and satellites at incredible speeds. AI can rapidly analyse satellite images and sensor data to identify patterns that indicate environmental change or human activities such as illegal fishing and oil spills that might harm these protected areas and marine ecosystem overall. The technology enhances our ability to maintain the health of marine protected areas (MPAs) allowing for timely interventions and more effective conservation efforts. *Ocean Mind* a non-profit organization based in the United Kingdom successfully aided identifying potential illegal unreported and unregulated (IUU) fishing risks within the Pitcairn Island Marine Reserve over a span of five years.

Eminent Professor Tainlu Wang (University of Hawaii) have tried to explore into the details of the transformative potential of soft robotics in matters of marine preservation. Fabricated with flexible materials, The soft robots are fabricated with flexible materials and it is an innovative approach for the benefit of the marine environments. The latest technology will emulate and track the fluid movements of the jellyfish, and offer measures and solution through their design and there will be minimal disturbances for the marine life. The latest technological revolutions offers promises for various applications in marine preservation efforts, across various domains including mechanical and electrical domains. Lastly, these robots offer the potential for safe interaction in the underwater environments

Coral bleaching have been seen in the numerous areas and places including the Caribbean Sea, South Atlantic, Red Sea, Gulf of Mexico, Western Indian Ocean and the waters of East Asia. As per the Global Coral Reef Monitoring Network, incidents of repeated bleaching of the corals are leading to their disappearance. As per the estimates, 14 percent of the coral reefs have been lost between the year 2009 to 2018. The sea and other water bodies especially the oceans are covering around 70% of the planet. The well-being and health of the marine life have degraded rapidly over the years and around 3.3 billion individuals have their employment connected with the marine life. The marine litter and contamination is reducing the employment prospects to a greater extent. It is estimated that predicted around $4-6 trillion comes from the ocean economy and it is expected to increase further by the year 2030. However, the restoration of the marine health is intertwined to the generation of value added and increase in employment opportunities for those dependent on the marine sector.

With the increase usage of the digital technology, ocean industries are capable of addressing the crucial challenges in the face of the quickly altering environment and can address one of the notable challenges in terms of addressing the well-being and restoration of the marine ecosystems and the associated living species in the form of survival of numerous aquatic lives. AI technology can assist in matters of decarbonizing the blue economy and alleviate the negative effects of environment change on the marine life and marine ecosystems. Effective digital apparatuses can be used to restore the damage done to the marine ecosystem and reduce contamination in the sea.

There is an increasing prospect to perform the fishing practices in a sustainable manner. By using the machine perception tools to feed the fish as per the requirements in terms of hunger and reduce the waste and carbon emissions. Fish feeding is causing substantial harm to by increasing the waste of unfed materials in the ocean. Newer technologies could minimise the ocean pollution and examine the data on fish populations, their health issues, the conditions of their breeding and the environment in which they are likely to breed more. The number of yields in terms of varieties of fish can be improved in a sustainable manner in the long run.

5. CHALLENGES OF ADOPTING AI IN BLUE ECONOMY

The biggest challenge of implementing AI into blue economy sectors is many organizations are unaware of its benefits. For this technology to become more widely used, sectors will need education on all its advantages. With more industries realizing all the upsides, it can help the blue economy create sustainable growth and preserve the ocean's ecosystem. With AI steadily becoming part of many industries, the sustainable future of the blue economy looks strong. AI has made great strides in the last decade and is now at a point where it can provide many advantages. This technology can help monitor the health of the ocean's ecosystems, restore them, aid with the design of renewable energy systems and provide valuable data regarding ocean health. Sectors can use this insight to propel the blue economy forward.

Around 40% of the ecosphere's inhabitants live nearby the coastal areas. Furthermore, 3 billion individuals use the oceans, sea and waterbodies for their living. Acidification, contamination, marine warming, and fishery ruin are prominent examples of the damage of the maritime ecosystems. The possible damages and hazards are harmful to earth and there will be long-term consequences if urgent action is not taken for the protection of the marine ecosystems.

But AI can only be as good as the information is given. If the data is flawed or biased, the AI's conclusion may be inaccurate or unfair. Imagine only showing AI pictures of cartoon Orcas and expecting them to recognize a real Orca. Every one of us is unique, as is the data fed to AI. For AI to be effective, it must understand and recognize the vast diversity in the data. It's like appreciating that every snowflake is different.

There are complexities in the ocean systems, stressing the importance of accurate pattern prediction. Oceanic predictions rely on ample robust data. Halifax, renowned for its sporadic hurricanes, serves as an ideal example. The unpredictability of these storms necessitates a wealth of multi-year data for precise forecasts. A comprehensive history of similar events is essential for understanding their patterns. AI comes to the fore by helping to manage and analyse the massive amounts of data that will be discovered and create the advanced ecosystems required. AI helps us to drive the data collection understand the data and deliver actionable insights around the data.

One of the challenges in attempting to manage the climate in the real world rather than just the digital world as human and we have evolved to solve problems case by case. However, from a climate and eco-system perspective when you treat a problem in isolation you end up making the whole ecosystem worse. There is a high probability of making the wrong choices repeatedly. We need to understand and measure the whole eco-system both on the business and nature side.

Recently, in aquaculture, they have developed an underwater autonomous robot that optimizes Salmon production in fish farms. Till date, more than 800 of these robots are operating along the Norwegian coastline inside fish pens. Each pen can contain up to 200,000 Salmon and the robots observe and if they are hungry, they automatically start feeding. They also look after the Salmon for such things as the presence of sea lice and integrity of the pen. Regarding fish farming, the feed is the biggest emitter of CO_2 emissions at around 60 percent, so ensuring that only the appropriate amount of food is dispensed is crucial to sustainability. The producers are able to produce more fish aided by AI projections.

Plastic pollution in the ocean is one of the major issues affecting marine life causing significant damage to habitats such as coral reefs and seagrass beds and harming marine animals. AI can play a crucial role in addressing this problem by rapidly identifying plastic objects in the ocean with high precision enabling timely clean-up operations. This proactive approach can help mitigate the environmental impact and protect the marine ecosystems more effectively.

AI optimizes fishing operations to reduce environmental impact by predicting optimal times and locations for fishing, thereby minimizing the catch and ensuring more targeted and sustainable harvests.

AI supports the creation of eco-friendly methods and technologies, encouraging efficient and responsible fishing practises. This includes developing precision aquaculture techniques and promoting habitat restoration efforts to maintain marine ecosystems.

The implementation of AI technology presents several significant challenges. These include-

Investment requirements- Implementing AI technology in the maritime industry requires substantial initial investment in hardware, software and training. The initial costs can be restrictive for smaller organization and developing countries limiting widespread adoption.

Establishing the necessary infrastructure for AI such as data collection systems, high-speed internet, and computing power adds to the financial burden. This can be a significant barrier especially in remote or underdeveloped regions.

Reliability Issues- The maritime industry's dependence on AI technology presents risks in association with the system sometimes malfunctioning in a quicker span of time. Technical glitches can lead to significant operational disruptions and financial losses.

Skill gaps- The implementation and maintenance of AI systems require specialized knowledge and skills. There is a need for continuous training and education to ensure that the workforce can efficiently be able to utilize the AI knowhows.

Adaptability- The rapid advancements and innovations in the use of AI technology makes the latest technology outdated in a shorter span of time. Continuous updates and adaptations are necessary to keep AI systems relevant and effective which can be challenging and resource intensive.

Data Privacy- The usage of the AI structures relies on vast quantities of data. Hence, there is a huge concern regarding the privacy and security of the delicate statistics. In the maritime industry, data privacy issues can arise from the collection and use of data related to vessel movements, fishing activities and environmental monitoring. Ensuring that data collection and usage comply with privacy laws and regulations is crucial to protect individuals, companies and proprietary information. While wildlife data may be public, operational data from vessels and specific fishing practices can be sensitive and require careful handling.

6. CONCLUSION

The term blue economy is widely well-defined as the maintainable, supportable and manageable use of marine resources for financial development, better livings and occupation creation. In addition, it also has prominent role in conserving the well-being of the marine ecosystems. In the year August 2022, the United Nation's Sustainable Blue Economy Investment Forum have seen different types of organization including the governments, banks and companies have come together at the *UN Ocean Conference.* They have committed to invest billions of dollars for the purpose of preserving the Blue Economy. The growing global interest in developing technologies that are responsible for the use of marine resources and fight climate change have become essential and a prominent firm in this direction is DeepSense. The firm is focused on adoption of AI in the ocean sector for the purpose of the growth of the Blue Economy. Deep Sense is working with other companies to classify and total fish passing done through the hydroelectric dams.

AI is also working in the direction of cleaning the ocean garbage. AI-powered autonomous floating garbage collecting system. The Ocean Cleanup invented by Dutch inventor Boyan Slat is cleaning up the Great Pacific Garbage Patch. The area has been found to contain the largest buildup of ocean plastic approximately (1.6 million square kilometres). It is located amid the regions Hawaii and California. AI and ML powered drones are being trained to recognize trash and plastic waste in oceans and pinpoint the location so that people can target their clean-up efforts accordingly. Satellite observation, image detection of ocean plastics will encourage tech enthusiasts to put in efforts in the direction. AI and related technologies are trying to resolve several pressing issues around ocean health. There is an abundance of data gathering through various initiatives. Deducing sensible results out of this data requires expertise in the field which is oceanography and physics. This is where ML comes as a rescue; it clubs the required expertise and data into meaningful results. We are all witnessing various awareness drives to stop plastic from entering our oceans. But it is also critical to make efforts to remove the plastic that is already inside and hence is entering the food chain and damaging ecosystems.

AI has the potential to revolutionize our understanding of oceans. For instance, numerous companies can employ AI to study different kelp species, using abundant footage and labelled data to track changes in kelp biodiversity and health, thereby contributing to a broader comprehensive of maritime ecosystems. In the shipping industry, AI effectively streamlines processes, particularly in managing extensive shipping logs through the use of Natural Language Processing, enhancing efficiency for shipping companies and border security. Additionally, AI aids in predictive maintenance by shifting through extensive records for many marine vessels, preventing potential malfunctions. Moreover, AI plays a pivotal role in shipping and port management, with sensor-equipped buoys collecting critical data that AI algorithms can analyze to provide valuable insights benefiting various industries from fisheries to shipping.

In the *Bay of Fundy*, tidal shifts occur dramatically making it a hub for tidal energy research. One innovative method is using an echosounder towed behind a boat to ping the ocean floor and detect entities in the water column, primarily fish. However, challenges arise from interferences caused by surface waves and air bubbles, complicating the task of extracting valuable data from the noise. Traditionally, this was a manual, tedious, and subjective process. Enter neural networks. By training these networks on historical data, the process of automating and streamlining data separation becomes more consistent and efficient.

Such technological advances aren't limited to data filtering. A project in Dartmouth at the *Center for Ocean Ventures (COVE)* uses computer vision to observe mussels. These aquatic creatures, with their environmentally responsive behaviours, potentially act as early indicators of water toxicity. By monitoring their opening and closing patterns, AI can detect anomalies faster than traditional sensors, aiding in quicker responses to potential hazards. Another integration of AI can be seen around hydro dams. While they represent a renewable energy source, they pose risks to migrating fish. In collaboration with Nova Scotia Power, Innova sea has developed a system using sonar and video data to detect, categorize, and count the fish, reducing the potential harm while maximizing the energy output.

Presently we are residing in an era of narrow AI, where AI excels at specific tasks based on human-fed data. It's about mimicking and not innovating. As we look towards the future especially in the realm of ocean sciences, we must consider the ethical implications of AI. It is a tool that holds great potential but equally great risks if not employed responsibly. Transparency in its creation, application and decision-making process is key. This responsibility falls not just on developers but on users and stakeholders too.

To ensure that oceans remain a vital, thriving and sustainable ecosystem, our strategies for exploration, research and conservation need to be as dynamic and interconnected as the waters themselves. The fusion of AI and oceanography provides a promising avenue, but like all tools, it requires wisdom in its use. According to the World Economic Forum, Scientists are increasingly using the innovative technology AI and underwater microphones. All these devices will be used to record the sounds and silence of different aquatic life including fish, whales and other marine lives.

Furthermore, they will be used to monitor the movements of dolphins and other sea life. There are eminent Scientists such as Mooney who want to conserve the marine ecosystem. In this direction, they recorded a health coral reef ten years back. To restore the dying coral reef, they broadcasted the sound to encourage coral larvae to return back to normalcy.

As per the Guardian, in the year 2023, some marine scientists researching on the Goa's coral reef recorded a reef and plugged the sounds into an algorithm that immediately identified the species. This identification is gradually opening new doors for further ocean discoveries. The seaside and maritime ecosystems in the forms of mangrove forestry, seagrass grazing land and salty marshlands have been acting as an extremely effective carbon sink. They are also habitats to diverse marine animals. Preserving and reinstating all these diverse ecosystems will alleviate the level of greenhouse gas emissions and help in further preserving the biodiversity. The marine sea life offers a significant source of renewable energy through various forms including wind, and tidal energy. The multiple use of different renewable energy sources can be particularly effective for nations such as India. Towards this direction, India has implemented a Coastal Regulation Zone announcement to categorize and be able to take care of the coastal regions and preserve environmentally complex seaside and maritime areas, including vital ecologies in the form of mangrove forestry, depleting coral reefs, and other aquatic forms of lives residing near the coastline.

The global population is quickly rising as per the UN estimates. The tentative figure suggests that we will be reaching a figure of around 8.5 billion by the year 2030. The aquaculture industry can support the growing global need. Nearly 17% of our present protein foundations come from fish and around 80% of the global fisheries are exhausted. There is a need for urgent action for minimizing the pollution in our oceans for the sustainable growth of the fisheries sector and preserving the marine ecosystem at the same time.

REFERENCES

Akhbarizadeh, , Dobaradaran, S., Nabipour, I., Tajbakhsh, S., Darabi, A. H., & Spitz, J. (2020). Abundance, composition, and potential intake of microplastics in canned fish. *Marine Pollution Bulletin*, 160, 111633. Advance online publication. DOI: 10.1016/j.marpolbul.2020.111633 PMID: 33181921

Bag, , Sabbir Rahman, M., Rogers, H., Srivastava, G., & Harm Christiaan Pretorius, J. (2023). Climate change adaptation and disaster risk reduction in the garment industry supply chain network. *Transportation Research Part E, Logistics and Transportation Review*, 171, 103031. DOI: 10.1016/j.tre.2023.103031

Barbier, E. B. (2023). Greening the ocean economy. *Front. Environ. Economics*, 2, 1096303. Advance online publication. DOI: 10.3389/frevc.2023.1096303

Betin, O., and Titova, G. (2023). The formation of the economy of the world ocean. Rybnoe Hozâ jstvo 2023, 18–23. DOI: 10.37663/0131-6184-2023-2-18-23

Chen, , Zou, X., Li, K., Li, K., Yang, X., & Chen, C. (2021). Multiple local 3D CNNs for region-based prediction in smart cities. *Information Sciences*, 542, 476–491. DOI: 10.1016/j.ins.2020.06.026

Doerr, . (2023). *"Marine entrepreneurship," in Routledge eBooks*. Routledge., DOI: 10.4324/9781003058151-18

Gaines, . (2019). *The expected impacts of climate change on the ocean economy*. World Resources Institute.

Gangisetty, S., & Rai, R. R. (2022). FloodNet: Underwater image restoration based on residual dense learning. *Signal Processing Image Communication*, 104, 116647. DOI: 10.1016/j.image.2022.116647

Gonzalez-Fernandez, D., Hanke, G., Pogojeva, M., Machitadze, N., Kotelnikova, Y., Tretiak, I., Savenko, O., Bilashvili, K., Gelashvili, N., Fedorov, A., Kulagin, D., Terentiev, A., & Slobodnik, J. (2022). Floating marine macro litter in the Black Sea: Toward baselines for large scale assessment. *Environmental Pollution*, 309, 119816. Advance online publication. DOI: 10.1016/j.envpol.2022.119816 PMID: 35872285

Häyhä, T., & Franzese, P. P. (2014). Ecosystem services assessment: A review under an ecological-economic and systems perspective. *Ecological Modelling*, 289, 124–132. DOI: 10.1016/j.ecolmodel.2014.07.002

Horton, A. A. (2022, January). Horton (2022). Plastic pollution: When do we know enough? *Journal of Hazardous Materials*, 422, 126885. Advance online publication. DOI: 10.1016/j.jhazmat.2021.126885

Howard, B. C. (2018). Blue growth: Stakeholder perspectives. *Marine Policy*, 87, 375–377. DOI: 10.1016/j.marpol.2017.11.002

IPCC. (2019). Technical Summary: IPCC Special Report on the Ocean and Cryosphere in a Changing Climate. Eds. H. O. Pörtner, D. C. Roberts, V. Masson-Delmotte, P. Zhai, E. Poloczanska, K. Mintenbeck, M. Tignor, A. Alegrıa, M. Nicolai, A. Okem, J. Petzold,́ B. Rama and N. M. Weyer (Geneva: IPCC).

Jain, H., Dhupper, R., Shrivastava, A., Kumar, D., & Kumari, M. (2023). AI-enabled strategies for climate change adaptation: Protecting communities, infrastructure, and businesses from the impacts of climate change. *Computational Urban Science*, 3(1), 25. DOI: 10.1007/s43762-023-00100-2

Jambeck, , Geyer, R., Wilcox, C., Siegler, T. R., Perryman, M., Andrady, A., Narayan, R., & Law, K. L. (2015). Marine pollution. Plastic waste inputs from land into the ocean. *Science*, 347(6223), 768–771. DOI: 10.1126/science.1260352 PMID: 25678662

Joshua, K Abbott & U Rashid Sumaila. (2019). Reducing Marine Plastic Pollution: Policy Insights from Economics. *Review of Environmental Economics and Policy, Association of Environmental and Resource Economists*, 13(2), 327–336.

Karanja, J. M., & Saito, O. (2018). Cost-benefit analysis of mangrove ecosystems in flood risk reduction: A case study of the Tana Delta, Kenya. *Sustainability Science*, 13, 503–516. DOI: 10.1007/s11625-017-0427-3

Kathijotes, N. (2013). Keynote: Blue economy - environmental and behavioural aspects towards sustainable coastal development. *Procedia: Social and Behavioral Sciences*, 101, 7–13. DOI: 10.1016/j.sbspro.2013.07.173

Kulp, S. A., & Strauss, B. H. (2018). CoastalDEM: A global coastal digital elevation model improved from SRTM using a neural network. *Remote Sensing of Environment*, 20, 231–239. DOI: 10.1016/j.rse.2017.12.026

Lillebø, A. I., Pita, C., Garcia Rodrigues, J., Ramos, S., & Villasante, S. (2017). How can marine ecosystem services support the blue growth agenda? *Marine Policy*, 81, 132–142. DOI: 10.1016/j.marpol.2017.03.008

Mihiretu, , Okoyo, E. N., & Lemma, T. (2023). Climate variability trendscommunity perspective-livelihood adaptation strategy nexus in the arid-tropics, Ethiopia. *Journal of Arid Environments*, 210, 104929. DOI: 10.1016/j.jaridenv.2022.104929

Mohabeer, R., & Roberts, J. L. (2020). Promoting the Blue Economy: The Challenge. In *Shaping the Future of Small Islands: Roadmap for Sustainable Development* (pp. 251–268). Springer Nature Singapore.

Mulazzani, , Trevisi, R., Manrique, R., & Malorgio, G. (2016). Blue growth and the relationship between ecosystem services and human activities: The Salento artisanal fisheries case study. *Ocean and Coastal Management*, 134, 120–128. DOI: 10.1016/j.ocecoaman.2016.09.019

Nost, E., & Colven, E. (2022). Earth for AI: A political ecology of data-driven climate initiatives. *Geoforum*, 130, 23–34. DOI: 10.1016/j.geoforum.2022.01.016

OECD. (2016). *The Ocean Economy in 2030*. OECD Publishing., DOI: 10.1787/9789264251724-

Rutenberg, . (2021). *"Use and impact of artificial intelligence on climate change adaptation in Africa," in African Handbook of Climate Change Adaptation* (Oguge, N., Ayal, D., Adeleke, L., & da Silva, I., Eds.). Springer International Publishing.

Sarker, , Bhuyan, M. A. H., Rahman, M. M., Islam, M. A., Hossain, M. S., Basak, S. C., & Islam, M. M. (2018). From science to action: Exploring the potentials of Blue Economy for enhancing economic sustainability in Bangladesh. *Ocean and Coastal Management*, 157, 180–192. DOI: 10.1016/j.ocecoaman.2018.03.001

Soma, , van den Burg, S. W. K., Hoefnagel, E. W. J., Stuiver, M., & van der Heide, C. M. (2018). Social innovation – a future pathway for blue growth? *Marine Policy*, 87, 363–370. DOI: 10.1016/j.marpol.2017.10.008

Son, X.. (2021). Vietnam's Regulations to Prevent Pollution from Plastic Waste: A Review Based on the Circular Economy Approach. *Journal of Environmental Law, Oxford University Press*, 33(1), 137–166.

UNCTAD. (2014). United Nations Conference on Trade and Development, The Ocean Economy: Opportunities and Challenges for Small Isalnd Developing States, available at https://unctad.org/en/publicationslibrary/ditcted2014d5_en.pdf

United Nations. (2014). Blue Economy Concept Paper Available at https://sustainabledevelopment.un.org/concent/documents/2978BEconcept.pdf

United Nations Conference on Trade and Development (UNCTAD). (2020). The COVID-19 Pandemic and the Blue Economy: New challenges and prospects for recovery and resilience (Geneva: UNCTAD).

Williams-Wynn, M. D., & Naidoo, P. (2020). A review of the treatment options for marine plastic waste in South Africa. *Marine Pollution Bulletin*, 161, 111785. Advance online publication. DOI: 10.1016/j.marpolbul.2020.111785 PMID: 33128983

World Bank. (2017). *The potential of the Blue Economy: Increasing long-term benefits of the sustainable use of marine resources for small island developing states and coastal least developed countries*. World Bank.

World Economic Forum. 2016. The New Plastics Economy: rethinking the future of plastics. https://www.ellenmacarthurfoundation.org/assets/downloads/EllenMacArthurFoundation_TheNewPlasticsEconomy_Pages.pdf

Zakeri, , Naji, A., Akbarzadeh, A., & Uddin, S. (2020). Microplastic ingestion in important commercial fish in the southern Caspian Sea. *Marine Pollution Bulletin*, 160, 111598. Advance online publication. DOI: 10.1016/j.marpolbul.2020.111598 PMID: 32871433

Zhang, , Ruiz-Menjivar, J., Luo, B., Liang, Z., & Swisher, M. E. (2020). Predicting climate change mitigation and adaptation behaviors in agricultural production: A comparison of the theory of planned behavior and the Value-Belief-Norm Theory. *Journal of Environmental Psychology*, 68, 101408. DOI: 10.1016/j.jenvp.2020.101408

Chapter 10
Uplifting Smart Sensors for Transforming Urban Transport and Parking:
Future of Urban Mobility Fostering Sustainability

Gunjan Srivastava
Babu Banarasi Das University, Lucknow, India

Purva Mange
Symbiosis International University, Pune, India

Tarun Kumar Kaushik
Sharda University, India

ABSTRACT

The growing complexity of urban life has led to the emergence of the notion of "smart cities." These pioneering urban areas are leading the way in terms of efficiency and sustainability, utilizing technology to enhance the overall standard of living. The integration of advanced IoT technology with wireless sensors is reasonable, as it serves as the basis for intelligent parking and transit systems. Real-time data analytics serves as the fundamental basis for intelligent transportation systems. With empowering communities to make decisions based on data, this trend reduces traffic and enhances the overall efficiency of transportation. The incorporation of wireless sensor technology with parking systems has led to a novel approach to urban mobility. This chapter examines the interaction between Highly Intelligent IoT and wireless sensor technologies in influencing the development of smart cities with a particular focus on intelligent and secure transportation and parking solutions.

DOI: 10.4018/979-8-3693-8684-2.ch010

1. INTRODUCTION

The smart city is demarcated as an urban infrastructure organization system that assimilates different information and communication technologies (ICT), including the Internet of Things (IoT). Urban infrastructure may interact with the public via ICT, track the development of the city and improve living conditions (Saranya & Amutha, 2024). Real-time data supplied by sensors must be processed and analyzed in order to use IoT systems in smart cities (Ullah et al., 2024). With the knowledge gained from the data, this study helps to resolve problems related to efficiency (Shahzad et al., 2021). This trajectory has the potential to completely transform parking for both locals and tourists (Varadharajan, 2021). The crucial factor to take into account is making sure smart parking systems are secure. Robust security measures such as encryption and authentication procedures are made possible by highly intelligent IoT (Ang et al., 2022). In the context of smart cities, the dynamic trajectories of wireless sensor technologies and the highly intelligent Internet of things (IoT) with a particular emphasis on the revolutionary effects these technologies have on parking solutions and safe, smart mobility. These sensors enable effective space management and lessen traffic congestion by providing real-time information on parking spot occupancy (Hanes et al., 2017). This adds to the continuing discussion about the creation of smart cities and sustainable urban ecosystems by providing a thorough examination of existing implementations, difficulties and future prospects (Baskar et al., 2024).

The Internet of Things (IoT) is a network that amalgamates diverse devices and technologies, thereby reducing reliance on human intervention (Vohra et al., 2024). This capability facilitates the creation of smart cities globally, enabling enhanced efficiency and functionality. with facilitating the coexistence of different technologies and facilitating their interaction, the IoT has propelled the evolution of smart city systems aimed at promoting sustainable living and enhancing citizens' comfort and productivity (Shen et al., 2018). Through a comprehensive exploration of literature, case studies, and real-world implementations, the paper aims to elucidate the potential of these technologies in contributing to urban sustainability. With examining challenges and proposing future directions, the research seeks to provide valuable insights for policymakers, urban planners and researchers engaged in the ongoing evolution of Smart Cities. The IoT for Smart Cities encompasses various domains and relies on a range of underlying systems to operate effectively (Giourka et al., 2019).

1.1 Background and Significance of Smart Urban Environments

There are many components that make up a smart city and the four main components of smart city applications are usually data collection, transmission/reception, storage and analysis (Mitieka et al., 2023). Data gathering is dependent on the requirements of the application and has been a major factor in the development of sensors in many different fields (Bagga et al., 2022). The second component is data interchange which is the transfer of data from data gathering devices to the cloud for analysis and storage (Nguyen et al., 2024). Several technologies are used to accomplish this goal such as citywide Wi-Fi networks, 4G and 5G technologies, and local networks that can transmit data both locally and globally 9Abi Sen et al., 2023). In order to prepare data for the fourth step, data analysis, several storage techniques are employed in the third stage, which involves cloud storage (Cardullo, 2020). The availability of cloud infrastructure makes it possible to collect, store, process and analyze heterogeneous data in real-time utilizing machine learning and deep learning algorithms as well as statistical techniques, for more complicated decision-making processes (Sani et al., 2024).

1.2 Intelligent IoT and Wireless Sensor Technologies: Overview and Relevance

With each new technological advancements the society is filled with more sophisticated tools that make our everyday lives easier and more productive (Bibri & Bibri, 2018). The Internet of Things (IoT), which connects numerous smart gadgets like smartphones, refrigerators, watches, fire alarms, door locks and more, enables smooth data exchange and communication between them is one of the most important technical innovations of our day (Miao & Phelps, 2019). The everyday activities, especially those involving transportation now heavily rely on IoT technology (Huda et al., 2024). They are particularly interested in smart transportation because of its potential to revolutionize the transportation of people and products (Almutairi et al., 2024). In smart cities, IoT provides drivers with several advantages, such as better logistics, safer roads, more effective parking systems, and better traffic control (Shackelford, 2020). Integrating these advantages into transportation applications is the goal of smart transportation (Chaloupkova & Janurova, 2019). Distributed ledgers, big data and machine learning are a few other technologies that have been investigated to augment the benefits offered by smart transportation (Silva et al., 2021). Road maintenance, anomalous traffic situation detection, parking places, street lighting, accident avoidance and route optimization are all areas where these technologies are put to use (Zhao et al., 2022).

1.3 Significance/ Importance of Transportation and Parking Systems in Urban Sustainability

Global urbanization has resulted in serious transportation-related issues for cities all over the world. Air pollution, traffic jams and rising greenhouse gas emissions are all consequences of the increasing number of automobiles on the road (Jain et al., 2021) (Lefebvre, 2021). Parking policies have become a critical area for change implementation as cities strive to lower their carbon footprint and improve the quality of life for its citizens (Jeong & Shin, 2020). A sustainable parking strategy is a collection of tactics and regulations intended to lessen the adverse effects of urban parking on the environment, society and economy (Yashi et al., 2024) (Oluwaseyi et al., 2024). These regulations encourage effective parking space usage to reduce traffic and pollution, as well as giving preference to environmentally friendly forms of mobility including walking, bicycling and public transportation over private vehicles (Brunetta & Caldarice, 2020).

A sustainable parking strategy must be included into urban planning in order to successfully solve these problems (Ullah et al., 2018). In order to practice sustainable urban planning, one must approach urban development holistically and take into account the long-term social, environmental and economic effects of growth and expansion (Leal Filho et al., 2020). With an emphasis on social justice, economic viability, and environmental sustainability, it seeks to build communities that are livable, dynamic and resilient (Liu et al., 2022). In order to create livable and sustainable communities that satisfy current demands while also protecting resources and possibilities for future generations, sustainable parking policies and sustainable urban planning are essential (Kumar et al., 2021).

1.4 Objectives of the Chapter

The chapter aims to contribute valuable insights to the ongoing discourse on shaping Smart Cities through the incorporation of highly intelligent IoT and wireless sensors, specifically focusing on the domains of smart and secure transportation and parking for a sustainable future (Bhardwaj et al., 2024) (Escap, 2022). It has the following objectives to:

- assess the various applications of IoT in shaping urban infrastructure with a particular focus on transportation and parking.
- evaluate the impact of these sensors on real-time monitoring, traffic management and predictive analytics for enhanced urban mobility.
- measure the effectiveness of real-time parking availability, reservation systems and enforcement mechanisms in optimizing urban parking spaces.

- highlight the potential societal and economic benefits of adopting highly intelligent IoT and wireless sensor technologies in urban environments.

Figure 1. Diverse Objectives of this Chapter (Source- Original)

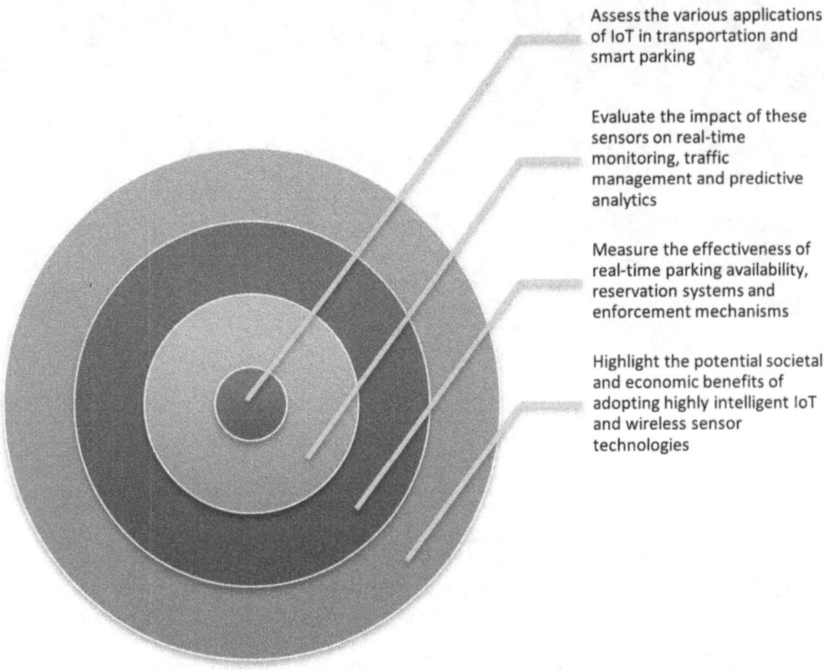

1.5 Structure of the Chapter

This chapter deeply dives in the diverse arena and use of Intelligent IoT and Wireless Sensors for Smart Urban Environments, Focusing on Secure Transportation and Parking for a Sustainable Tomorrow. Section 2 elaborates the Smart Transportation Systems: Intelligent IoT and Wireless Networks. Section 3 expresses the Parking Management Solutions: Orchestrating Intelligent IoT and Wireless Sensors: Section 4 highlights the Sustainability and Environmental Impact. Finally, Section 5 Conclude the Chapter with Future Scope.

Figure 2. charting Structure of the Chapter (Source- Original)

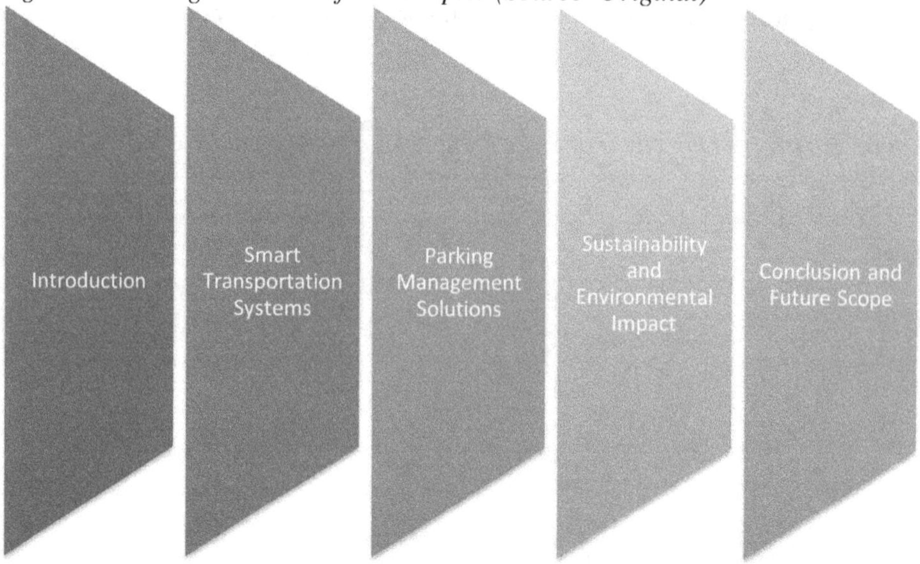

2. SMART TRANSPORTATION SYSTEMS: INTELLIGENT IOT AND WIRELESS NETWORKS

"Smart transportation" refers to an advanced technical approach that aims to link all infrastructure systems and vehicles together (Abdelhady et al., 2022). With utilizing GPS signals and data from cell providers, this communication network makes it possible to control traffic flow and conduct a thorough assessment of road segments (Mukhopadhyay et al., 2022). Globally, urban regions are facing unchecked increases in transportation volumes, which result in traffic jams, delays, detours, higher CO2 emissions, mishaps, pressure on emergency services, and disturbances to public life. For example, traffic congestion cost the Texas Transportation Institute an astounding $160 billion in electricity. As urbanization trends throughout the world point to an ongoing migration of people into cities, it is critical to make sure transportation is both safe and efficient (Devi & Kishore, 2024).

The development of ICT (information and communication technology) presents viable answers to these problems. Intelligent Transport Systems (ITS) which rely on data collection, processing, integration and distribution, are an example of how ICT is integrated with transportation infrastructure. ITS operates surreptitiously using wireless communication or buried wires, facilitating quick coordination of road users and dispatching activities. The potential of Intelligent Transportation

Systems (ITS) to enhance transportation is underestimated since many people are not aware of its various uses, despite its pervasive presence in daily life. Because of resource constraints, energy consumption, and interference, these systems frequently rely on wired networks or short-range communication technologies like Bluetooth, ZigBee, Wi-Fi and RFID, which pose reliability problems in indoor and outdoor parking situations (Sharma & Singh, 2022). The introduction of long-range technologies such as Long-Range Wide Area Network (LoRaWAN) has been one of the driving forces behind the advancement of smart parking. LoRaWAN, a low-power wide-area networking (LPWAN) protocol and LoRa, a proprietary IoT wireless communication technology, make it easier to integrate LoRa technology into traditional infrastructure and enable battery-operated, low-data-rate, low-cost IoT applications (Marr, 2020).

Figure 3. The genesis of the Smart Transportation Systems (Source- Original)

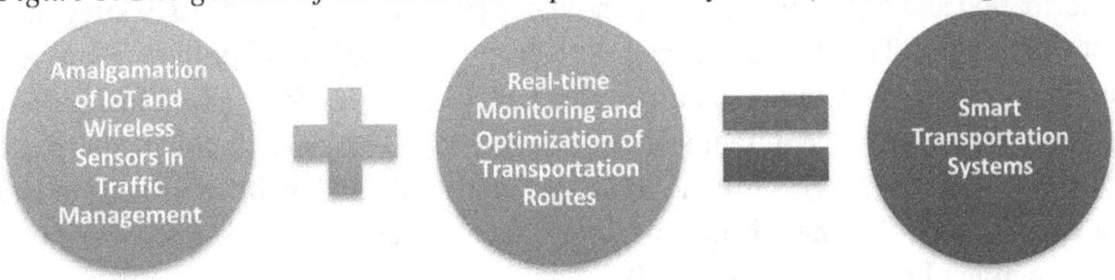

2.1 Amalgamation of IoT and Wireless Sensors in Traffic Management

Within the framework of ITS development and research, a number of clearly defined directions have emerged, including automated diagnostics, autonomous delivery control systems, intelligent navigation systems, automated payment systems, traveler information systems, traffic control systems, automated stationary vehicle control systems, automated vehicle support systems and traffic accident monitoring systems (Gaur & Sahoo, 2022). Through the combination of these methodologies, an all-encompassing ITS complex may be developed to improve transportation efficiency by utilizing novel ideas in sensory networks, controllers, and information system optimization techniques (Patil et al., 2024). Elevating road safety standards, guaranteeing efficient response to accidents and disasters, optimizing the capacity of the current transport infrastructure, improving the quality of public transportation services, establishing information infrastructure to support planning systems and infrastructure development, managing the technical state of transportation networks, enforcing traffic regulations, automating electronic payment processes, increasing road user awareness, and minimizing the ecological impact of transportation are some of the key modern objectives within the ITS framework (Sartal et al., 2020).

2.2 Real-time Monitoring and Optimization of Transportation Routes

Increasing the use of information technology is necessary to improve the operational efficiency of transportation networks (Obiako, 2022). Smart Transportation, also known as Intelligent Transportation Systems (ITS) is the application of modern sensor, computer, electronics and communication technologies (Ardakani & Cheshmehzangi, 2023). Its goal is to integrate these technologies with management methods to improve surface transportation's efficiency and safety. With ITS, travel times and fuel consumption are decreased while traffic flow and safety are optimized (Ahuja et al., 2021). To create a transportation system that is both intelligent and sustainable, information and communication technologies (ICT) must be seamlessly integrated into Internet of Things infrastructures (Knaap et al., 2023). With the LoRaWAN protocol's integrated authentication and encryption for security, LoRa technology enables safe, two-way communication for Internet of Things applications that scale to millions of devices (Mohring, 2018). With the ability to handle thousands of end devices or nodes, a single LoRa gateway device makes LoRaWAN appropriate for huge parking lots (Kataray et al., 2023).

3. PARKING MANAGEMENT SOLUTIONS: ORCHESTRATING INTELLIGENT IOT AND WIRELESS SENSORS

An advanced parking management system that uses data and technology to improve efficacy and efficiency is referred to as smart parking (Ashraf, 2021). In order to lead cars to open parking spots, offer real-time information on parking availability and expedite the parking procedure overall, it makes use of a variety of technologies, including sensors, cameras and mobile applications (Malini et al., 2021). A graphical user interface (GUI) allows end users to examine the status of parking spots and make reservations based on that information (Lv et al., 2019). Sensors are used in an Internet of Things (IoT)-based parking system (Janani et al., 2021). Many IoT-based Smart Parking Systems (SPSs) have been developed for the purpose of improving urban life by taking into account different social, financial and environmental factors (Wang et al., 2020). However, many of these systems have limitations when it comes to the sensitivity of the detection method, system interface, and microcontroller processing capabilities (Ponnusamy & Alagarsamy, 2023). The majority of parking monitoring systems in use are local in nature, meaning that vehicles are not provided with up-to-date information on available parking spots (Ramirez-Moreno et al., 2021). Among the main attributes and advantages of smart parking systems are:

Real-time parking availability: Smart parking systems monitor parking spots in real-time and gather information on occupancy status by using sensors or cameras (Nourildean et al., 2022). With using digital signs or smartphone apps, this information is communicated to drivers, saving them time and allowing them to find vacant parking spaces more rapidly 9Jha et al., 2020).

Payment and reservation systems: There are frequently integrated with smart parking, enabling vehicles to reserve parking spots in advance (Sharma & Arya, 2023). This guarantees a place for when you get there. It also provides easy ways to pay, such as cashless transactions or mobile payment applications which streamline parking procedures and do away with the requirement for cash payments at meters (Ghazal et al., 2023).

Traffic management: Smart parking systems can reduce traffic congestion brought on by circling cars by giving real-time information on parking availability (Oladimeji et al., 2023). As a result, there is less pollution, more efficient traffic flow, and increased road safety (Rao & Deebak, 2023).

Analytics and optimization of data: Intelligent parking systems collect and examine information on parking trends, occupancy percentages and user conduct (Vadivel et al., 2023). With the use of this data, parking operations may be optimized, trends can be recognized, and well-informed choices about new infrastructure or price changes can be made (Balasubramanian et al., 2023).

Sustainability and environmental benefits: Smart parking systems can reduce fuel usage and carbon emissions by cutting down on the amount of time spent looking for space (Fan, 2023). Also, they maximize the use of available parking spots, which may lessen the demand for additional parking buildings and preserve green spaces in metropolitan settings (Pramanik et al., 2023).

Figure 4. The Attributes and Advantages of Smart Parking Systems (Source- Original)

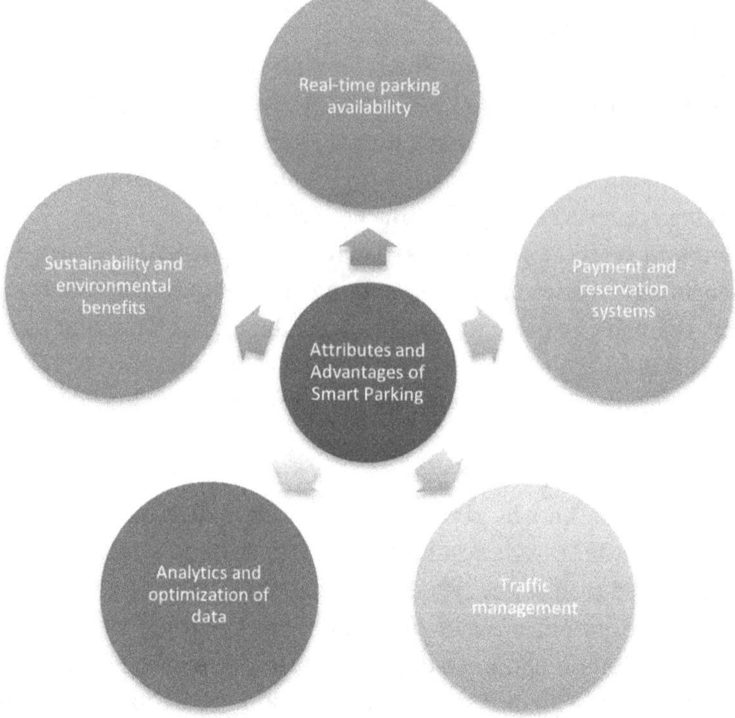

Smart parking uses data-driven solutions and technology to improve the parking experience for drivers and operators alike (Kanellopoulos et al., 2023). It maximizes the use of parking spaces, lessens traffic, enhances driving convenience and encourages environmentally friendly urban growth (Alahi et al., 2023).

3.1 Challenges in Urban Parking and Congestion

The complexity of transportation in cities is expanding due to increased urbanization. Public transportation access is still restricted and parking for private automobiles has grown even more difficult and costly (Abbasi et al., 2023). As a result, the need for Internet of Things (IoT)-enabled parking systems is rising in order to mitigate persistent traffic jams and deal with the volatility of parking availability (Alharbi et al., 2023). The proliferation of multiple linked devices has made the construction of an Internet of Things (IoT)-based smart parking system increasingly viable (Vinodhini et al., 2023). With giving drivers and parking facility managers access to advanced information such as parking spots that are available and the current cost of parking, this infrastructure gives them more power and efficiency

(Almalki et al., 2023). Because of data unreliability, limited transmission resilience, high prices for smart sensing units, complexity and energy consumption, conventional parking systems have not been able to deliver accurate, real-time information (Kumar & Chhillar, 2023). Consequently, any suggested remedies need to take these limitations into account and create smart parking units that are IoT-enabled, more dependable, affordable, maintenance-free and simple to install (Saini et al., 2023). Smart parking applications are increasingly using LPWANs such as LoRa, NB-IoT, and SigFox which makes the creation of smart sensing units based on these cutting-edge wireless technologies necessary (Almagrabi, 2023).

3.2 IoT-based Smart Parking Systems for Efficient Utilization of Parking Spaces

The idea of a "smart road" unifies several components onto a single platform (Rosayyan et al., 2023). These components, even when used alone, may handle a wide range of regional issues (Rani & Sharma, 2023). Adaptive traffic signals improve traffic flow and lessen congestion by adjusting in response to current traffic circumstances (Hussain et al., 2024). Automated infraction detection increases driving responsibility and reduces the risk of collisions. Energy savings are possible with intelligent regulation of street lights (Kolhe et al., 2023). The effect of developing a nation's road infrastructure on its economy clearly illustrates an increase in tax receipts, public goods, road services, tourism, accessibility to transportation, demand for the transportation of people and commodities, production of consumer products, expenditures associated with connected services, and a reduction in journey time (Salman & Hasar, 2023).

4. SUSTAINABILITY AND ENVIRONMENTAL IMPACT

There are noteworthy advantages linked to environmentally friendly parking regulations (Arshi & Mondal, 2023). The creation of additional car-free and environmentally friendly zones may significantly reduce traffic, improve the quality of the air and make areas safer and more pedestrian-friendly (Kwon et al., 2023). As centralized sites where people can park their automobiles and use several transportation alternatives such as public transit, car-sharing, bike-sharing, and park-and-ride facilities, mobility hubs provide another important possibility within sustainable urban development (Zakutynskyi et al., 2023). With arranging parking facilities strategically and integrating them with well-thought-out public transit systems, communities may lessen pollution and traffic in their downtown areas, increase accessibility, and cut down on travel times (Amlan et al., 2023). The implementation of intelligent transport systems holds strategic significance, influencing the competitive positioning of each country in the global market. Given the substantial capital requirements involved, state participation is indispensable (Siddiqui et al., 2023). Intelligent road ecosystem solutions include those that gather process and evaluate vehicle and road infrastructure data for decision-making purposes (Husain et al., 2023). They consist of GPS/GLONASS systems, automated road weather stations, traffic flow detectors, adaptive (smart) traffic lights, automated violation detection mechanisms, electronic nonstop fare payment systems, parking meters, connected information boards, automated lighting control systems and other connected objects (Hashem et al., 2023).

A key component of upgrading and enabling sustainable urban planning is defining efficient parking rules such as- mobility hubs and car-free zones which will ultimately improve living conditions in cities (Mohapatra & Rath, 2023). But putting such proposals into action necessitates a careful analysis of the risks as well as the opportunities (Venckauskas et al., 2024). Modifications such as the adoption of sustainable parking regulations, present both opportunities and risks that need to be properly evaluated (Jyothi et al., 2023). People may be discouraged from visiting specific parts of the city by unduly stringent parking regulations and inadequate connection between parking facilities and public/shared transit choices (Demertzi et al., 2023). The implementation of car-free and environmental zones may result in the displacement of individuals and companies that depend on private automobiles for transit (Hassan et al., 2023). Also, individuals may find it more difficult to go around the city without a car if there are insufficient or unreliable public transit choices (Xu et al., 2023). Thus, the key to a successful implementation is striking a balance between preserving accessibility and cutting down on the number of cars on the road (Das et al., 2023).

Figure 5. The Applications of IoT and Sensors: Sustainability and Environmental Impact (Source- Original)

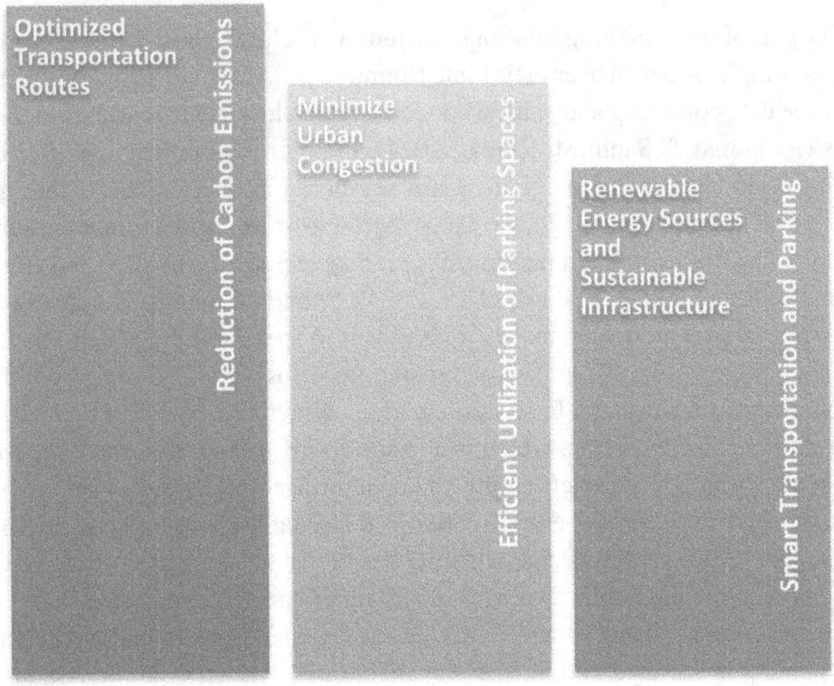

4.1 Reduction of Carbon Emissions through Optimized Transportation Routes

When looking for parking, drivers may use their cellphones to remotely search for and book places using an IoT-based parking system which acts as a centralized administration platform (Rao et al., 2024). This is a simple way to prevent any traffic jams (Adnan et al., 2024). The hardware sensors of the system identify parking spots that are available and notify nearby cars in real-time, removing the anxiety about finding an open place and guaranteeing convenient travel (Nassereddine & Khang, 2024). The linked gadget also notifies drivers of busy hours and surcharges, saving them from having to struggle to find parking or from incurring unforeseen increased expenses (Malik et al., 2023). With the convenience of a mobile app, smart parking technology improves parking operations efficiency, optimizes the use of currently available parking spots and promotes better traffic flow (Iftikhar et al., 2023).

4.2 Efficient Utilization of Parking Spaces to Minimize Urban Congestion

The goal of smart parking solutions is to remove the headache of parking by giving drivers complete control over their route (Sumit et al., 2024). IoT technology makes real-time data collecting and analysis possible which lowers costs and shortens trip times (Rajkumar & Santhosh Kumar, 2024). In order to optimize operations, the parking ecosystem's numerous gadgets and sensors are connected via the Internet of Things (IoT) (Salunke et al., 2024). The integration of IoT with autonomous cars is critical to the future of smart parking since it may increase road space and enhance commuter accessibility (Prakash et al., 2024). Suburban regions frequently have a shortage of parking which makes it necessary to spend a lot of time looking for available spaces (Aoki, 2024). This adds to the problems of traffic jams, air pollution, fuel waste and irate drivers. In addition, parking searches account for up to 30% of inner-city traffic congestion, which is a result of urban growth (Rao et al., 2024). But, some parking lots are left empty and underutilized because cars are unable to reach them because real-time information is unavailable (Nassereddine & Khang, 2024). As a result, parking slot availability has grown to be a major worry in urban development, raising public awareness of its significance (Adnanet al., 2024). It is imperative to put in place a smart parking management system that collects real-time parking space data and makes it available to the general public (Daousis et al., 20240. With effectively matching cars with open parking spaces, this kind of technology may minimize traffic congestion, save time, maximize parking spot use and save parking management expenses (Lepakshi & Ranga, 2024).

4.3 Assimilation of Smart Transportation and Parking with Renewable Energy Sources and Sustainable Infrastructure

The idea of "smart cities" has developed in modern times from being a catch-phrase to an essential component of urban development and planning (Gonzalez et al., 2024). Information and communication technologies (ICT) are the mainstay of smart cities' efforts to encourage sustainable urban life (Bahrepour & Maleki, 2024). Integrating renewable energy sources such as- solar, wind and other alternatives into the city's infrastructure is a crucial component of this sustainability strategy (Van Hoang, 2024). This integration guarantees long-term social and economic sustainability in addition to providing benefits for the environment. In the field of renewable energy, solar energy has become a front-runner, especially when considering smart cities (Asaju, 2024). Solar panels can be installed on rooftops of residential and commercial buildings, integrated into public infrastructure like street lighting and public transportation systems or deployed in a variety of other ways in urban settings (Faliagka et al., 2024). The creation of solar highways is a creative example of the use of solar energy in smart cities (Hussain et al., 2024). These roadways are equipped with solar cells, which allow them to produce electricity for traffic control and street lighting. The development of smart buildings structures outfitted with automated systems and energy-efficient technologies where solar panels greatly lower carbon emissions is greatly aided by solar energy (Amutha, 2024).

5. CONCLUSION AND FUTURE SCOPE

The intellectualization of transportation systems is the primary goal of Intelligent Transport Systems (ITS). Globally, the application of telematics in the context of ITS offers a chance to tackle issues related to social and economic transportation, increasing the effectiveness of public transit, reducing traffic accidents, boosting environmental sustainability and guaranteeing general safety of transportation. Putting into practice cutting-edge electrical, computer, and communication capabilities makes managing transportation networks, traffic flow control, and information transfer easier. Four essential characteristics are prioritized when adopting and implementing new technology in transportation systems: safety, responsiveness, integration and sustainability. These ideas are essential to accomplishing the main objectives of smart transportation, which are economic growth, environmental sustainability and access and mobility. It navigate the complexities of urban growth, the integration of Highly Intelligent IoT and wireless sensors stands as a beacon guiding towards a more connected, secure and sustainable urban future.

REFERENCES

Abbasi, R., Bashir, A. K., Mateen, A., Amin, F., Ge, Y., & Omar, M. (2023). Efficient Security and Privacy of Lossless Secure Communication for Sensor-based Urban Cities. *IEEE Sensors Journal*.

Abdelhady, A. M., Amin, O., Alouini, M. S., & Shihada, B. (2022). Revolutionizing optical wireless communications via smart optics. *IEEE Open Journal of the Communications Society*, 3, 654–669. DOI: 10.1109/OJCOMS.2022.3164790

Abi Sen, A. A., Aljohani, Y. M., Hakami, A. A. H., Bahbouh, N. M., Alkhodre, A., & Alghamdi, M. S. (2023, March). A Framework for Moving from Traditional to Smart University. In *2023 10th International Conference on Computing for Sustainable Global Development (INDIACom)* (pp. 562-569). IEEE.

Adnan, M., Haque, A., Sarwar, M., & Siddiqui, A. S. (2024). Fundamentals of internet of things (IoT) for smart cities. In *Smart Cities: Power Electronics, Renewable Energy, and Internet of Things* (pp. 50-65). CRC Press.

Adnan, M., Haque, A., Sarwar, M., & Siddiqui, A. S. (2024). Fundamentals of internet of things (IoT) for smart cities. In *Smart Cities: Power Electronics, Renewable Energy, and Internet of Things* (pp. 50-65). CRC Press.

. Ahuja, N. J., Srikanth, P., Konstantinou, C., & KAUST, C. (2021). Blockchain and Autonomous Vehicles: Recent Advances and Future Directions.

Alahi, M. E. E., Sukkuea, A., Tina, F. W., Nag, A., Kurdthongmee, W., Suwannarat, K., & Mukhopadhyay, S. C. (2023). Integration of IoT-enabled technologies and artificial intelligence (AI) for smart city scenario: Recent advancements and future trends. *Sensors (Basel)*, 23(11), 5206. DOI: 10.3390/s23115206 PMID: 37299934

Alharbi, F., Zakariah, M., Alshahrani, R., Albakri, A., Viriyasitavat, W., & Alghamdi, A. A. (2023). Intelligent transportation using wireless sensor networks blockchain and license plate recognition. *Sensors (Basel)*, 23(5), 2670. DOI: 10.3390/s23052670 PMID: 36904873

Almagrabi, A. O. (2023). Challenges and vulnerability evaluation of smart cities in IoT device based on cybersecurity mechanism. *Expert Systems: International Journal of Knowledge Engineering and Neural Networks*, 40(4), e13113. DOI: 10.1111/exsy.13113

Almalki, F. A., Alsamhi, S. H., Sahal, R., Hassan, J., Hawbani, A., Rajput, N. S., Saif, A., Morgan, J., & Breslin, J. (2023). Green IoT for eco-friendly and sustainable smart cities: Future directions and opportunities. *Mobile Networks and Applications*, 28(1), 178–202. DOI: 10.1007/s11036-021-01790-w

Almutairi, R., Bergami, G., & Morgan, G. (2024). Advancements and Challenges in IoT Simulators: A Comprehensive Review. *Sensors (Basel)*, 24(5), 1511. DOI: 10.3390/s24051511 PMID: 38475047

Amlan, K. N. H., Uddin, M. S., Mahmud, T., & Riyan, N. B. (2023). IoT, Cloud Computing, and Sensing Technology for Smart Cities. In *Intelligent Techniques for Cyber-Physical Systems* (pp. 267–291). CRC Press. DOI: 10.1201/9781003438588-16

. Amutha, B. (2024). Velocious: A Resilient IoT Architecture for 6G based Intelligent Transportation System with Expeditious Movement Mechanism.

Ang, K. L. M., Seng, J. K. P., & Ngharamike, E. (2022). Towards crowdsourcing internet of things (crowd-iot): Architectures, security and applications. *Future Internet*, 14(2), 49. DOI: 10.3390/fi14020049

Aoki, S. (2024). Privacy, Security, and Reliability for Urban Ambient Sensing with Sensor-Rich Autonomous Vehicles. *IEEE Internet of Things Magazine*, 7(2), 114–118. DOI: 10.1109/IOTM.001.2200225

Ardakani, S. P., & Cheshmehzangi, A. (2023). *Big Data Analytics for Smart Urban Systems*. Springer Nature. DOI: 10.1007/978-981-99-5543-5

Arshi, O., & Mondal, S. (2023). Advancements in sensors and actuators technologies for smart cities: A comprehensive review. *Smart Construction and Sustainable Cities*, 1(1), 18. DOI: 10.1007/s44268-023-00022-2

Asaju, B. J. (2024). Cybersecurity Frameworks for Autonomous Vehicle Systems: Safeguarding Onboard Systems, Communication Networks, and Data Privacy in Smart City Ecosystems. *Internet of Things and Edge Computing Journal*, 4(1), 27–48.

Ashraf, S. (2021). A proactive role of IoT devices in building smart cities. *Internet of Things and Cyber-Physical Systems*, 1, 8–13. DOI: 10.1016/j.iotcps.2021.08.001

Bagga, P., Das, A. K., Chamola, V., & Guizani, M. (2022). Blockchain-envisioned access control for internet of things applications: A comprehensive survey and future directions. *Telecommunication Systems*, 81(1), 125–173. DOI: 10.1007/s11235-022-00938-7 PMID: 35879968

Bahrepour, D., & Maleki, R. (2024). Benefit and limitation of using blockchain in smart cities to improve citizen services. *GeoJournal*, 89(2), 57. DOI: 10.1007/s10708-024-11040-7

Balasubramanian, S. B., Balaji, P., Munshi, A., Almukadi, W., Prabhu, T. N., Venkatachalam, K., & Abouhawwash, M. (2023). Machine learning based IoT system for secure traffic management and accident detection in smart cities. *PeerJ. Computer Science*, 9, e1259. DOI: 10.7717/peerj-cs.1259 PMID: 37346697

Baskar, P., Khang, A., Annadurai, S., Amritha, S. S., & Annadurai, P. (2024). Quantum Computing Smart City: Fostering Sustainability and Green Infrastructure With Smart City. In *Applications and Principles of Quantum Computing* (pp. 162-213). IGI Global.

Bhardwaj, A., Bharany, S., Abulfaraj, A. W., Ibrahim, A. O., & Nagmeldin, W. (2024). Fortifying home IoT security: A framework for comprehensive examination of vulnerabilities and intrusion detection strategies for smart cities. *Egyptian Informatics Journal*, 25, 100443. DOI: 10.1016/j.eij.2024.100443

Bibri, S. E., & Bibri, S. E. (2018). Conceptual, theoretical, disciplinary, and discursive foundations: A multidimensional framework. *Smart Sustainable Cities of the Future: The Untapped Potential of Big Data Analytics and Context–Aware Computing for Advancing Sustainability*, 39-131.

Brunetta, G., & Caldarice, O. (2020). Spatial resilience in planning: Meanings, challenges, and perspectives for urban transition. *Sustainable Cities and Communities*, 628-640.

Cardullo, P. (2020). *Citizens in the'smart City': Participation, Co-production, Governance*. Routledge. DOI: 10.4324/9780429438806

Chaloupková, M., & Janurová, M. (2019, June). Applications of the Smart Cities Concept. In *Proceedings of FEB Zagreb International Odyssey Conference on Economics and Business* (Vol. 1, No. 1, pp. 532-546). University of Zagreb, Faculty of Economics and Business.

Daousis, S., Peladarinos, N., Cheimaras, V., Papageorgas, P., Piromalis, D. D., & Munteanu, R. A. (2024). Overview of Protocols and Standards for Wireless Sensor Networks in Critical Infrastructures. *Future Internet*, 16(1), 33. DOI: 10.3390/fi16010033

Das, D., Banerjee, S., Chatterjee, P., Ghosh, U., & Biswas, U. (2023). Blockchain for intelligent transportation systems: Applications, challenges, and opportunities. *IEEE Internet of Things Journal*, 10(21), 18961–18970. DOI: 10.1109/JIOT.2023.3277923

Demertzi, V., Demertzis, S., & Demertzis, K. (2023). An overview of cyber threats, attacks and countermeasures on the primary domains of smart cities. *Applied Sciences (Basel, Switzerland)*, 13(2), 790. DOI: 10.3390/app13020790

Devi, N., & Kishore, N. (2024). IoT Optimization for Smart Cities and Mobility in Smart Cities. In *Applications of Optimization and Machine Learning in Image Processing and IoT* (pp. 54–66). Chapman and Hall/CRC.

. ESCAP, U. (2022). Asia-Pacific digital transformation report 2022: shaping our digital future.

Faliagka, E., Christopoulou, E., Ringas, D., Politi, T., Kostis, N., Leonardos, D., Tranoris, C., Antonopoulos, C. P., Denazis, S., & Voros, N. (2024). Trends in Digital Twin Framework Architectures for Smart Cities: A Case Study in Smart Mobility. *Sensors (Basel)*, 24(5), 1665. DOI: 10.3390/s24051665 PMID: 38475201

Fan, J., Yang, W., Liu, Z., Kang, J., Niyato, D., Lam, K. Y., & Du, H. (2023). Understanding security in smart city domains from the ANT-centric perspective. *IEEE Internet of Things Journal*, 10(13), 11199–11223. DOI: 10.1109/JIOT.2023.3252040

Gaur, L., & Sahoo, B. M. (2022). *Explainable Artificial Intelligence for Intelligent Transportation Systems: Ethics and Applications*. Springer Nature. DOI: 10.1007/978-3-031-09644-0

Ghazal, T. M., Hasan, M. K., Alzoubi, H. M., Alshurideh, M., Ahmad, M., & Akbar, S. S. (2023). Internet of things connected wireless sensor networks for smart cities. In *The Effect of Information Technology on Business and Marketing Intelligence Systems* (pp. 1953–1968). Springer International Publishing. DOI: 10.1007/978-3-031-12382-5_107

Giourka, P., Sanders, M. W., Angelakoglou, K., Pramangioulis, D., Nikolopoulos, N., Rakopoulos, D., Tryferidis, A., & Tzovaras, D. (2019). The smart city business model canvas—A smart city business modeling framework and practical tool. *Energies*, 12(24), 4798. DOI: 10.3390/en12244798

González, V., Martín, L., Santana, J. R., Sotres, P., Lanza, J., & Sánchez, L. (2024). Reshaping Smart Cities through NGSI-LD Enrichment. *Sensors (Basel)*, 24(6), 1858. DOI: 10.3390/s24061858 PMID: 38544121

Hanes, D., Salgueiro, G., Grossetete, P., Barton, R., & Henry, J. (2017). *IoT fundamentals: Networking technologies, protocols, and use cases for the internet of things*. Cisco Press.

Hashem, I. A. T., Usmani, R. S. A., Almutairi, M. S., Ibrahim, A. O., Zakari, A., Alotaibi, F., Alhashmi, S. M., & Chiroma, H. (2023). Urban computing for sustainable smart cities: Recent advances, taxonomy, and open research challenges. *Sustainability (Basel)*, 15(5), 3916. DOI: 10.3390/su15053916

Hassan, M. A., Javed, R., Granelli, F., Gen, X., Rizwan, M., Ali, S. H., & Ullah, S. (2023, March). Intelligent transportation systems in smart city: a systematic survey. In *2023 International Conference on Robotics and Automation in Industry (ICRAI)* (pp. 1-9). IEEE. DOI: 10.1109/ICRAI57502.2023.10089543

Huda, N. U., Ahmed, I., Adnan, M., Ali, M., & Naeem, F. (2024). Experts and intelligent systems for smart homes' Transformation to Sustainable Smart Cities: A comprehensive review. *Expert Systems with Applications*, 238, 122380. DOI: 10.1016/j.eswa.2023.122380

Husain, S. F., Tutumluer, E., Mechitov, K. A., Qamhia, I. I., Spencer, B., & Edwards, J. R. (2023). Towards a wireless sensing infrastructure for smart mobility. *Transportation Geotechnics*, 40, 100985. DOI: 10.1016/j.trgeo.2023.100985

Hussain, I., Elomri, A., Kerbache, L., & El Omri, A. (2024). Smart city solutions: Comparative analysis of waste management models in IoT-enabled environments using multiagent simulation. *Sustainable Cities and Society*, 103, 105247. DOI: 10.1016/j.scs.2024.105247

Hussain, I., Elomri, A., Kerbache, L., & El Omri, A. (2024). Smart city solutions: Comparative analysis of waste management models in IoT-enabled environments using multiagent simulation. *Sustainable Cities and Society*, 103, 105247. DOI: 10.1016/j.scs.2024.105247

Iftikhar, A. A., Qureshi, K. N., Altalbe, A. A., & Javeed, K. (2023). Security Provision by Using Detection and Prevention Methods to Ensure Trust in Edge-Based Smart City Networks. *IEEE Access : Practical Innovations, Open Solutions*, 11, 137529–137547. DOI: 10.1109/ACCESS.2023.3338550

Jain, S., Ahuja, N. J., Srikanth, P., Bhadane, K. V., Nagaiah, B., Kumar, A., & Konstantinou, C. (2021). Blockchain and autonomous vehicles: Recent advances and future directions. *IEEE Access : Practical Innovations, Open Solutions*, 9, 130264–130328. DOI: 10.1109/ACCESS.2021.3113649

Janani, R. P., Renuka, K., & Aruna, A. (2021). IoT in smart cities: A contemporary survey. *Global Transitions Proceedings*, 2(2), 187–193. DOI: 10.1016/j.gltp.2021.08.069

Jeong, M., & Shin, H. H. (2020). Tourists' experiences with smart tourism technology at smart destinations and their behavior intentions. *Journal of Travel Research*, 59(8), 1464–1477. DOI: 10.1177/0047287519883034

Jha, S., Nkenyereye, L., Joshi, G. P., & Yang, E. (2020). Mitigating and monitoring smart city using internet of things. *Computers, Materials & Continua*, 65(2), 1059–1079. DOI: 10.32604/cmc.2020.011754

Jyothi, A. P., Shankar, A., & Shukla, A., Jr., A. N., Kumar, S. S., & Yashwanth, A. (2023, October). Intelligent Systems and Methods for Smart City in Multiple Domains. In *2023 International Conference on Evolutionary Algorithms and Soft Computing Techniques (EASCT)* (pp. 1-7). IEEE.

Kanellopoulos, D., Sharma, V. K., Panagiotakopoulos, T., & Kameas, A. (2023). Networking architectures and protocols for IoT applications in smart cities: Recent developments and perspectives. *Electronics (Basel)*, 12(11), 2490. DOI: 10.3390/electronics12112490

Kataray, T., Nitesh, B., Yarram, B., Sinha, S., Cuce, E., Shaik, S., & Roy, A. (2023). Integration of smart grid with renewable energy sources: Opportunities and challenges–A comprehensive review. *Sustainable Energy Technologies and Assessments*, 58, 103363. DOI: 10.1016/j.seta.2023.103363

. Knaap, G. J., Lewis, R., Chakraborty, A., & June-Friesen, K. (2022). 18. Toward a "Smart Growth 2.0". *Handbook on Smart Growth: Promise, Principles, and Prospects for Planning*, 324.

Kolhe, R. V., William, P., Yawalkar, P. M., Paithankar, D. N., & Pabale, A. R. (2023). Smart city implementation based on Internet of Things integrated with optimization technology. *Measurement. Sensors*, 27, 100789. DOI: 10.1016/j.measen.2023.100789

Kumar, R. L., Pham, Q. V., Khan, F., Piran, M. J., & Dev, K. (2021). Blockchain for securing aerial communications: Potentials, solutions, and research directions. *Physical Communication*, 47, 101390. DOI: 10.1016/j.phycom.2021.101390

Kumar, S., & Chhillar, R. S. (2023, December). Comparative analysis of existing security techniques in intelligent transportation system for smart cities. In *AIP Conference Proceedings* (Vol. 2916, No. 1). AIP Publishing. DOI: 10.1063/5.0177444

Kwon, H. J., Salim, M. M., & Park, J. H. (2023). Recent trends on Smart City security: A comprehensive overview. *Journal of Information Processing Systems*, 19(1), 118–129.

Leal Filho, W., Marisa Azul, A., Brandli, L., Gökçin Özuyar, P., & Wall, T. (Eds.). (2020). *Sustainable cities and communities*. Springer International Publishing.

Lefebvre, M. (2021). *Reimagining Smart Cities: reconsidering pluralism and data subjectivities* (Doctoral dissertation, University of Oxford).

Lepakshi, V. A., & Ranga, P. C. (2024). Artificial Intelligence for Smart City Development: Review/Survey of Existing Systems in Smart Cities. *Handbook of Artificial Intelligence for Smart City Development*, 60-75.

Liu, F., Fan, H. Y., & Qi, J. Y. (2022). Blockchain technology, cryptocurrency: Entropy-based perspective. *Entropy (Basel, Switzerland)*, 24(4), 557. DOI: 10.3390/e24040557 PMID: 35455220

Lv, Z., Hu, B., & Lv, H. (2019). Infrastructure monitoring and operation for smart cities based on IoT system. *IEEE Transactions on Industrial Informatics*, 16(3), 1957–1962. DOI: 10.1109/TII.2019.2913535

Malik, P. K., Roges, R., Tiwari, P., Malik, P., Kumar, V., & Gehlot, A. (2023, July). Smart Cities Monitoring using Internet of Things: Opportunities and Challenges. In 2023 4th International Conference on Electronics and Sustainable Communication Systems (ICESC) (pp. 450-455). IEEE.

Malini, P., Gowthaman, N., Gautami, A., & Thillaiarasu, N. (2021). Internet of Everything (IoE) in smart city paradigm using advanced sensors for handheld devices and equipment. In *IoT and IoE Driven Smart Cities* (pp. 121–141). Springer International Publishing.

Marr, B. (2020). *Tech Trends in Practice: The 25 technologies that are driving the 4th Industrial Revolution.* John Wiley & Sons.

Miao, J. T., & Phelps, N. (2019). Smart City Report 1: The evolution of smart city: case studies of Barcelona, Spain and Helsinki, Finland.

Mitieka, D., Luke, R., Twinomurinzi, H., & Mageto, J. (2023). Smart Mobility in Urban Areas: A Bibliometric Review and Research Agenda. *Sustainability (Basel)*, 15(8), 6754. DOI: 10.3390/su15086754

Mohapatra, H., & Rath, A. K. (2023). Designing of fault-tolerant models for wireless sensor network-assisted smart city applications. In *Intelligent Technologies: Concepts, Applications, and Future Directions* (Vol. 2, pp. 25–43). Springer Nature Singapore. DOI: 10.1007/978-981-99-1482-1_2

Mohring, K. A. (2018). *Smart streetlights: a feasibility study* (Doctoral dissertation, James Cook University).

Mukhopadhyay, D., Bhattacharyya, S., Krishnan, B., & Roy, S. (Eds.). (2022). *Blockchain for IoT*. CRC Press. DOI: 10.1201/9781003188247

Nassereddine, M., & Khang, A. (2024). Applications of Internet of Things (IoT) in smart cities. In *Advanced IoT Technologies and Applications in the Industry 4.0 Digital Economy* (pp. 109–136). CRC Press. DOI: 10.1201/9781003434269-6

Nassereddine, M., & Khang, A. (2024). Applications of Internet of Things (IoT) in smart cities. In *Advanced IoT Technologies and Applications in the Industry 4.0 Digital Economy* (pp. 109–136). CRC Press. DOI: 10.1201/9781003434269-6

. Nguyen, H., Nawara, D., & Kashef, R. (2024). Connecting the Indispensable Roles of IoT and Artificial Intelligence in Smart Cities: A Survey. *Journal of Information and Intelligence*.

Nourildean, S. W., Hassib, M. D., & Mohammed, Y. A. (2022). Internet of things based wireless sensor network: A review. *Indonesian Journal of Electrical Engineering and Computer Science*, 27(1), 246–261. DOI: 10.11591/ijeecs.v27.i1.pp246-261

Obiako, I. V. (2022). *Toward a Bio-Inspired System Architecting Framework: Simulation of the Integration of Autonomous Bus Fleets and Alternative Fueling Infrastructures in Closed Sociotechnical Environments* (Doctoral dissertation, University of South Alabama).

Oladimeji, D., Gupta, K., Kose, N. A., Gundogan, K., Ge, L., & Liang, F. (2023). Smart transportation: An overview of technologies and applications. *Sensors (Basel)*, 23(8), 3880. DOI: 10.3390/s23083880 PMID: 37112221

Oluwaseyi, J., Odu, A., & Vincent, J. (2024). Explore how this system integrates with broader smart city initiatives, such as traffic management, IoT (Internet of Things) sensors, and data-driven decision-making for urban planning.

Patil, S. C., & Gidde, M. R. (2024). An IoT-based waste management approach for environmental sustainability. *Journal of Integrated Science and Technology*, 12(3), 769–769. DOI: 10.62110/sciencein.jist.2024.v12.769

Ponnusamy, M., & Alagarsamy, A. (2023). Traffic monitoring in smart cities using internet of things assisted robotics. *Materials Today: Proceedings*, 81, 290–294. DOI: 10.1016/j.matpr.2021.03.192

Prakash, J., Murali, L., Manikandan, N., Nagaprasad, N., & Ramaswamy, K. (2024). A vehicular network based intelligent transport system for smart cities using machine learning algorithms. *Scientific Reports*, 14(1), 468. DOI: 10.1038/s41598-023-50906-7 PMID: 38172149

Pramanik, S., Pandey, D., Joardar, S., Niranjanamurthy, M., Pandey, B. K., & Kaur, J. (2023, October). An overview of IoT privacy and security in smart cities. In *AIP Conference Proceedings* (Vol. 2495, No. 1). AIP Publishing. DOI: 10.1063/5.0123511

Rajkumar, Y., & Santhosh Kumar, S. V. N. (2024). A comprehensive survey on communication techniques for the realization of intelligent transportation systems in IoT based smart cities. *Peer-to-Peer Networking and Applications*, 17(3), 1–46. DOI: 10.1007/s12083-024-01627-9

Ramírez-Moreno, M. A., Keshtkar, S., Padilla-Reyes, D. A., Ramos-López, E., García-Martínez, M., Hernández-Luna, M. C., & Lozoya-Santos, J. D. J. (2021). Sensors for sustainable smart cities: A review. *Applied Sciences (Basel, Switzerland)*, 11(17), 8198. DOI: 10.3390/app11178198

Rani, P., & Sharma, R. (2023). Intelligent transportation system for internet of vehicles based vehicular networks for smart cities. *Computers & Electrical Engineering*, 105, 108543. DOI: 10.1016/j.compeleceng.2022.108543

Rao, P. M., & Deebak, B. D. (2023). Security and privacy issues in smart cities/industries: Technologies, applications, and challenges. *Journal of Ambient Intelligence and Humanized Computing*, 14(8), 10517–10553. DOI: 10.1007/s12652-022-03707-1

Rao, P. M., Pedada, S., Jangirala, S., Das, A. K., & Rodrigues, J. J. (2024). Role of IoT in the ages of digital to smart cities: Security challenges and countermeasures. *IEEE Internet of Things Magazine*, 7(1), 56–64. DOI: 10.1109/IOTM.001.2300124

Rao, P. M., Pedada, S., Jangirala, S., Das, A. K., & Rodrigues, J. J. (2024). Role of IoT in the ages of digital to smart cities: Security challenges and countermeasures. *IEEE Internet of Things Magazine*, 7(1), 56–64. DOI: 10.1109/IOTM.001.2300124

Rosayyan, P., Paul, J., Subramaniam, S., & Ganesan, S. I. (2023). An optimal control strategy for emergency vehicle priority system in smart cities using edge computing and IOT sensors. *Measurement. Sensors*, 26, 100697. DOI: 10.1016/j.measen.2023.100697

Saini, S., Chauhan, A., Thakur, G., & Sapra, L. (2023). Challenges and opportunities in secure smart cities for enhancing the security and privacy. *Enabling Technologies for Effective Planning and Management in Sustainable Smart Cities*, 1-27.

Salman, M. Y., & Hasar, H. (2023). Review on environmental aspects in smart city concept: Water, waste, air pollution and transportation smart applications using IoT techniques. *Sustainable Cities and Society*, 94, 104567. DOI: 10.1016/j.scs.2023.104567

Salunke, M. B., Kulkarni, S. V., Ovale, S., Manjre, B. M., Limkar, S., Shaikh, F., & Sai, V. (2024). A Sustainable IoT-Based Smart Transportation System for Urban Mobility. In *WSN and IoT* (pp. 144–164). CRC Press. DOI: 10.1201/9781003437079-6

Sani, M. S., Iranmanesh, S., Salarian, H., Raad, R., & Jamalipour, A. (2024). BIDS: Blockchain-Enabled Intrusion Detection System in Smart Cities. *IEEE Internet of Things Magazine*, 7(2), 107–113. DOI: 10.1109/IOTM.001.2300191

Saranya, M., & Amutha, B. (2024). A Survey of Innovative Machine Learning Approaches in Smart City Applications. In *Innovative Machine Learning Applications for Cryptography* (pp. 231–241). IGI Global. DOI: 10.4018/979-8-3693-1642-9.ch013

Sartal, A., Carou, D., & Davim, J. P. (Eds.). (2020). *Enabling technologies for the successful deployment of industry 4.0*. CRC Press. DOI: 10.1201/9780429055621

Shackelford, S. J. (2020). *The Internet of Things: What Everyone Needs to Know*. Oxford University Press. DOI: 10.1093/wentk/9780190943813.001.0001

Shahzad, F., Javed, A. R., Zikria, Y. B., Rehman, S., & Jalil, Z. (2021). Future smart cities: requirements, emerging technologies, applications, challenges, and future aspects. *TechRxiv*.

Sharma, A., & Singh, B. (2022). Measuring Impact of E-commerce on Small Scale Business: A Systematic Review. *Journal of Corporate Governance and International Business Law*, 5(1).

Sharma, R., & Arya, R. (2023). Security threats and measures in the Internet of Things for smart city infrastructure: A state of art. *Transactions on Emerging Telecommunications Technologies*, 34(11), e4571. DOI: 10.1002/ett.4571

Shen, S., Wei, Z. Q., Sun, L. J., Su, Y. Q., Wang, R. C., & Jiang, H. M. (2018). The shared bicycle and its network—internet of shared bicycle (IOSB): A review and survey. *Sensors (Basel)*, 18(8), 2581. DOI: 10.3390/s18082581 PMID: 30087263

Siddiqui, S., Hameed, S., Shah, S. A., Khan, A. K., & Aneiba, A. (2023). Smart contract-based security architecture for collaborative services in municipal smart cities. *Journal of Systems Architecture*, 135, 102802. DOI: 10.1016/j.sysarc.2022.102802

. Silva, L., Magaia, N., Sousa, B., Kobusińska, A., Casimiro, A., Mavromoustakis, C. X., ... & De Albuquerque, V. H. C. (2021). Computing paradigms in emerging vehicular environments: A review. *IEEE/CAA Journal of Automatica Sinica, 8*(3), 491-511.

Sumit, C., Chhillar, R. S., Dalal, S., Dalal, S., Lilhore, U. K., & Samiya, S. (2024). A dynamic and optimized routing approach for VANET communication in smart cities to secure intelligent transportation system via a chaotic multi-verse optimization algorithm. *Cluster Computing*, 27(5), 1–26. DOI: 10.1007/s10586-024-04322-9

Ullah, A., Qi, G., Hussain, S., Ullah, I., & Ali, Z. (2024). The Role of LLMs in Sustainable Smart Cities: Applications, Challenges, and Future Directions. *arXiv preprint arXiv:2402.14596*.

Ullah, F., Sepasgozar, S. M., & Wang, C. (2018). A systematic review of smart real estate technology: Drivers of, and barriers to, the use of digital disruptive technologies and online platforms. *Sustainability (Basel)*, 10(9), 3142. DOI: 10.3390/su10093142

Vadivel, G., Hussain, M. J. M., & Sangeetha, S. T. (2023). SMART TRANSPORTATION SYSTEMS: IOT-CONNECTED WIRELESS SENSOR NETWORKS FOR TRAFFIC CONGESTION MANAGEMENT. *INTERNATIONAL JOURNAL OF ADVANCES IN SIGNAL AND IMAGE SCIENCES*, 9(1), 40–49. DOI: 10.29284/IJASIS.9.1.2023.40-49

Van Hoang, T. (2024). Impact of Integrated Artificial Intelligence and Internet of Things Technologies on Smart City Transformation. *Journal of Technical Education Science*, 19(1), 64–73. DOI: 10.54644/jte.2024.1532

Varadharajan, V. (2021). Next Generation of Intelligent Cities: Case Studies From Europe. In *Developing and Monitoring Smart Environments for Intelligent Cities* (pp. 87-111). IGI Global.

Venčkauskas, A., Taparauskas, M., Grigaliūnas, Š., & Brūzgienė, R. (2024). Enhancing Communication Security an In-Vehicle Wireless Sensor Network. *Electronics (Basel)*, 13(6), 1003. DOI: 10.3390/electronics13061003

Vinodhini, M., Rajkumar, S., & Subramaniam, S. K. (2023). Real-time Internet of LoRa Things (IoLT)-based accident detection and prevention system in vehicular networks towards smart city. *International Journal of Communication Systems*.

Vohra, S. K., Kumar, V. S., Krishnamoorthy, R., Mahesh, P., Haralayya, B., Soni, N., & Gupta, S. K. (2024). Original Research Article The 5G revolution: Tackling challenges in smart cities and intelligent transportation systems. *Journal of Autonomous Intelligence*, 7(2). Advance online publication. DOI: 10.32629/jai.v7i2.1342

Wang, A., Wang, P., Miao, X., Li, X., Ye, N., & Liu, Y. (2020). A review on non-terrestrial wireless technologies for Smart City Internet of Things. *International Journal of Distributed Sensor Networks*, 16(6), 1550147720936824. DOI: 10.1177/1550147720936824

Xu, J., Li, M., He, Z., & Anwlnkom, T. (2023). Security and privacy protection communication protocol for Internet of vehicles in smart cities. *Computers & Electrical Engineering*, 109, 108778. DOI: 10.1016/j.compeleceng.2023.108778

Yashi, Y., Nijhawan, R., & Bhargava, D. (2024). Intelligent Communication Systems for Urban Things. In *Computational Intelligence in Urban Infrastructure* (pp. 155–168). CRC Press.

Zakutynskyi, I., Sibruk, L., & Kokarieva, A. (2023). IoT System for Monitoring and Managing Public Transport Data. *WSEAS Transactions on Systems*, 22, 242–248. DOI: 10.37394/23202.2023.22.25

Zhao, W., Liang, Z., & Li, B. (2022). Realizing a Rural Sustainable Development through a Digital Village Construction: Experiences from China. *Sustainability (Basel)*, 14(21), 14199. DOI: 10.3390/su142114199

Compilation of References

Aayog, N. (n.d.). National strategy for artificial intelligence, 2018. *URL:* Https://Niti. Gov. *in/Sites/Default/Files/2019-01/NationalStrategy-for-AI-Discussion-Paper. Pdf.*

Abbasi, R., Bashir, A. K., Mateen, A., Amin, F., Ge, Y., & Omar, M. (2023). Efficient Security and Privacy of Lossless Secure Communication for Sensor-based Urban Cities. *IEEE Sensors Journal.*

Abdelhady, A. M., Amin, O., Alouini, M. S., & Shihada, B. (2022). Revolutionizing optical wireless communications via smart optics. *IEEE Open Journal of the Communications Society*, 3, 654–669. DOI: 10.1109/OJCOMS.2022.3164790

Abedinia, O., Ghasemi-Marzbali, A., Gouran-Orimi, S., & Bagheri, M. (2023). Presence of renewable resources in a smart city for supplying clean and sustainable energy. In *Decision making using AI in energy and sustainability: methods and models for policy and practice* (pp. 233–251). Springer International Publishing. DOI: 10.1007/978-3-031-38387-8_14

Abi Sen, A. A., Aljohani, Y. M., Hakami, A. A. H., Bahbouh, N. M., Alkhodre, A., & Alghamdi, M. S. (2023, March). A Framework for Moving from Traditional to Smart University. In *2023 10th International Conference on Computing for Sustainable Global Development (INDIACom)* (pp. 562-569). IEEE.

Adeleye, R. A., Awonuga, K. F., Asuzu, O. F., Ndubuisi, N. L., & Tubokirifuruar, T. S. (2024). Digital marketing analytics: A review of strategies in the age of big data and AI. *World Journal of Advanced Research and Reviews, 21*(2), 073-084.

Adesina, A. A., Iyelolu, T. V., & Paul, P. O. (2024). Leveraging predictive analytics for strategic decision-making: Enhancing business performance through data-driven insights. *World Journal of Advanced Research and Reviews*, 22(3), 1927–1934. DOI: 10.30574/wjarr.2024.22.3.1961

Adnan, M., Haque, A., Sarwar, M., & Siddiqui, A. S. (2024). Fundamentals of internet of things (IoT) for smart cities. In *Smart Cities: Power Electronics, Renewable Energy, and Internet of Things* (pp. 50-65). CRC Press.

Adnan, M., Xiao, B., Ali, M. U., Bibi, S., Yu, H., Xiao, P., Zhao, P., Wang, H., & An, X. (2024). Human inventions and its environmental challenges, especially artificial intelligence: New challenges require new thinking. *Environmental Challenges*, 16, 100976. DOI: 10.1016/j.envc.2024.100976

Afzal, S., Mubeen, M., Hussain, S., Ali, M., Javeed, H. M. R., Al-Ashkar, I., . . . Jatoi, W. N. (2023). Modern breeding approaches for climate change. In *climate change impacts on agriculture: Concepts, Issues and Policies for Developing Countries* (pp. 299-313). Cham: Springer International Publishing. DOI: 10.1007/978-3-031-26692-8_17

Ahmed, Z., Gui, D., Murtaza, G., Yunfei, L., & Ali, S. (2023). An overview of smart irrigation management for improving water productivity under climate change in drylands. *Agronomy (Basel)*, 13(8), 2113. DOI: 10.3390/agronomy13082113

AI and the governance of sustainable development. An idea analysis of the European Union, the United Nations, and the World Economic Forum—ScienceDirect. (n.d.). Retrieved August 15, 2024, from https://www.sciencedirect.com/science/article/pii/S1462901123002393

AI-Based Modeling: Techniques, Applications and Research Issues Towards Automation, Intelligent and Smart Systems | SN Computer Science. (n.d.). Retrieved August 15, 2024, from https://link.springer.com/article/10.1007/s42979-022-01043-x

Ajiga, D. I., Ndubuisi, N. L., Asuzu, O. F., Owolabi, O. R., Tubokirifuruar, T. S., & Adeleye, R. A. (2024). AI-driven predictive analytics in retail: A review of emerging trends and customer engagement strategies. *International Journal of Management & Entrepreneurship Research*, 6(2), 307–321. DOI: 10.51594/ijmer.v6i2.772

Akhbarizadeh, , Dobaradaran, S., Nabipour, I., Tajbakhsh, S., Darabi, A. H., & Spitz, J. (2020). Abundance, composition, and potential intake of microplastics in canned fish. *Marine Pollution Bulletin*, 160, 111633. Advance online publication. DOI: 10.1016/j.marpolbul.2020.111633 PMID: 33181921

Akter, S., Babu, M. M., Hani, U., Sultana, S., Bandara, R., & Grant, D. (2024). Unleashing the power of artificial intelligence for climate action in industrial markets. *Industrial Marketing Management*, 117, 92–113. DOI: 10.1016/j.indmarman.2023.12.011

Al Hajri, N. H., Al Harthi, R. N., Pasam, G. K., & Natarajan, R. (2024). Machine Learning and Machine Learning based Green Energy Generation using Hybrid Renewable Energy Sources of Solar, Wind and Hydrogen Fuel Cells. In *E3S Web of Conferences* (Vol. 472, p. 01008). EDP Sciences. DOI: 10.1051/e3sconf/202447201008

Al Mahri, S., & Bouchama, A. (2018). Heatstroke. *Handbook of Clinical Neurology*, 157, 531–545. DOI: 10.1016/B978-0-444-64074-1.00032-X PMID: 30459024

Alahi, M. E. E., Sukkuea, A., Tina, F. W., Nag, A., Kurdthongmee, W., Suwannarat, K., & Mukhopadhyay, S. C. (2023). Integration of Machine Learning-enabled technologies and artificial intelligence (AI) for smart city scenario: Recent advancements and future trends. *Sensors (Basel)*, 23(11), 5206. DOI: 10.3390/s23115206 PMID: 37299934

Aldoseri, A., Al-Khalifa, K. N., & Hamouda, A. M. (2024). Methodological approach to assessing the current state of organizations for AI-Based digital transformation. *Applied System Innovation*, 7(1), 14. DOI: 10.3390/asi7010014

Alharbi, F., Zakariah, M., Alshahrani, R., Albakri, A., Viriyasitavat, W., & Alghamdi, A. A. (2023). Intelligent transportation using wireless sensor networks blockchain and license plate recognition. *Sensors (Basel)*, 23(5), 2670. DOI: 10.3390/s23052670 PMID: 36904873

Alhosani, K., & Alhashmi, S. M. (2024). Opportunities, challenges, and benefits of AI innovation in government services: A review. *Discover Artificial Intelligence*, 4(1), 18. DOI: 10.1007/s44163-024-00111-w

Ali, A. H., & Thakkar, R. (2023). Climate changes through data science: Understanding and mitigating environmental crisis. *Mesopotamian Journal of Big Data*, 2023, 125–137. DOI: 10.58496/MJBD/2023/017

Ali, G., Mijwil, M. M., Buruga, B. A., Abotaleb, M., & Adamopoulos, I. (2024). A Survey on Artificial Intelligence in Cybersecurity for Smart Agriculture: State-of-the-Art, Cyber Threats, Artificial Intelligence Applications, and Ethical Concerns. *Mesopotamian Journal of Computer Science*, 2024, 71–121. DOI: 10.58496/MJCSC/2024/007

Ali, M. (2023). Synergizing Solar Cell Innovation, Radio Wave Technology, and AI for Sustainable Business Growth: A Comprehensive Review. *International Journal of Social. Humanities and Life Sciences*, 1(1), 19–28.

Ali, S., Abuhmed, T., El-Sappagh, S., Muhammad, K., Alonso-Moral, J. M., Confalonieri, R., Guidotti, R., Del Ser, J., Díaz-Rodríguez, N., & Herrera, F. (2023). Explainable Artificial Intelligence (XAI): What we know and what is left to attain Trustworthy Artificial Intelligence. *Information Fusion*, 99, 101805. DOI: 10.1016/j.inffus.2023.101805

Allal-Chérif, O., Simón-Moya, V., & Ballester, A. C. C. (2021). Intelligent purchasing: How artificial intelligence can redefine the purchasing function. *Journal of Business Research*, 124, 69–76. DOI: 10.1016/j.jbusres.2020.11.050

Alliance, A. G. (2024). *Briefing Paper Series*. https://www3.weforum.org/docs/WEF_AI_Governance_Alliance_Briefing_Paper_Series_2024.pdf

Almagrabi, A. O. (2023). Challenges and vulnerability evaluation of smart cities in IoT device based on cybersecurity mechanism. *Expert Systems: International Journal of Knowledge Engineering and Neural Networks*, 40(4), e13113. DOI: 10.1111/exsy.13113

Almalki, F. A., Alsamhi, S. H., Sahal, R., Hassan, J., Hawbani, A., Rajput, N. S., Saif, A., Morgan, J., & Breslin, J. (2023). Green Machine Learning for eco-friendly and sustainable smart cities: Future directions and opportunities. *Mobile Networks and Applications*, 28(1), 178–202. DOI: 10.1007/s11036-021-01790-w

Almutairi, R., Bergami, G., & Morgan, G. (2024). Advancements and Challenges in IoT Simulators: A Comprehensive Review. *Sensors (Basel)*, 24(5), 1511. DOI: 10.3390/s24051511 PMID: 38475047

Alordiah, C. O. (2023). Proliferation of Artificial Intelligence Tools: Adaptation Strategies in the Higher Education Sector. *Propellers Journal of Education*, 2(1), 53–65.

Alsamhi, S. H., Rowan, N. J., Hawbani, A., Nashwan, F., & O'Brolchain, N. (2024, August). AI-Empowered Peatland Management for Climate-Resilient and Sustainable Future: Challenges and Opportunities. In *2024 4th International Conference on Emerging Smart Technologies and Applications (eSmarTA)* (pp. 1-8). IEEE.

Alshmeel, G. H. A., Al-Doori, A. S. B., Ahmed, S. R., Abrahim, Z. A., Ghaffoori, A. J., & Hussain, A. S. T. (2024, May). Self-Sustaining Buoy System: Harnessing Water Wave Energy for Smart, Wireless Sensing and Data Transmission. In *Proceedings of the Cognitive Models and Artificial Intelligence Conference* (pp. 349-356). DOI: 10.1145/3660853.3660940

Ameen, N., Tarhini, A., Reppel, A., & Anand, A. (2021). Customer experiences in the age of artificial intelligence. *Computers in Human Behavior*, 114, 106548. DOI: 10.1016/j.chb.2020.106548 PMID: 32905175

Amlan, K. N. H., Uddin, M. S., Mahmud, T., & Riyan, N. B. (2023). IoT, Cloud Computing, and Sensing Technology for Smart Cities. In *Intelligent Techniques for Cyber-Physical Systems* (pp. 267–291). CRC Press. DOI: 10.1201/9781003438588-16

Andrae, A. (2017). Total consumer power consumption forecast. Conference: Nordic Digital Business Summit. Available online: https://www.researchgate.net/publication/320225452_Total_Consumer_Power_Consumption_Forecast [Accessed on 21 May 2024].

Andrae, A., & Edler, T. (2015). On global electricity usage of communication technology: Trends to 2030. *Challenges*, 6(1), 117–157. DOI: 10.3390/challe6010117

Ang, K. L. M., Seng, J. K. P., & Ngharamike, E. (2022). Towards crowdsourcing internet of things (crowd-iot): Architectures, security and applications. *Future Internet*, 14(2), 49. DOI: 10.3390/fi14020049

Aoki, S. (2024). Privacy, Security, and Reliability for Urban Ambient Sensing with Sensor-Rich Autonomous Vehicles. *IEEE Internet of Things Magazine*, 7(2), 114–118. DOI: 10.1109/IOTM.001.2200225

Ara, A., Maraj, M. A. A., Rahman, M. A., & Bari, M. H. (2024). The Impact Of Machine Learning On Prescriptive Analytics For Optimized Business Decision-Making. *International Journal of Management Information Systems and Data Science*, 1(1), 7–18.

Ardakani, S. P., & Cheshmehzangi, A. (2023). *Big Data Analytics for Smart Urban Systems*. Springer Nature. DOI: 10.1007/978-981-99-5543-5

Arshi, O., & Mondal, S. (2023). Advancements in sensors and actuators technologies for smart cities: A comprehensive review. *Smart Construction and Sustainable Cities*, 1(1), 18. DOI: 10.1007/s44268-023-00022-2

Artificial intelligence technology, public trust, and effective governance—Penn State. (n.d.). Retrieved August 15, 2024, from https://pure.psu.edu/en/publications/artificial-intelligence-technology-public-trust-and-effective-gov

Asaduz-Zaman, M., Ongsakul, W., & Hossain, M. J. (2023). Microgrid Energy Management for Smart City Planning on Saint Martin's Island in Bangladesh. *Energies*, 16(10), 4088. DOI: 10.3390/en16104088

Asaju, B. J. (2024). Cybersecurity Frameworks for Autonomous Vehicle Systems: Safeguarding Onboard Systems, Communication Networks, and Data Privacy in Smart City Ecosystems. *Internet of Things and Edge Computing Journal*, 4(1), 27–48.

Ashraf, S. (2021). A proactive role of IoT devices in building smart cities. *Internet of Things and Cyber-Physical Systems*, 1, 8–13. DOI: 10.1016/j.iotcps.2021.08.001

Aslam, R., & Rasheed, H. (2023). *Integrating AI.* Cloud Computing, and IoT for Robust Cybersecurity Ecosystems and Adaptive Threat Mitigation.

Autodesk. (2016). Reimagining the future of air travel. Available online: https://www.autodesk.com/customer-stories/airbus [Accessed on 21 May 2024].

Awasthi, A., Vishwakarma, K., & Pattnayak, K. C. (2021). Retrospection of heatwave and heat index: a case study using ERA5 dataset.

Awogbemi, O., Von Kallon, D. V., & Kumar, K. S. (2024). Contributions of Artificial Intelligence and Digitization in Achieving Clean and Affordable Energy. *Intelligent Systems with Applications*, 22, 200389. DOI: 10.1016/j.iswa.2024.200389

Babatunde, S. O., Odejide, O. A., Edunjobi, T. E., & Ogundipe, D. O. (2024). The role of AI in marketing personalization: A theoretical exploration of consumer engagement strategies. *International Journal of Management & Entrepreneurship Research*, 6(3), 936–949. DOI: 10.51594/ijmer.v6i3.964

Babu, C. S. (2024). Adaptive AI for Dynamic Cybersecurity Systems: Enhancing Protection in a Rapidly Evolving Digital Landscap. In *Principles and Applications of Adaptive Artificial Intelligence* (pp. 52-72). IGI Global.

Bacelar, E., Pinto, T., Anjos, R., Morais, M. C., Oliveira, I., Vilela, A., & Cosme, F. (2024). Impacts of climate change and mitigation strategies for some abiotic and biotic constraints influencing fruit growth and quality. *Plants*, 13(14), 1942. DOI: 10.3390/plants13141942 PMID: 39065469

Bagga, P., Das, A. K., Chamola, V., & Guizani, M. (2022). Blockchain-envisioned access control for internet of things applications: A comprehensive survey and future directions. *Telecommunication Systems*, 81(1), 125–173. DOI: 10.1007/s11235-022-00938-7 PMID: 35879968

Bag, S., Srivastava, G., Bashir, M. M. A., Kumari, S., Giannakis, M., & Chowdhury, A. H. (2022). Journey of customers in this digital era: Understanding the role of artificial intelligence technologies in user engagement and conversion. *Benchmarking*, 29(7), 2074–2098. DOI: 10.1108/BIJ-07-2021-0415

Bag, , Sabbir Rahman, M., Rogers, H., Srivastava, G., & Harm Christiaan Pretorius, J. (2023). Climate change adaptation and disaster risk reduction in the garment industry supply chain network. *Transportation Research Part E, Logistics and Transportation Review*, 171, 103031. DOI: 10.1016/j.tre.2023.103031

Bahrepour, D., & Maleki, R. (2024). Benefit and limitation of using blockchain in smart cities to improve citizen services. *GeoJournal*, 89(2), 57. DOI: 10.1007/s10708-024-11040-7

Bakker, K. (2024). *Gaia's Web: How Digital Environmentalism Can Combat Climate Change, Restore Biodiversity, Cultivate Empathy, and Regenerate the Earth.* MIT Press. DOI: 10.7551/mitpress/14912.001.0001

Balasubramanian, S. B., Balaji, P., Munshi, A., Almukadi, W., Prabhu, T. N., Venkatachalam, K., & Abouhawwash, M. (2023). Machine learning based IoT system for secure traffic management and accident detection in smart cities. *PeerJ. Computer Science*, 9, e1259. DOI: 10.7717/peerj-cs.1259 PMID: 37346697

Balyan, S., Jangir, H., Tripathi, S. N., Tripathi, A., Jhang, T., & Pandey, P. (2024). Seeding a Sustainable Future: Navigating the Digital Horizon of Smart Agriculture. *Sustainability (Basel)*, 16(2), 475. DOI: 10.3390/su16020475

Barbier, E. B. (2023). Greening the ocean economy. *Front. Environ. Economics*, 2, 1096303. Advance online publication. DOI: 10.3389/frevc.2023.1096303

Baskar, P., Khang, A., Annadurai, S., Amritha, S. S., & Annadurai, P. (2024). Quantum Computing Smart City: Fostering Sustainability and Green Infrastructure With Smart City. In *Applications and Principles of Quantum Computing* (pp. 162-213). IGI Global.

Bayomi, N., & Fernandez, J. E. (2023). Eyes in the sky: Drones applications in the built environment under climate change challenges. *Drones (Basel)*, 7(10), 637. DOI: 10.3390/drones7100637

Behera, C. K., & Dadra, R. (2023). Understanding young consumers' attitude formation for new-age fintech credit products: An SOR framework perspective. *Journal of Financial Services Marketing*, •••, 1–15.

Beitelmal, W. H., Nwokolo, S. C., Meyer, E. L., & Ahia, C. C. (2024). Exploring Adaptation Strategies to Mitigate Climate Threats to Transportation Infrastructure in Nigeria: Lagos City, as a Case Study. *Climate (Basel)*, 12(8), 117. DOI: 10.3390/cli12080117

Belkhir, L., & Elmeligi, A. (2018). Assessing ICT global emissions footprint: Trends to 2040 & recommendations. *Journal of Cleaner Production*, 177, 448–463. DOI: 10.1016/j.jclepro.2017.12.239

Berry, I. (2021, November 19). *10 ways AI can be used in Smart Cities*. https://aimagazine.com/top10/10-ways-ai-can-be-used-smart-cities

Betin, O., and Titova, G. (2023). The formation of the economy of the world ocean. Rybnoe Hozâ jstvo 2023, 18–23. DOI: 10.37663/0131-6184-2023-2-18-23

Beyari, H., & Garamoun, H. (2022). The Effect of Artificial Intelligence on End-User Online Purchasing Decisions: Toward an Integrated Conceptual Framework. *Sustainability (Basel)*, 14(15), 9637. DOI: 10.3390/su14159637

Bhardwaj, A., Bharany, S., Abulfaraj, A. W., Ibrahim, A. O., & Nagmeldin, W. (2024). Fortifying home IoT security: A framework for comprehensive examination of vulnerabilities and intrusion detection strategies for smart cities. *Egyptian Informatics Journal*, 25, 100443. DOI: 10.1016/j.eij.2024.100443

Bibri, S. E., & Bibri, S. E. (2018). Conceptual, theoretical, disciplinary, and discursive foundations: A multidimensional framework. *Smart Sustainable Cities of the Future: The Untapped Potential of Big Data Analytics and Context–Aware Computing for Advancing Sustainability*, 39-131.

Bibri, S. E., Krogstie, J., Kaboli, A., & Alahi, A. (2024). Smarter eco-cities and their leading-edge artificial intelligence of things solutions for environmental sustainability: A comprehensive systematic review. *Environmental Science and Ecotechnology*, 19, 100330. DOI: 10.1016/j.ese.2023.100330 PMID: 38021367

Bin Abu Sofian, A. D. A., Lim, H. R., Siti Halimatul Munawaroh, H., Ma, Z., Chew, K. W., & Show, P. L. (2024). Machine learning and the renewable energy revolution: Exploring solar and wind energy solutions for a sustainable future including innovations in energy storage. *Sustainable Development (Bradford)*, 32(4), 3953–3978. DOI: 10.1002/sd.2885

Birchall, S. J., & Kehler, S. (2023). Denial and discretion as a governance process: How actor perceptions of risk and responsibility hinder adaptation to climate change. *Environmental Science & Policy*, 147, 1–10. DOI: 10.1016/j.envsci.2023.05.017

Bonatti, P. A., Decker, S., Polleres, A., & Presutti, V. (2019). Machine learning: New directions for knowledge representation on the semantic web (dagstuhl seminar 18371). [). Schloss Dagstuhl-Leibniz-Zentrum fuer Informatik.]. *Dagstuhl Reports*, 8(9), •••.

Brickson, L., Zhang, L., Vollrath, F., Douglas-Hamilton, I., & Titus, A. J. (2023). Elephants and algorithms: A review of the current and future role of AI in elephant monitoring. *Journal of the Royal Society, Interface*, 20(208), 20230367. DOI: 10.1098/rsif.2023.0367 PMID: 37963556

Brunetta, G., & Caldarice, O. (2020). Spatial resilience in planning: Meanings, challenges, and perspectives for urban transition. *Sustainable Cities and Communities*, 628-640.

Büchner, M., Zürn, J., Todoran, I. G., Valada, A., & Burgard, W. (2023). Learning and aggregating lane MLs for urban automated driving. In *Proceedings of the IEEE/CVF Conference on Computer Vision and Pattern Recognition* (pp. 13415-13424).

Bui, T., & Son, N. T. (2023). The survey of digital transformation in education: A systematic review. *International Journal of TESOL & Education*, 3(4), 32–51. DOI: 10.54855/ijte.23343

Busman, S. A., & Ananda, N. A. (2022). Artificial Intelligence and Digital Marketing Role in Increasing Consumer Purchase Intention. [AIJBM]. *American International Journal of Business Management*, 5(01), 63–68.

Butt, A., Imran, F., Helo, P., & Kantola, J. (2024). Strategic design of culture for digital transformation. *Long Range Planning*, 57(2), 102415. DOI: 10.1016/j.lrp.2024.102415

Cai, P., Wang, H., Sun, Y., & Liu, M. (2022). DQ-GAT: Towards safe and efficient autonomous driving with deep Q-learning and ML attention networks. *IEEE Transactions on Intelligent Transportation Systems*, 23(11), 21102–21112. DOI: 10.1109/TITS.2022.3184990

Caissie, D. (2006). The thermal regime of rivers: A review. *Freshwater Biology*, 51(8), 1389–1406. DOI: 10.1111/j.1365-2427.2006.01597.x

Camacho, J. D. J., Aguirre, B., Ponce, P., Anthony, B., & Molina, A. (2024). Leveraging Artificial Intelligence to Bolster the Energy Sector in Smart Cities: A Literature Review. *Energies*, 17(2), 353. DOI: 10.3390/en17020353

Campbell, S., Remenyi, T. A., White, C. J., & Johnston, F. H. (2018). Heatwave and health impact research: A global review. *Health & Place*, 53, 210–218. DOI: 10.1016/j.healthplace.2018.08.017 PMID: 30189362

Cao, L. (2023). AI and data science for smart emergency, crisis and disaster resilience. *International Journal of Data Science and Analytics*, 15(3), 231–246. DOI: 10.1007/s41060-023-00393-w PMID: 37035277

Capraro, V., Lentsch, A., Acemoglu, D., Akgun, S., Akhmedova, A., Bilancini, E., Bonnefon, J.-F., Brañas-Garza, P., Butera, L., Douglas, K. M., Everett, J. A. C., Gigerenzer, G., Greenhow, C., Hashimoto, D. A., Holt-Lunstad, J., Jetten, J., Johnson, S., Kunz, W. H., Longoni, C., & Viale, R. (2024). The impact of generative artificial intelligence on socioeconomic inequalities and policy making. *PNAS Nexus*, 3(6), pgae191. Advance online publication. DOI: 10.1093/pnasnexus/pgae191 PMID: 38864006

Cardullo, P. (2020). *Citizens in the'smart City': Participation, Co-production, Governance*. Routledge. DOI: 10.4324/9780429438806

Cavanaugh, C., McConney, A., Maor, D., & McCarthy, A. (2023). Digital transformation in education: Critical components for leaders of system change. *Computers in Human Behavior*, 145, 107751. Advance online publication. DOI: 10.1016/j.chb.2023.107751

Chaloupková, M., & Janurová, M. (2019, June). Applications of the Smart Cities Concept. In *Proceedings of FEB Zagreb International Odyssey Conference on Economics and Business* (Vol. 1, No. 1, pp. 532-546). University of Zagreb, Faculty of Economics and Business.

Cheng, Y., & Jiang, H. (2022). Customer–brand relationship in the era of artificial intelligence: Understanding the role of chatbot marketing efforts. *Journal of Product and Brand Management*, 31(2), 252–264. DOI: 10.1108/JPBM-05-2020-2907

Chen, L., Chen, Z., Zhang, Y., Liu, Y., Osman, A. I., Farghali, M., Hua, J., Al-Fatesh, A., Ihara, I., Rooney, D. W., & Yap, P.-S. (2023). Artificial intelligence-based solutions for climate change: A review. *Environmental Chemistry Letters*, 21(5), 2525–2557. DOI: 10.1007/s10311-023-01617-y

Chen, Y., & Jin, S. (2023). Artificial intelligence and carbon emissions in manufacturing firms: The moderating role of green innovation. *Processes (Basel, Switzerland)*, 11(9), 2705. DOI: 10.3390/pr11092705

Chen, , Zou, X., Li, K., Li, K., Yang, X., & Chen, C. (2021). Multiple local 3D CNNs for region-based prediction in smart cities. *Information Sciences*, 542, 476–491. DOI: 10.1016/j.ins.2020.06.026

Cheong, S., Sankaran, K., & Bastani, H. (2022). Artificial intelligence for climate change adaptation. *Wiley Interdisciplinary Reviews. Data Mining and Knowledge Discovery*, 12(5), e1459. Advance online publication. DOI: 10.1002/widm.1459

Ciais, P., Reichstein, M., Viovy, N., Granier, A., Ogée, J., Allard, V., Aubinet, M., Buchmann, N., Bernhofer, C., Carrara, A., Chevallier, F., De Noblet, N., Friend, A. D., Friedlingstein, P., Grünwald, T., Heinesch, B., Keronen, P., Knohl, A., Krinner, G., & Valentini, R. (2005). Europe-wide reduction in primary productivity caused by the heat and drought in 2003. *Nature*, 437(7058), 529–533. DOI: 10.1038/nature03972 PMID: 16177786

Clancy, M. (2021). *Reflections on the financial and ethical implications of music generated by artificial intelligence* (Doctoral dissertation, PhD Thesis. Trinity College, Dublin).

Clemente, D., Cabral, T., Rosa-Santos, P., & Taveira-Pinto, F. (2023). Blue seaports: The smart, sustainable and electrified ports of the future. *Smart Cities*, 6(3), 1560–1588. DOI: 10.3390/smartcities6030074

Coeckelbergh, M. (2020a). *AI ethics*. Cambridge, MA: MIT Press Essential Knowledge series, 167-183.

Coeckelbergh, M. (2020b). AI for climate: Freedom, justice, and other ethical and political challenges. *AI and Ethics*, 1(1), 67–72. DOI: 10.1007/s43681-020-00007-2

COMEST. (2010). The ethical implications of global climate change. Available online: http://www.gci.org.uk/Documents/UNESCO_COMEST_.pdf [Accessed on 21 May 2024].

Cook, G., Lee, J., Tsai, T., Kong, A., Deans, J., Johnson, B., & Jardim, E. (2017). Clicking clean: Who is winning the race to build a green internet. Greenpeace Report. Available online: https://www.actu-environnement.com/media/pdf/news-28245-clicking-clean-2017.pdf [Accessed on 21 May 2024].

Copernicus Climate Data Store | Copernicus Climate Data Store. (n.d.-b). https://cds.climate.copernicus.eu/cdsapp#!/dataset/derived-utci-historical?tab=overview

Cornell University. (2019). AI helps reduce Amazon hydropower dams' carbon footprint. *Science Daily,* September 19, 2019. Available online: www.sciencedaily.com/releases/2019/09/190919134703.htm [Accessed on 21 May 2024].

Cortez, F. (2023). Artificial Intelligence, Climate Change and Innovative Democratic Governance. *European Journal of Risk Regulation*, 14(3), 484–503. DOI: 10.1017/err.2023.60

Costello, A., Abbas, M. M. E., Allen, A., Ball, S., Bell, S., Bellamy, R., Friel, S., Groce, N., Johnson, A. M., Kett, M., Lee, M., Lévy, C., Maslin, M., McCoy, D., McGuire, B., Montgomery, H., Napier, D., Pagel, C., Patel, J., & Patterson, C. (2009, May 1). Managing the health effects of climate change. *Lancet*, 373(9676), 1693–1733. Advance online publication. DOI: 10.1016/S0140-6736(09)60935-1 PMID: 19447250

Coulombel, N., Boutueil, V., Liu, L., Viguié, V., & Yin, B. (2019). Substantial rebound effects in urban ridesharing: Simulating travel decisions in Paris, France. *Transportation Research Part D, Transport and Environment*, 71, 110–126. DOI: 10.1016/j.trd.2018.12.006

Crawford, K., & Joler, V. (2018). Anatomy of an AI system: The Amazon Echo as an anatomical map of human labor, data and planetary resources. Available online: https://anatomyof.ai [Accessed on 21 May 2024].

Crutzen, P. J. (2006). The "Anthropocene". In Ehlers, E., & Krafft, T. (Eds.), *Earth system science in the Anthropocene*. Springer., DOI: 10.1007/3-540-26590-2_3

Cruz, L., Gutierrez, X. F., & Martínez-Fernández, S. (2024). Innovating for Tomorrow: The Convergence of SE and Green AI. *arXiv preprint arXiv:2406.18142*.

Cuntz, A., Fink, C., & Stamm, H. (2024). Artificial Intelligence and Intellectual Property: An Economic Perspective. *World Intellectual Property Organization (WIPO) Economic Research Working Paper Series, 77*. https://papers.ssrn.com/sol3/papers.cfm?abstract_id=4757971

Dai, S., Liang, Y., Liu, S., Wang, Y., Shao, W., Lin, X., & Feng, X. (2018, March). Learning entity and relation embeddings with entity description for AI completion. In *2018 2nd International Conference on Artificial Intelligence: Technologies and Applications (ICAITA 2018)* (pp. 194-197). Atlantis Press.

Dai, A. (2013). Increasing drought under global warming in observations and models. *Nature Climate Change*, 3(1), 52–58. DOI: 10.1038/nclimate1633

Danish, M. S. S. (2023). AI and expert insights for sustainable energy future. *Energies*, 16(8), 3309. DOI: 10.3390/en16083309

Daousis, S., Peladarinos, N., Cheimaras, V., Papageorgas, P., Piromalis, D. D., & Munteanu, R. A. (2024). Overview of Protocols and Standards for Wireless Sensor Networks in Critical Infrastructures. *Future Internet*, 16(1), 33. DOI: 10.3390/fi16010033

Darjee, K. B., Neupane, P. R., & Köhl, M. (2023). Proactive Adaptation Responses by Vulnerable Communities to Climate Change Impacts. *Sustainability (Basel)*, 15(14), 10952. DOI: 10.3390/su151410952

Das, D., Banerjee, S., Chatterjee, P., Ghosh, U., & Biswas, U. (2023). Blockchain for intelligent transportation systems: Applications, challenges, and opportunities. *IEEE Internet ofThings Journal*, 10(21), 18961–18970. DOI: 10.1109/JIOT.2023.3277923

De Almeida, P. G. R., Dos Santos, C. D., & Farias, J. S. (2021). Artificial Intelligence Regulation: A framework for governance. *Ethics and Information Technology*, 23(3), 505–525. DOI: 10.1007/s10676-021-09593-z

de Marcellis-Warin, N., Marty, F., Thelisson, E., & Warin, T. (2022). Artificial intelligence and consumer manipulations: From consumer's counter algorithms to firm's self-regulation tools. *AI and Ethics*, 2(2), 259–268. DOI: 10.1007/s43681-022-00149-5

December 2018 COM (2018) 795 final. Available online: https://eur-lex.europa.eu/legal-content/EN/TXT/?uri=CELEX%3A52018DC0795 [Accessed on 22 May 2024].

Dellaert, B. G., Shu, S. B., Arentze, T. A., Baker, T., Diehl, K., Donkers, B., & Steffel, M. (2020). Consumer decisions with artificially intelligent voice assistants. *Marketing Letters*, 31(4), 335–347. DOI: 10.1007/s11002-020-09537-5

Deloitte. (n.d.). *Deloitte Survey: Analytics and Data-driven Culture Help Companies Outperform Business Goals in the "Age of With."* Retrieved August 15, 2024, from https://www.prnewswire.com/news-releases/deloitte-survey-analytics-and-data-driven-culture-help-companies-outperform-business-goals-in-the-age-of-with-300889703.html

Demertzi, V., Demertzis, S., & Demertzis, K. (2023). An overview of cyber threats, attacks and countermeasures on the primary domains of smart cities. *Applied Sciences (Basel, Switzerland)*, 13(2), 790. DOI: 10.3390/app13020790

Devi, N., & Kishore, N. (2024). IoT Optimization for Smart Cities and Mobility in Smart Cities. In *Applications of Optimization and Machine Learning in Image Processing and IoT* (pp. 54–66). Chapman and Hall/CRC.

Diamandis, P. H., & Kotler, S. (2020). *The future is faster than you think: How converging technologies are transforming business, industries, and our lives*. Simon & Schuster.

Díaz-Rodríguez, N., Del Ser, J., Coeckelbergh, M., López de Prado, M., Herrera-Viedma, E., & Herrera, F. (2023). Connecting the dots in trustworthy Artificial Intelligence: From AI principles, ethics, and key requirements to responsible AI systems and regulation. *Information Fusion*, 99, 101896. DOI: 10.1016/j.inffus.2023.101896

Dignum, V. (2019). Responsible artificial intelligence: How to develop and use AI in a responsible way. Artificial Intelligence: Foundations, Theory, and Algorithms. Cham.: Springer. https://doi.org/DOI: 10.1007/978-3-030-30371-6

Dimitrieska, S., Stankovska, A., & Efremova, T. (2018). Artificial intelligence and marketing. *Entrepreneurship*, 6(2), 298–304.

Dobbe, R., & Whittake, M. (2019). AI and climate change: How they're connected, and what we can do about it. AI Now Institute, October 17, 2019. Available online: https://medium.com/@AINowInstitute/ai-and-climate-change-how-theyre-connected-and-what-we-can-do-about-it-6aa8d0f5b32c [Accessed on 21 May 2024].

Doerr, . (2023). *"Marine entrepreneurship," in Routledge eBooks*. Routledge., DOI: 10.4324/9781003058151-18

Dong, X. L., He, X., Kan, A., Li, X., Liang, Y., Ma, J., & Han, J. (2020, August). Autoknow: Self-driving knowledge collection for products of thousands of types. In *Proceedings of the 26th ACM SIGKDD International Conference on Knowledge Discovery & Data Mining* (pp. 2724-2734). DOI: 10.1145/3394486.3403323

Duff, A. S. (2015). *Information society. International Encyclopedia of the Social & Behavioral Sciences* (2nd ed.). Elsevier., DOI: 10.1016/B978-0-08-097086-8.95017-7

Dwivedi, D. N., Mahanty, G., & nath Dwivedi, V. (2024). Intelligent Conservation: A Comprehensive Study on AI-Enhanced Environmental Monitoring and Preservation. In *The Convergence of Self-Sustaining Systems With AI and IoT* (pp. 215-226). IGI Global.

Ebi, K. L., & Meehl, G. A. (2007). The Heat is On: Climate Change & Heatwaves in the Midwest. In Regional Impacts of Climate Change: Four Case Studies in the United States. Pew Center on Global Climate Change.

Eckert, E., & Kovalevska, O. (2021). Sustainability in the European Union: Analyzing the Discourse of the European Green Deal. *Journal of Risk and Financial Management*, 14(2), 2. Advance online publication. DOI: 10.3390/jrfm14020080

Efremova, N., Foley, C., Unagaev, A., Karimi, R., Averkin, A., & Gutiérrez-Romero, R. (2022). Monitoring large-scale regenerative grazing using artificial intelligence. https://doi.org/DOI: 10.21203/rs.3.rs-2136218/v1

Elahi, M. F., Luo, X., & Tian, R. (2020, July). A framework for modeling Machine learning via processing natural descriptions of vehicle-pedestrian interactions. In *International Conference on Human-Computer Interaction* (pp. 40-50). Cham: Springer International Publishing.

El-Hussieny, E. (2023). Emerging the concept of Sustainable Architecture in forming a Smart City. *Resourceedings*, 3(1), 22–32. DOI: 10.21625/resourceedings.v3i1.948

Elwahsh, H., Allakany, A., Alsabaan, M., Ibrahem, M. I., & El-Shafeiy, E. (2023). A deep learning technique to improve road maintenance systems based on climate change. *Applied Sciences (Basel, Switzerland)*, 13(15), 8899. DOI: 10.3390/app13158899

Environment, U. N. (2022, February 10). *Frontiers 2022: Noise, Blazes and Mismatches*. https://www.unep.org/resources/frontiers-2022-noise-blazes-and-mismatches

Eom, T., Woo, C., & Chun, D. (2024). Predicting an ICT business process innovation as a digital transformation with machine learning techniques. *Technology Analysis and Strategic Management*, 36(9), 2271–2283. DOI: 10.1080/09537325.2022.2132927

Espinosa, L. A., Portela, M. M., Moreira Freitas, L. M., & Gharbia, S. (2023). Addressing the Spatiotemporal Patterns of Heatwaves in Portugal with a Validated ERA5-Land Dataset (1980–2021). *Water (Basel)*, 15(17), 3102. DOI: 10.3390/w15173102

European Commission. (2018a). EU declaration on cooperation on artificial intelligence. Available online: https://ec.europa.eu/jrc/communities/en/node/1286/document/eu-declaration-cooperation-artificial-intelligence [Accessed on 22 May 2024].

European Commission. (2018b). Communication from the commission to the European Parliament, the European Council, the Council, the European Economic and Social Committee and the Committee of the Regions. Artificial intelligence for Europe. Brussels, 25 April 2018. Available online: https://eur-lex.europa.eu/legal-content/EN/TXT/?uri=COM%3A2018%3A237%3AFIN [Accessed on 22 May 2024].

European Commission. (2018c). Communication from the commission to the European Parliament, the European Council, the Council, the European Economic and Social Committee and the Committee of the Regions. Coordinated plan on artificial intelligence, Brussels, 7

European Commission. (2019). A definition of artificial intelligence: Main capabilities and scientific disciplines. Available online: https://ec.europa.eu/digital-single-market/en/news/definition-artificial-intelligence-main-capabilities-and-scientific-disciplines [Accessed on 21 May 2024].

European Commission. (2019a). Ethics guidelines for trustworthy AI. Available online: https://digital-strategy.ec.europa.eu/en/library/ethics-guidelines-trustworthy-ai [Accessed 22 May 2024].

European Commission. (2020a). Robotics and artificial intelligence (Unit A.1). Available online: https://ec.europa.eu/digital-single-market/en/content/robotics-and-artificial-intelligence-innovation-and-excellence-unit-a1 [Accessed 22 May 2024].

European Commission. (2020b). White paper on artificial intelligence: A European approach to excellence and trust of 19 February 2020. Available online: https://ec.europa.eu/info/publications/white-paper-artificial-intelligence-european-approach-excellence-and-trust_en [Accessed on 21 May 2024].

European Commission. (2020c). Commission report on safety and liability implications of AI, the Internet of Things and robotics of 19 February 2020. Available online: https://ec.europa.eu/info/publications/commission-report-safety-and-liability-implications-ai-internet-things-and-robotics-0_en [Accessed on 22 May 2024].

European Parliament. (2017). Resolution of 16 February 2017 with recommendations to the commission on civil law rules on robotics. Available online: https://www.europarl.europa.eu/doceo/document/TA-8-2017-0051_EN.html [Accessed on 21 May 2024].

European Parliament. (2020a). Resolution of 20 October 2020 on intellectual property rights for the development of artificial intelligence technologies. Available online: https://www.europarl.europa.eu/doceo/document/TA-9-2020-0277_EN.html [Accessed on 22 May 2024].

European Parliament. (2020b). Resolution of 20 October 2020 with recommendations to the commission on a framework of ethical aspects of artificial intelligence, robotics and related technologies. Available online: https://www.europarl.europa.eu/doceo/document/TA-9-2020-0275_EN.html [Accessed on 22 May 2024].

European Parliament. (2020c). Resolution of 20 October 2020 with recommendations to the commission on a civil liability regime for artificial intelligence. Available online: https://www.europarl.europa.eu/doceo/document/TA-9-2020-0276_EN.html [Accessed on 22 May 2024].

Faliagka, E., Christopoulou, E., Ringas, D., Politi, T., Kostis, N., Leonardos, D., Tranoris, C., Antonopoulos, C. P., Denazis, S., & Voros, N. (2024). Trends in Digital Twin Framework Architectures for Smart Cities: A Case Study in Smart Mobility. *Sensors (Basel)*, 24(5), 1665. DOI: 10.3390/s24051665 PMID: 38475201

Fan, J., Yang, W., Liu, Z., Kang, J., Niyato, D., Lam, K. Y., & Du, H. (2023). Understanding security in smart city domains from the ANT-centric perspective. *IEEE Internet of Things Journal*, 10(13), 11199–11223. DOI: 10.1109/JIOT.2023.3252040

Fan, Z., Yan, Z., & Wen, S. (2023). Deep learning and artificial intelligence in sustainability: A review of SDGs, renewable energy, and environmental health. *Sustainability (Basel)*, 15(18), 13493. DOI: 10.3390/su151813493

Fedorov, M. (2020). By developing the sphere of artificial intelligence, we ensure Ukraine's competitiveness on the international market. December 2, 2020. Available online: https://www.kmu.gov.ua/news/mihajlo-fedorov-rozvivayuchi-sferu-shtuchnogo-intelektu-mi-zabezpechuyemo-konkurentospromozhnist-ukrayini-na-mizhnarodnomu-rinku [Accessed on 21 May 2024].

Feijóo, C., Kwon, Y., Bauer, J. M., Bohlin, E., Howell, B., Jain, R., Potgieter, P., Vu, K., Whalley, J., & Xia, J. (2020). Harnessing artificial intelligence (AI) to increase wellbeing for all: The case for a new technology diplomacy. *Telecommunications Policy*, 44(6), 101988. DOI: 10.1016/j.telpol.2020.101988 PMID: 32377031

Feindor-Schmidt, U. (2020). Regulation of artificial intelligence in Europe - What's in the pipeline? Lexology, December 1, 2020. Available online: https://www.lexology.com/library/detail.aspx?g=d9f74ab9-139c-49e1-9d82-70de718af80f [Accessed on 22 May 2024].

Felzmann, H., Fosch-Villaronga, E., Lutz, C., & Tamò-Larrieux, A. (2020). Towards Transparency by Design for Artificial Intelligence. *Science and Engineering Ethics*, 26(6), 3333–3361. DOI: 10.1007/s11948-020-00276-4 PMID: 33196975

Feng, S., & Liu, S. (2023). Does AI Application Matter in Promoting Carbon Productivity? Fresh Evidence from 30 Provinces in China. *Sustainability (Basel)*, 15(23), 23. Advance online publication. DOI: 10.3390/su152316261

Filippova, A. (2021). Current security issues in the information society. *SHS Web of Conferences, 109,* n. 01014. https://doi.org/DOI: 10.1051/shsconf/202110901014

Floridi, L., Cowls, J., Beltrametti, M., Chatila, R., Chazerand, P., Dignum, V., Luetge, C., Madelin, R., Pagallo, U., Rossi, F., Schafer, B., Valcke, P., & Vayena, E. (2018). AI4People—An ethical framework for a good AI society: Opportunities, risks, principles, and recommendations. *Minds and Machines*, 28(4), 689–707. DOI: 10.1007/s11023-018-9482-5 PMID: 30930541

Foss-Solbrekk, K. (2021). Three routes to protecting AI systems and their algorithms under IP law: The good, the bad and the ugly. *Journal of Intellectual Property Law & Practice*, 16(3), 247–258. DOI: 10.1093/jiplp/jpab033

Francisco, M., & Linnér, B.-O. (2023). AI and the governance of sustainable development. An idea analysis of the European Union, the United Nations, and the World Economic Forum. *Environmental Science & Policy*, 150, 103590. DOI: 10.1016/j.envsci.2023.103590

Full article: Governance of artificial intelligence. (n.d.). Retrieved August 15, 2024, from https://www.tandfonline.com/doi/full/10.1080/14494035.2021.1928377

Füller, J., Hutter, K., Wahl, J., Bilgram, V., & Tekic, Z. (2022). How AI revolutionizes innovation management – Perceptions and implementation preferences of AI-based innovators. *Technological Forecasting and Social Change*, 178, 121598. DOI: 10.1016/j.techfore.2022.121598

Gaines, . (2019). *The expected impacts of climate change on the ocean economy.* World Resources Institute.

Galaz, V., Centeno, M. A., Callahan, P. W., Causevic, A., Patterson, T., Brass, I., Baum, S., Farber, D., Fischer, J., Garcia, D., McPhearson, T., Jimenez, D., King, B., Larcey, P., & Levy, K. (2021). Artificial intelligence, systemic risks, and sustainability. *Technology in Society*, 67, 101741. DOI: 10.1016/j.techsoc.2021.101741

Gangisetty, S., & Rai, R. R. (2022). FloodNet: Underwater image restoration based on residual dense learning. *Signal Processing Image Communication*, 104, 116647. DOI: 10.1016/j.image.2022.116647

Gardner, A. V. D. L. (1984). An artificial intelligence approach to legal reasoning. *Thesis (Ph.D.),* Stanford University. The MIT Press, p. 239.

Garg, A. (2024). AI for a Better World: Sustainability and Technology. *Shodh Sagar Journal of Artificial Intelligence and Machine Learning*, 1(1), 33–38.

Garzón, J., & Acevedo, M. (2019). Augmented reality applications in education: A systematic review of the literature. *Journal of Educational Technology & Society*, 22(1), 1–12.

Gates, B. (2020). COVID-19 is awful. Climate change could be worse. The blog of Bill Gates, August 04, 2020. Available online: https://www.gatesnotes.com/Energy/Climate-and-COVID-19 [Accessed on 22 May 2024].

Gaur, L., Afaq, A., Arora, G. K., & Khan, N. (2023). Artificial intelligence for carbon emissions using system of systems theory. *Ecological Informatics*, 76, 102165. DOI: 10.1016/j.ecoinf.2023.102165

Gaur, L., & Sahoo, B. M. (2022). *Explainable Artificial Intelligence for Intelligent Transportation Systems: Ethics and Applications*. Springer Nature. DOI: 10.1007/978-3-031-09644-0

Genish, T., & Boopathiraja, S. (2023). Biomass Renewable Energy: Introduction and Application of AI and Machine Learning. In *AI-Powered Machine Learning in the Energy Industry: Digital Technology and Sustainable Energy Systems* (pp. 165–187). Springer International Publishing.

Gent, E. (2020). This 'once-for-all' neural network could slash AI's carbon footprint. SingularityHub, May 4, 2020. Available online: https://singularityhub.com/2020/05/04/this-once-for-all-neural-network-could-slash-ais-carbon-footprint/ [Accessed on 21 May 2024].

Ghatrehsamani, S., Jha, G., Dutta, W., Molaei, F., Nazrul, F., Fortin, M., Bansal, S., Debangshi, U., & Neupane, J. (2023). Artificial intelligence tools and techniques to combat herbicide resistant weeds—A review. *Sustainability (Basel)*, 15(3), 1843. DOI: 10.3390/su15031843

Ghazal, T. M., Hasan, M. K., Alzoubi, H. M., Alshurideh, M., Ahmad, M., & Akbar, S. S. (2023). Internet of things connected wireless sensor networks for smart cities. In *The Effect of Information Technology on Business and Marketing Intelligence Systems* (pp. 1953–1968). Springer International Publishing. DOI: 10.1007/978-3-031-12382-5_107

Giles, M. (2019). Is AI the next big climate-change threat? We haven't a clue. MIT Technology Review, July 29, 2019. Available online: https://www.technologyreview.com/2019/07/29/663/ai-computing-cloud-computing-microchips/ [Accessed on 21 May 2024].

Giourka, P., Sanders, M. W., Angelakoglou, K., Pramangioulis, D., Nikolopoulos, N., Rakopoulos, D., Tryferidis, A., & Tzovaras, D. (2019). The smart city business model canvas—A smart city business modeling framework and practical tool. *Energies*, 12(24), 4798. DOI: 10.3390/en12244798

Gochhait, S., Mazumdar, O., Chahal, S., Kanwat, P., Gupta, S., Sharma, R., & Sachan, R. (2020, May). Role of Machine leaning (ML) in understanding the behavior pattern: a study on e-commerce. In *ICDSMLA 2019: Proceedings of the 1st International Conference on Data Science, Machine Learning and Applications* (pp. 1600-1606). Singapore: Springer Singapore. DOI: 10.1007/978-981-15-1420-3_166

Gonzalez-Fernandez, D., Hanke, G., Pogojeva, M., Machitadze, N., Kotelnikova, Y., Tretiak, I., Savenko, O., Bilashvili, K., Gelashvili, N., Fedorov, A., Kulagin, D., Terentiev, A., & Slobodnik, J. (2022). Floating marine macro litter in the Black Sea: Toward baselines for large scale assessment. *Environmental Pollution*, 309, 119816. Advance online publication. DOI: 10.1016/j.envpol.2022.119816 PMID: 35872285

González, V., Martín, L., Santana, J. R., Sotres, P., Lanza, J., & Sánchez, L. (2024). Reshaping Smart Cities through NGSI-LD Enrichment. *Sensors (Basel)*, 24(6), 1858. DOI: 10.3390/s24061858 PMID: 38544121

Govindan, K. (2024). Unlocking the potential of quality as a core marketing strategy in remanufactured circular products: A machine learning enabled multi-theoretical perspective. *International Journal of Production Economics*, 269, 109123. DOI: 10.1016/j.ijpe.2023.109123

Grigorieva, E., Livenets, A., & Stelmakh, E. (2023). Adaptation of agriculture to climate change: A scoping review. *Climate (Basel)*, 11(10), 202. DOI: 10.3390/cli11100202

Guidance, W. H. O. (2021). Ethics and governance of artificial intelligence for health. *World Health Organization*. https://iris.who.int/rest/bitstreams/1352854/retrieve?ref=thinkml.ai

Gumaelius, L., Skogh, I. B., Matthíasdóttir, Á., & Pantzos, P. (2024). Engineering education in change. A case study on the impact of digital transformation on content and teaching methods in different engineering disciplines. *European Journal of Engineering Education*, 49(1), 70–93. DOI: 10.1080/03043797.2023.2285794

Guo, R., Hong, Z., & Xue, X. (2021). *Reward Function Design via Human AI and Inverse Reinforcement Learning for Intelligent Driving* (No. 2021-01-0180). SAE Technical Paper.

Gustafsson, S., Hermelin, B., & Smas, L. (2019). Integrating environmental sustainability into strategic spatial planning: The importance of management. *Journal of Environmental Planning and Management*, 62(8), 1321–1338. DOI: 10.1080/09640568.2018.1495620

Guzman, A. (2021). Race after technology: Abolitionist tools for the new Jim code. *Information Communication and Society*, 24(13), 1989–1990. DOI: 10.1080/1369118X.2020.1844269

Hadeed, M. Z., Ali, A., Malik, A., Raza, A., & Shoaib, M. (2024). Harnessing AI and GIS Technologies for Climate-Resilient Agriculture and Environmental Sustainability. In *Maintaining a Sustainable World in the Nexus of Environmental Science and AI* (pp. 333–364). IGI Global. DOI: 10.4018/979-8-3693-6336-2.ch013

Haldorai, A., Babitha Lincy, R., Suriya, M., & Balakrishnan, M. (2024). Climate Resilience Via Smart Technologies Over Natural Disaster. In *Artificial Intelligence for Sustainable Development* (pp. 383–406). Springer Nature Switzerland. DOI: 10.1007/978-3-031-53972-5_20

Haleem, A., Javaid, M., Khan, I. H., & Mohan, S. (2023). Significant Applications of Artificial Intelligence Towards Attaining Sustainability. *Journal of Industrial Integration & Management*, 8(4), 489–520. DOI: 10.1142/S2424862223500331

Halilaj, L., Dindorkar, I., Lüttin, J., & Rothermel, S. (2021). A AI-based approach for situation comprehension in driving scenarios. In *The Semantic Web: 18th International Conference, ESWC 2021, Virtual Event, June 6–10, 2021* [Springer International Publishing.]. *Proceedings*, 18, 699–716.

Halilaj, L., Luettin, J., Henson, C., & Monka, S. (2022, September). Machine learning for automated driving. In *2022 IEEE Fifth International Conference on Artificial Intelligence and Knowledge Engineering (AIKE)* (pp. 98-105). IEEE.

Halilaj, L., Luettin, J., Monka, S., Henson, C., & Schmid, S. (2023). AI-Based Integration of Autonomous Driving Datasets. *International Journal of Semantic Computing*, 17(02), 249–271. DOI: 10.1142/S1793351X23600048

Hanes, D., Salgueiro, G., Grossetete, P., Barton, R., & Henry, J. (2017). *IoT fundamentals: Networking technologies, protocols, and use cases for the internet of things*. Cisco Press.

Han, H., Liu, Z., Li, J., & Zeng, Z. (2024). Challenges in remote sensing based climate and crop monitoring: Navigating the complexities using AI. *Journal of Cloud Computing (Heidelberg, Germany)*, 13(1), 34. DOI: 10.1186/s13677-023-00583-8

Han, Y., Li, Z., Feng, T., Qiu, S., Hu, J., Yadav, K. K., & Obaidullah, A. J. (2024). Unraveling the impact of digital transformation on green innovation through microdata and machine learning. *Journal of Environmental Management*, 354, 120271. DOI: 10.1016/j.jenvman.2024.120271 PMID: 38354610

Hao, K. (2019). Here are 10 ways AI could help fight climate change. MIT Technology Review, June 20, 2019. Available online: https://www.technologyreview.com/2019/06/20/134864/ai-climate-change-machine-learning/ [Accessed 21 May 2024].

Harrer, S., Menard, J., Rivers, M., Green, D. V., Karpiak, J., Jeliazkov, J. R., & Sternke, M. C. (2024). Artificial intelligence drives the digital transformation of pharma. In *Artificial intelligence in clinical practice* (pp. 345–372). Academic Press. DOI: 10.1016/B978-0-443-15688-5.00049-8

Hasan, R., Weaven, S., & Thaichon, P. (2021). Blurring the line between physical and digital environment: The impact of artificial intelligence on customers' relationship and customer experience. In *Developing digital marketing* (pp. 135–153). Emerald Publishing Limited. DOI: 10.1108/978-1-80071-348-220211008

Hashem, I. A. T., Usmani, R. S. A., Almutairi, M. S., Ibrahim, A. O., Zakari, A., Alotaibi, F., Alhashmi, S. M., & Chiroma, H. (2023). Urban computing for sustainable smart cities: Recent advances, taxonomy, and open research challenges. *Sustainability (Basel)*, 15(5), 3916. DOI: 10.3390/su15053916

Hassan, M. A., Javed, R., Granelli, F., Gen, X., Rizwan, M., Ali, S. H., & Ullah, S. (2023, March). Intelligent transportation systems in smart city: a systematic survey. In *2023 International Conference on Robotics and Automation in Industry (ICRAI)* (pp. 1-9). IEEE. DOI: 10.1109/ICRAI57502.2023.10089543

Häyhä, T., & Franzese, P. P. (2014). Ecosystem services assessment: A review under an ecological-economic and systems perspective. *Ecological Modelling*, 289, 124–132. DOI: 10.1016/j.ecolmodel.2014.07.002

Hazas, M., Morley, J., Bates, O., & Friday, A. (2016). Are there limits to growth in data traffic?: On time use, data generation and speed. *Proceedings of the Second Workshop on Computing within Limits, 14,* 1–5. https://doi.org/DOI: 10.1145/2926676.2926690

Heat wave. (2024, May 7). Wikipedia. https://en.wikipedia.org/wiki/Heat_wave

Heat Waves and Climate Change - Center for Climate and Energy Solutions. (2023, July 14). Center for Climate and Energy Solutions. https://www.c2es.org/content/heat-waves-and-climate-change/

He, Q., Zhu, Z., Zhao, D., Song, W., & Huang, D. (2024). An Interpretable Deep Learning Approach for Detecting Marine Heatwaves Patterns. *Applied Sciences (Basel, Switzerland)*, 14(2), 601. DOI: 10.3390/app14020601

Herath, H. M. K. K. M. B., & Mittal, M. (2022). Adoption of artificial intelligence in smart cities: A comprehensive review. *International Journal of Information Management Data Insights*, 2(1), 100076. DOI: 10.1016/j.jjimei.2022.100076

Herhausen, D., Bernritter, S. F., Ngai, E. W., Kumar, A., & Delen, D. (2024). Machine learning in marketing: Recent progress and future research directions. *Journal of Business Research*, 170, 114254. DOI: 10.1016/j.jbusres.2023.114254

Hern, A. (2018). Bitcoin's energy usage is huge – We can't afford to ignore it. *The Guardian,* January 17, 2018. Available online: https://perma.cc/2X2H-CF9V [Accessed on 21 May 2024].

Hoegh-Guldberg, O., Mumby, P. J., Hooten, A. J., Steneck, R. S., Greenfield, P., Gomez, E., ... & Hatziolos, M. (2007). Coral reefs under rapid climate change and ocean acidification. science, 318(5857), 1737-1742.

Holzinger, A., Saranti, A., Hauschild, A. C., Beinecke, J., Heider, D., Roettger, R., . . . Pfeifer, B. (2023, August). Human-in-the-loop integration with domain-Machine learning for explainable federated deep learning. In *International Cross-Domain Conference for Machine Learning and Knowledge Extraction* (pp. 45-64). Cham: Springer Nature Switzerland. DOI: 10.1007/978-3-031-40837-3_4

Holzinger, A., Keiblinger, K., Holub, P., Zatloukal, K., & Müller, H. (2023). AI for life: Trends in artificial intelligence for biotechnology. *New Biotechnology*, 74, 16–24. DOI: 10.1016/j.nbt.2023.02.001 PMID: 36754147

Horton, A. A. (2022, January). Horton (2022). Plastic pollution: When do we know enough? *Journal of Hazardous Materials*, 422, 126885. Advance online publication. DOI: 10.1016/j.jhazmat.2021.126885

Ho, S. P. S., & Chow, M. Y. C. (2023). The role of artificial intelligence in consumers' brand preference for retail banks in Hong Kong. *Journal of Financial Services Marketing*, •••, 1–14.

Hossain, M., Nijhum, I. R., Shad, M. T. M., Sadi, A. A., Peyal, M. M. K., & Rahman, R. M. (2023). An end-to-end pollution analysis and detection system using artificial intelligence and object detection algorithms. *Decision Analytics Journal*, 8, 100283. DOI: 10.1016/j.dajour.2023.100283

How AI Can Help Save Endangered Species | Scientific American. (n.d.). Retrieved August 15, 2024, from https://www.scientificamerican.com/article/how-ai-can-help-save-endangered-species/

Howard, B. C. (2018). Blue growth: Stakeholder perspectives. *Marine Policy*, 87, 375–377. DOI: 10.1016/j.marpol.2017.11.002

Huang, H., Sun, L., Du, B., Liu, C., Lv, W., & Xiong, H. (2021, August). Representation learning on Machine learning for node importance estimation. In *Proceedings of the 27th ACM SIGKDD Conference on Knowledge Discovery & Data Mining* (pp. 646-655).

Huda, N. U., Ahmed, I., Adnan, M., Ali, M., & Naeem, F. (2024). Experts and intelligent systems for smart homes' Transformation to Sustainable Smart Cities: A comprehensive review. *Expert Systems with Applications*, 238, 122380. DOI: 10.1016/j.eswa.2023.122380

Hughes, B. a GE company and C3.ai. (2019). Announce joint venture, June 24, 2019. Available online: https://c3.ai/baker-hughes-and-c3-ai-announce-joint-venture-to-deliver-ai-solutions [Accessed on 21 May 2024].

Hui, C. X., Dan, G., Alamri, S., & Toghraie, D. (2023). Greening smart cities: An investigation of the integration of urban natural resources and smart city technologies for promoting environmental sustainability. *Sustainable Cities and Society*, 99, 104985. DOI: 10.1016/j.scs.2023.104985

Husain, S. F., Tutumluer, E., Mechitov, K. A., Qamhia, I. I., Spencer, B., & Edwards, J. R. (2023). Towards a wireless sensing infrastructure for smart mobility. *Transportation Geotechnics*, 40, 100985. DOI: 10.1016/j.trgeo.2023.100985

Hussain, I., Elomri, A., Kerbache, L., & El Omri, A. (2024). Smart city solutions: Comparative analysis of waste management models in IoT-enabled environments using multiagent simulation. *Sustainable Cities and Society*, 103, 105247. DOI: 10.1016/j.scs.2024.105247

Hu, X., Liu, Y., Tang, B., Yan, J., & Chen, L. (2023). Learning dynamic ML for overtaking strategy in autonomous driving. *IEEE Transactions on Intelligent Transportation Systems*, 24(11), 11921–11933. DOI: 10.1109/TITS.2023.3287223

Huynh-The, T., Pham, Q. V., Pham, X. Q., Nguyen, T. T., Han, Z., & Kim, D. S. (2023). Artificial intelligence for the metaverse: A survey. *Engineering Applications of Artificial Intelligence*, 117, 105581. DOI: 10.1016/j.engappai.2022.105581

Iftikhar, A. A., Qureshi, K. N., Altalbe, A. A., & Javeed, K. (2023). Security Provision by Using Detection and Prevention Methods to Ensure Trust in Edge-Based Smart City Networks. *IEEE Access : Practical Innovations, Open Solutions*, 11, 137529–137547. DOI: 10.1109/ACCESS.2023.3338550

Ijiga, A. C., Peace, A. E., Idoko, I. P., Ezebuka, C. I., Harry, K. D., Ukatu, I. E., & Agbo, D. O.Amina Catherine IjigaAbutu Ehi PeaceIdoko Peter IdokoChijioke Charles EzebukaKimberly D. HarryIfeanyi Emmanuel UkatuDaniel Obekpa Agbo. (2024). Technological innovations in mitigating winter health challenges in New York City, USA. *International Journal of Science and Research Archive*, 11(1), 535–551. DOI: 10.30574/ijsra.2024.11.1.0078

Impact of artificial intelligence on human loss in decision making, laziness and safety in education—PMC. (n.d.). Retrieved August 15, 2024, from https://www.ncbi.nlm.nih.gov/pmc/articles/PMC10251321/

Information privacy, impact assessment, and the place of ethics—ScienceDirect. (n.d.). Retrieved August 15, 2024, from https://www.sciencedirect.com/science/article/pii/S0267364920300091

IPCC (Intergovernmental Panel on Climate Change). (2018). An IPCC special report on the impacts of global warming of 1.5°C. Available online: https://www.ipcc.ch/sr15/ [Accessed on 21 May 2024].

IPCC. (2019). Technical Summary: IPCC Special Report on the Ocean and Cryosphere in a Changing Climate. Eds. H. O. Pörtner, D. C. Roberts, V. Masson-Delmotte, P. Zhai, E. Poloczanska, K. Mintenbeck, M. Tignor, A. Alegrıa, M. Nicolai, A. Okem, J. Petzold,́ B. Rama and N. M. Weyer (Geneva: IPCC).

Ise, T., & Oba, Y. (2019). Forecasting climatic trends using neural networks: An experimental study using global historical data. *Frontiers in Robotics and AI*, 6, 32. DOI: 10.3389/frobt.2019.00032 PMID: 33501048

Issa Zadeh, S. B., Esteban Perez, M. D., López-Gutiérrez, J. S., & Fernández-Sánchez, G. (2023). Optimizing Smart Energy Infrastructure in Smart Ports: A Systematic Scoping Review of Carbon Footprint Reduction. *Journal of Marine Science and Engineering*, 11(10), 1921. DOI: 10.3390/jmse11101921

Ivanov, S. H., & Webster, C. (2017, June). The robot as a consumer: a research agenda. In *Marketing: experience and perspectives* (pp. 29–30). Conference.

Jacques-Dumas, V., Ragone, F., Borgnat, P., Abry, P., & Bouchet, F. (2022). Deep learning-based extreme heatwave forecast. *Frontiers in Climate*, •••, 4.

Jain, A., Sharma, A., Jately, V., & Azzopardi, B. (Eds.). (2023). *Sustainable Energy Solutions with Artificial Intelligence, Blockchain Technology, and Internet of Things*. CRC Press. DOI: 10.1201/9781003356639

Jain, H., Dhupper, R., Shrivastava, A., Kumar, D., & Kumari, M. (2023). AI-enabled strategies for climate change adaptation: Protecting communities, infrastructure, and businesses from the impacts of climate change. *Computational Urban Science*, 3(1), 25. DOI: 10.1007/s43762-023-00100-2

Jain, S., Ahuja, N. J., Srikanth, P., Bhadane, K. V., Nagaiah, B., Kumar, A., & Konstantinou, C. (2021). Blockchain and autonomous vehicles: Recent advances and future directions. *IEEE Access : Practical Innovations, Open Solutions*, 9, 130264–130328. DOI: 10.1109/ACCESS.2021.3113649

Jain, S., & Gandhi, A. V. (2021). Impact of artificial intelligence on impulse buying behaviour of Indian shoppers in fashion retail outlets. *International Journal of Innovation Science*, 13(2), 193–204. DOI: 10.1108/IJIS-10-2020-0181

Jambeck, , Geyer, R., Wilcox, C., Siegler, T. R., Perryman, M., Andrady, A., Narayan, R., & Law, K. L. (2015). Marine pollution. Plastic waste inputs from land into the ocean. *Science*, 347(6223), 768–771. DOI: 10.1126/science.1260352 PMID: 25678662

Janani, R. P., Renuka, K., & Aruna, A. (2021). IoT in smart cities: A contemporary survey. *Global Transitions Proceedings*, 2(2), 187–193. DOI: 10.1016/j.gltp.2021.08.069

Javaid, M., Haleem, A., Khan, I. H., & Suman, R. (2023). Understanding the potential applications of Artificial Intelligence in Agriculture Sector. *Advanced Agrochem*, 2(1), 15–30. DOI: 10.1016/j.aac.2022.10.001

Jeong, M., & Shin, H. H. (2020). Tourists' experiences with smart tourism technology at smart destinations and their behavior intentions. *Journal of Travel Research*, 59(8), 1464–1477. DOI: 10.1177/0047287519883034

Jha, S., Nkenyereye, L., Joshi, G. P., & Yang, E. (2020). Mitigating and monitoring smart city using internet of things. *Computers, Materials & Continua*, 65(2), 1059–1079. DOI: 10.32604/cmc.2020.011754

Jia, J., Xu, Y., & Li, W. (2024). A study on the strategic momentum of SMEs' digital transformation: Evidence from China. *Technological Forecasting and Social Change*, 200, 123038. DOI: 10.1016/j.techfore.2023.123038

Jiao, D., Xu, N., Yang, F., & Xu, K. (2021). Evaluation of spatial-temporal variation performance of ERA5 precipitation data in China. *Scientific Reports*, 11(1), 17956. DOI: 10.1038/s41598-021-97432-y PMID: 34504211

Jin, K., Zhong, Z. Z., & Zhao, E. Y. (2024). Sustainable digital marketing under big data: An AI random forest model approach. *IEEE Transactions on Engineering Management*, 71, 3566–3579. DOI: 10.1109/TEM.2023.3348991

Ji, S., Pan, S., Cambria, E., Marttinen, P., & Philip, S. Y. (2021). A survey on Machine learning: Representation, acquisition, and applications. *IEEE Transactions on Neural Networks and Learning Systems*, 33(2), 494–514. DOI: 10.1109/TNNLS.2021.3070843 PMID: 33900922

Joshua, P. Meltzer, & Cameron F. Kerry. (2021). *Strengthening international cooperation on artificial intelligence*. Brookings. https://www.brookings.edu/articles/strengthening-international-cooperation-on-artificial-intelligence/

Joshua, K Abbott & U Rashid Sumaila. (2019). Reducing Marine Plastic Pollution: Policy Insights from Economics. *Review of Environmental Economics and Policy, Association of Environmental and Resource Economists*, 13(2), 327–336.

Jyothi, A. P., Shankar, A., & Shukla, A., Jr., A. N., Kumar, S. S., & Yashwanth, A. (2023, October). Intelligent Systems and Methods for Smart City in Multiple Domains. In *2023 International Conference on Evolutionary Algorithms and Soft Computing Techniques (EASCT)* (pp. 1-7). IEEE.

Kaack, L., Donti, P., Strubell, E., Kamiya, G., Creutzig, F., & Rolnick, D. (2022). Aligning artificial intelligence with climate change mitigation. *Nature Climate Change*, 12(6), 518–527. DOI: 10.1038/s41558-022-01377-7

Kalemkuş, F., & Kalemkuş, E. (2022). The effects of augmented reality on students' learning outcomes: A meta-analysis study. *Journal of Educational Technology & Society*, 25(3), 1–15.

Kalogiannidis, S., Kalfas, D., Patitsa, C., & Chalaris, M. (2024). Relationship between climate change and business risk: strategies for adaptation and mitigation: evidence from a Mediterranean country. *WSEAS Transactions on Environment and Development*, 20, 276–294. DOI: 10.37394/232015.2024.20.28

Kaminski, M. E. (2022). Regulating the Risks of AI. SSRN *Electronic Journal*. DOI: 10.2139/ssrn.4195066

Kanellopoulos, D., Sharma, V. K., Panagiotakopoulos, T., & Kameas, A. (2023). Networking architectures and protocols for IoT applications in smart cities: Recent developments and perspectives. *Electronics (Basel)*, 12(11), 2490. DOI: 10.3390/electronics12112490

Kang, B. G., & Nam, Y. (2024). Responsible Artificial Intelligence for Climate Action: A Theoretical Framework for Sustainable Development. *Sustainable Machine Intelligence Journal*, 8, 1–1. DOI: 10.61356/SMIJ.2024.88101

Kao, L. J., Chiu, C. C., Lin, H. T., Hung, Y. W., & Lu, C. C. (2024). Unveiling the dimensions of digital transformation: A comprehensive taxonomy and assessment model for business. *Journal of Business Research*, 176, 114595. DOI: 10.1016/j.jbusres.2024.114595

Kaperonis, S. (2024). How artificial intelligence (AI) is transforming the user experience in digital marketing. In *The use of artificial intelligence in digital marketing: Competitive strategies and tactics* (pp. 117–141). IGI Global.

Karanja, J. M., & Saito, O. (2018). Cost-benefit analysis of mangrove ecosystems in flood risk reduction: A case study of the Tana Delta, Kenya. *Sustainability Science*, 13, 503–516. DOI: 10.1007/s11625-017-0427-3

Kataray, T., Nitesh, B., Yarram, B., Sinha, S., Cuce, E., Shaik, S., & Roy, A. (2023). Integration of smart grid with renewable energy sources: Opportunities and challenges–A comprehensive review. *Sustainable Energy Technologies and Assessments*, 58, 103363. DOI: 10.1016/j.seta.2023.103363

Kates-Harbeck, J., Svyatkovskiy, A., & Tang, W. (2019). Predicting disruptive instabilities in controlled fusion plasmas through deep learning. *Nature*, 568(7753), 526–531. DOI: 10.1038/s41586-019-1116-4 PMID: 30996348

Kathijotes, N. (2013). Keynote: Blue economy - environmental and behavioural aspects towards sustainable coastal development. *Procedia: Social and Behavioral Sciences*, 101, 7–13. DOI: 10.1016/j.sbspro.2013.07.173

Katsamakas, E., Pavlov, O. V., & Saklad, R. (2024). Artificial intelligence and the transformation of higher education institutions. *arXiv preprint arXiv:2402.08143.*

Kazu, I., & Yalçin, M. (2022). The role of mobile devices in education: A systematic review of the literature from 2010 to 2020. *Journal of Educational Computing Research*, 60(1), 1–27.

Kedi, W. E., Ejimuda, C., Idemudia, C., & Ijomah, T. I. (2024). AI software for personalized marketing automation in SMEs: Enhancing customer experience and sales. *World Journal of Advanced Research and Reviews*, 23(1), 1981–1990. DOI: 10.30574/wjarr.2024.23.1.2159

Kedi, W. E., Ejimuda, C., Idemudia, C., & Ijomah, T. I. (2024). Machine learning software for optimizing SME social media marketing campaigns. *Computer Science & IT Research Journal*, 5(7), 1634–1647. DOI: 10.51594/csitrj.v5i7.1349

Kenny, G. P., Yardley, J., Brown, C., Sigal, R. J., & Jay, O. (2010). Heat stress in older individuals and patients with common chronic diseases. *Canadian Medical Association Journal*, 182(10), 1053–1060. DOI: 10.1503/cmaj.081050 PMID: 19703915

Khalid, M. (2024). Energy 4.0: AI-enabled digital transformation for sustainable power networks. *Computers & Industrial Engineering*, 193, 110253. DOI: 10.1016/j.cie.2024.110253

Khan, M. R., Haider, Z. M., Malik, F. H., Almasoudi, F. M., Alatawi, K. S. S., & Bhutta, M. S. (2024). A comprehensive review of microgrid energy management strategies considering electric vehicles, energy storage systems, and AI techniques. *Processes (Basel, Switzerland)*, 12(2), 270. DOI: 10.3390/pr12020270

Khan, M. W., Saad, S., Ammad, S., Rasheed, K., & Jamal, Q. (2024). Smart Infrastructure and AI. In *AI in Material Science* (pp. 193–215). CRC Press. DOI: 10.1201/9781003438489-9

Khaspuria, G., Ranjan, A., Soni, P., & Dadhich, K. (2024). Natural Disaster Mitigation Strategies: A Comprehensive Review. *Journal of Scientific Research and Reports*, 30(8), 20–34. DOI: 10.9734/jsrr/2024/v30i82221

Khunakornbodintr, S. (2024). Examining the impact of green technological specialization and the integration of AI technologies on green innovation performance: Evidence from China. *Frontiers in Artificial Intelligence*, 6, 1237285. Advance online publication. DOI: 10.3389/frai.2023.1237285 PMID: 38725486

Kim, J. E., Henson, C., Huang, K., Tran, T. A., & Lin, W. Y. (2021). Accelerating road sign ground truth construction with AI and machine learning. In *Intelligent Computing:Proceedings of the 2021 Computing Conference,* Volume 2 (pp. 325-340). Springer International Publishing.

Ki-moon, B. (2011). *Address to the 66th General Assembly*. We the Peoples.

Kirchgaessner, S. (2019). Revealed: Google made large contributions to climate change deniers. *The Guardian,* October 11, 2019. Available online: https://amp.theguardian.com/environment/2019/oct/11/google-contributions-climate-change-deniers?twitter_impression=true [Accessed on 21 May 2024].

Knowlton, K., Rotkin-Ellman, M., King, G., Margolis, H. G., Smith, D., Solomon, G., Trent, R., & English, P. (2009). The 2006 California heat wave: Impacts on hospitalizations and emergency department visits. *Environmental Health Perspectives*, 117(1), 61–67. DOI: 10.1289/ehp.11594 PMID: 19165388

Kolhe, R. V., William, P., Yawalkar, P. M., Paithankar, D. N., & Pabale, A. R. (2023). Smart city implementation based on Internet of Things integrated with optimization technology. *Measurement. Sensors*, 27, 100789. DOI: 10.1016/j.measen.2023.100789

Konya, A., & Nematzadeh, P. (2024). Recent applications of AI to environmental disciplines: A review. *The Science of the Total Environment*, 906, 167705. DOI: 10.1016/j.scitotenv.2023.167705 PMID: 37820816

Kopalle, P. K., Gangwar, M., Kaplan, A., Ramachandran, D., Reinartz, W., & Rindfleisch, A. (2022). Examining Machine leaning (ML) technologies in marketing via a global lens: Current trends and future research opportunities. *International Journal of Research in Marketing*, 39(2), 522–540. DOI: 10.1016/j.ijresmar.2021.11.002

Koppe, C., Kovats, S., Jendritzky, G., & Menne, B. (2004). *Heat-waves: risks and responses (No. EUR/03/5036810)*. World Health Organization. Regional Office for Europe.

Kosatsky, T. (2005). The 2003 European heat waves. *Eurosurveillance*, 10(7), 3–4. DOI: 10.2807/esm.10.07.00552-en PMID: 29208081

Kovats, R. S., & Koppe, C. (2005). Heatwaves: past and future impacts on health. Integration of public health with adaptation to climate change: lessons learned and new directions, 136-160.

Kraft, M., Bai, J., Mosbach, S., Taylor, C., Karan, D., Lee, K. F., ... & Lapkin, A. (2023). From Platform to AI: Distributed Self-Driving Laboratories.

Kulkarni, A. V., Joseph, S., & Patil, K. P. (2024). Artificial intelligence technology readiness for social sustainability and business ethics: Evidence from MSMEs in developing nations. *International Journal of Information Management Data Insights*, 4(2), 100250. DOI: 10.1016/j.jjimei.2024.100250

Kulp, S. A., & Strauss, B. H. (2018). CoastalDEM: A global coastal digital elevation model improved from SRTM using a neural network. *Remote Sensing of Environment*, 20, 231–239. DOI: 10.1016/j.rse.2017.12.026

Kumar, S., & Chhillar, R. S. (2023, December). Comparative analysis of existing security techniques in intelligent transportation system for smart cities. In *AIP Conference Proceedings* (Vol. 2916, No. 1). AIP Publishing. DOI: 10.1063/5.0177444

Kumar, D. (2023). Concept of Sustainable Energy System for Smart Cities. In *Renewable Energy Scenarios in Future Indian Smart Cities: A Geospatial Technology Perspective* (pp. 1–20). Springer Nature Singapore. DOI: 10.1007/978-981-19-8456-3_1

Kumar, R. L., Pham, Q. V., Khan, F., Piran, M. J., & Dev, K. (2021). Blockchain for securing aerial communications: Potentials, solutions, and research directions. *Physical Communication*, 47, 101390. DOI: 10.1016/j.phycom.2021.101390

Kwiatkowski, C. A., Pawłowska, M., Harasim, E., & Pawłowski, L. (2023). Strategies of Climate Change Mitigation in Agriculture Plant Production—A Critical Review. *Energies*, 16(10), 4225. DOI: 10.3390/en16104225

Kwon, H. J., Salim, M. M., & Park, J. H. (2023). Recent trends on Smart City security: A comprehensive overview. *Journal of Information Processing Systems*, 19(1), 118–129.

Larsson, S., & Heintz, F. (2020). Transparency in artificial intelligence. *Internet Policy Review*, 9(2). Advance online publication. https://policyreview.info/concepts/transparency-artificial-intelligence. DOI: 10.14763/2020.2.1469

Laux, J., Wachter, S., & Mittelstadt, B. (2024). Three pathways for standardisation and ethical disclosure by default under the European union artificial intelligence act. *Computer Law & Security Report*, 53, 105957. DOI: 10.1016/j.clsr.2024.105957

Lawyer, C., An, L., & Goharian, E. (2023). A Review of Climate Adaptation Impacts and Strategies in Coastal Communities: From Agent-Based Modeling towards a System of Systems Approach. *Water (Basel)*, 15(14), 2635. DOI: 10.3390/w15142635

Leal Filho, W., Marisa Azul, A., Brandli, L., Gökçin Özuyar, P., & Wall, T. (Eds.). (2020). *Sustainable cities and communities*. Springer International Publishing.

Lecue, F. (2020). On the role of Machine learning in explainable AI. *Semantic Web*, 11(1), 41–51. DOI: 10.3233/SW-190374

Lee, S., Park, Y., & Park, G. (2024). Using AI chatbots in climate change mitigation: A moderated serial mediation model. *Behaviour & Information Technology*, 1–17. DOI: 10.1080/0144929X.2024.2402532

Lefebvre, M. (2021). *Reimagining Smart Cities: reconsidering pluralism and data subjectivities* (Doctoral dissertation, University of Oxford).

Lepakshi, V. A., & Ranga, P. C. (2024). Artificial Intelligence for Smart City Development: Review/Survey of Existing Systems in Smart Cities. *Handbook of Artificial Intelligence for Smart City Development*, 60-75.

Leso, B. H., Cortimiglia, M. N., Ghezzi, A., & Minatogawa, V. (2024). Exploring digital transformation capability via a blended perspective of dynamic capabilities and digital maturity: A pattern matching approach. *Review of Managerial Science*, 18(4), 1149–1187. DOI: 10.1007/s11846-023-00692-3

Leyer, M., Oberlaender, A., Dootson, P., & Kowalkiewicz, M. (2020, June). Decision-making with artificial intelligence: Towards a novel conceptualization of patterns. In *PACIS* (p. 224).

Li, X., Liu, J., Li, J., Yu, W., Cao, Z., Qiu, S., . . . Jiao, X. (2023, August). ML Structure-Based Implicit Risk Reasoning for Long-Tail Scenarios of Automated Driving. In *2023 4th International Conference on Big Data, Artificial Intelligence and Internet of Things Engineering (ICBAIE)* (pp. 415-420). IEEE.

Li, J., & Zhou, T. (2023). Multiagent deep meta reinforcement learning for sea computing-based energy management of interconnected grids considering renewable energy sources in sustainable cities. *Sustainable Cities and Society*, 99, 104917. DOI: 10.1016/j.scs.2023.104917

Lillebø, A. I., Pita, C., Garcia Rodrigues, J., Ramos, S., & Villasante, S. (2017). How can marine ecosystem services support the blue growth agenda? *Marine Policy*, 81, 132–142. DOI: 10.1016/j.marpol.2017.03.008

Lima, M. A., Mendes, L. F. R., Mothé, G. A., Linhares, F. G., de Castro, M. P. P., da Silva, M. G., & Sthel, M. S. (2020). Renewable energy in reducing greenhouse gas emissions: Reaching the goals of the Paris agreement in Brazil. *Environmental Development*, 33, 100504. DOI: 10.1016/j.envdev.2020.100504

Lingam, Y. K. (2018). The role of Machine leaning (ML) in making accurate stock decisions in E-commerce industry. *Int. J. Adv. Res. Ideas Innov. Technol*, 4(3), 2281–2286.

Liu, F., Fan, H. Y., & Qi, J. Y. (2022). Blockchain technology, cryptocurrency: Entropy-based perspective. *Entropy (Basel, Switzerland)*, 24(4), 557. DOI: 10.3390/e24040557 PMID: 35455220

Liu, H., Li, Y., Hong, R., Li, Z., Li, M., Pan, W., Glowacz, A., & He, H. (2020). AI analysis and visualization of research trends on driver behavior. *Journal of Intelligent & Fuzzy Systems*, 38(1), 495–511. DOI: 10.3233/JIFS-179424

Liu, J., Zhang, X., Li, Y., Wang, J., & Kim, H. J. (2019). Deep learning-based reasoning with multi-ontology for IoT applications. *IEEE Access : Practical Innovations, Open Solutions*, 7, 124688–124701. DOI: 10.1109/ACCESS.2019.2937353

Liu, Q., Li, X., Tang, Y., Gao, X., Yang, F., & Li, Z. (2023). ML Reinforcement Learning-Based Decision-Making Technology for Connected and Autonomous Vehicles: Framework, Review, and Future Trends. *Sensors (Basel)*, 23(19), 8229. DOI: 10.3390/s23198229 PMID: 37837063

Liu, W., McKibbin, W. J., Morris, A. C., & Wilcoxen, P. J. (2020). Global economic and environmental outcomes of the Paris Agreement. [i]. *Energy Economics*, 90, 104838. DOI: 10.1016/j.eneco.2020.104838

Li, W., Yigitcanlar, T., Browne, W., & Nili, A. (2023). The Making of Responsible Innovation and Technology: An Overview and Framework. *Smart Cities*, 6(4), 1996–2034. DOI: 10.3390/smartcities6040093

Li, X., Zhao, H., Feng, Y., Li, J., Zhao, Y., & Wang, X. (2024). Research on key technologies of high energy efficiency and low power consumption of new data acquisition equipment of power Internet of Things based on artificial intelligence. *International Journal of Thermofluids*, 21, 100575. DOI: 10.1016/j.ijft.2024.100575

Lobell, D. B., Schlenker, W., & Costa-Roberts, J. (2011). Climate trends and global crop production since 1980. *Science*, 333(6042), 616–620. DOI: 10.1126/science.1204531 PMID: 21551030

Lopez-Gomez, I., McGovern, A., Agrawal, S., & Hickey, J. (2023). Global Extreme Heat Forecasting Using Neural Weather Models. *Artificial Intelligence for the Earth Systems*, 2(1), e220035. DOI: 10.1175/AIES-D-22-0035.1

López, O. L., Rosabal, O. M., Ruiz-Guirola, D. E., Raghuwanshi, P., Mikhaylov, K., Lovén, L., & Iyer, S. (2023). Energy-sustainable Machine Learning connectivity: Vision, technological enablers, challenges, and future directions. *IEEE Open Journal of the Communications Society*, 4, 2609–2666. DOI: 10.1109/OJCOMS.2023.3323832

Lücking, H. (2023). *The use of AI technology for sustainability: The envisioned meaning and role of sustainable AI in climate political discourses in Denmark and Germany* (Bachelor's thesis, University of Twente).

Luettin, J., Monka, S., Henson, C., & Halilaj, L. (2022, November). A survey on AI-based methods for automated driving. In *Iberoamerican Machine learning and Semantic Web Conference* (pp. 16–31). Springer International Publishing.

Luo, X., Tong, S., Fang, Z., & Qu, Z. (2019). Frontiers: Machines vs. humans: The impact of artificial intelligence chatbot disclosure on customer purchases. *Marketing Science*, 38(6), 937–947. DOI: 10.1287/mksc.2019.1192

Lv, Z., Hu, B., & Lv, H. (2019). Infrastructure monitoring and operation for smart cities based on IoT system. *IEEE Transactions on Industrial Informatics*, 16(3), 1957–1962. DOI: 10.1109/TII.2019.2913535

Macenaite, M., & Kosta, E. (2017). Consent for processing children's personal data in the EU: Following in US footsteps? *Information & Communications Technology Law*, 26(2), 146–197. DOI: 10.1080/13600834.2017.1321096

Maddireddy, B. R., & Maddireddy, B. R. (2023). Adaptive Cyber Defense: Using Machine Learning to Counter Advanced Persistent Threats. *International Journal of Advanced Engineering Technologies and Innovations*, 1(03), 305–324.

Madiega, T. (2019). EU guidelines on ethics in artificial intelligence: Context and implementation. *European Parliamentary Research Service,* 1-13. Available online: https://www.europarl.europa.eu/RegData/etudes/BRIE/2019/640163/EPRS_BRI(2019)640163_EN.pdf [Accessed on 22 May 2024].

Mahmoud, A. B., Tehseen, S., & Fuxman, L. (2020). The dark side of artificial intelligence in retail innovation. In *Retail Futures* (pp. 165–180). Emerald Publishing Limited. DOI: 10.1108/978-1-83867-663-620201019

Mahon, L., Giunchiglia, E., Li, B., & Lukasiewicz, T. (2020, December). AI extraction from videos. In *2020 19th IEEE International Conference on Machine Learning and Applications (ICMLA)* (pp. 25-32). IEEE.

Malik, P. K., Roges, R., Tiwari, P., Malik, P., Kumar, V., & Gehlot, A. (2023, July). Smart Cities Monitoring using Internet of Things: Opportunities and Challenges. In 2023 4th International Conference on Electronics and Sustainable Communication Systems (ICESC) (pp. 450-455). IEEE.

Malini, P., Gowthaman, N., Gautami, A., & Thillaiarasu, N. (2021). Internet of Everything (IoE) in smart city paradigm using advanced sensors for handheld devices and equipment. In *IoT and IoE Driven Smart Cities* (pp. 121–141). Springer International Publishing.

Maor, D., McCarthy, A., McConney, A., & Cavanaugh, C. (2023). Leading digital transformation in education: Insights from educators and stakeholders across various contexts worldwide. *Educational Management Administration & Leadership,* 51(2), 215–234.

Marda, V. (2018). Artificial intelligence policy in India: A framework for engaging the limits of data-driven decision-making. *Philosophical Transactions. Series A, Mathematical, Physical, and Engineering Sciences*, 376(2133), 20180087. DOI: 10.1098/rsta.2018.0087 PMID: 30323001

Marr, B. (2020). *Tech Trends in Practice: The 25 technologies that are driving the 4th Industrial Revolution.* John Wiley & Sons.

Martins, N. R. B., Angelica, A., Chakravarthy, K., Svidinenko, Y., Boehm, F. J., Opris, I., Lebedev, M. A., Swan, M., Garan, S. A., Rosenfeld, J. V., Hogg, T., & Freitas, R. A.Jr. (2019). Human brain/cloud interface. *Frontiers in Neuroscience,* 13, 112. DOI: 10.3389/fnins.2019.00112 PMID: 30983948

Martsenko, N. (2019). Legal regime of artificial intelligence in civil law. *Aktual ni problemy pravoznavstva, 4,* 91-98. Available online: http://dspace.wunu.edu.ua/handle/316497/38382 [Accessed on 22 May 2024].

Matheson, R. (2020). Reducing the carbon footprint of artificial intelligence. MIT News, April 23, 2020. Available online: https://news.mit.edu/2020/artificial-intelligence-ai-carbon-footprint-0423 [Accessed on 21 May 2024].

McGlade, C., & Ekins, P. (2015). The geographical distribution of fossil fuels unused when limiting global warming to 2°C. *Nature*, 517(7533), 187–190. DOI: 10.1038/nature14016 PMID: 25567285

McGovern, A., Elmore, K., Gagne, D., Haupt, S., Karstens, C., Lagerquist, R., Smith, T., & Williams, J. (2017). Using artificial intelligence to improve real-time decision-making for high-impact weather. *Bulletin of the American Meteorological Society*, 98(10), 2073–2090. DOI: 10.1175/BAMS-D-16-0123.1

McKinsey & Company. (2023). *The state of AI in 2023: Generative AI's breakout year | McKinsey*. https://www.mckinsey.com/capabilities/quantumblack/our-insights/the-state-of-ai-in-2023-generative-ais-breakout-year

McQuade, S., & Monteleoni, C. (2012). Global climate model tracking using geospatial neighborhoods. *Proceedings of the AAAI Conference on Artificial Intelligence*, 26(1), 335–341. DOI: 10.1609/aaai.v26i1.8178

Mehta, P., Dhaliwal, L. K., Baweja, P. K., Jangra, M. S., & Bhardwaj, S. K. (2023). Concept of climate smart villages using artificial intelligence/machine learning. In *Visualization Techniques for Climate Change with Machine Learning and Artificial Intelligence* (pp. 359–377). Elsevier. DOI: 10.1016/B978-0-323-99714-0.00010-8

Mennella, C., Maniscalco, U., De Pietro, G., & Esposito, M. (2024). Ethical and regulatory challenges of AI technologies in healthcare: A narrative review. *Heliyon*, 10(4), e26297. DOI: 10.1016/j.heliyon.2024.e26297 PMID: 38384518

Miao, J. T., & Phelps, N. (2019). Smart City Report 1: The evolution of smart city: case studies of Barcelona, Spain and Helsinki, Finland.

Microsoft News Center. (2018). Microsoft demonstrates the power of AI and Cloud to Oil and Gas players, at ADIPEC 2018. November 12, 2018. Available online: https://news.microsoft.com/en-xm/2018/11/12/microsoft-demonstrates-the-power-of-ai-and-cloud-to-oil-and-gas-players-at-adipec-2018/ [Accessed on 21 May 2024].

Mihiretu, , Okoyo, E. N., & Lemma, T. (2023). Climate variability trendscommunity perspective-livelihood adaptation strategy nexus in the arid-tropics, Ethiopia. *Journal of Arid Environments*, 210, 104929. DOI: 10.1016/j.jaridenv.2022.104929

Milojevic-Dupont, N., & Creutzig, F. (2021). Machine learning for geographically differentiated climate change mitigation in urban areas. *Sustainable Cities and Society*, 64, 102526. DOI: 10.1016/j.scs.2020.102526

Mirshafiee, F., Shahbazi, E., Safi, M., & Rituraj, R. (2023). Predicting power and hydrogen generation of a renewable energy converter utilizing data-driven methods: A sustainable smart grid case study. *Energies*, 16(1), 502. DOI: 10.3390/en16010502

Mitchell, A., Dokei, T., Hickman, T., & Albagli, D. (2020). Regulation of artificial intelligence in Europe and Japan. White & Case LLP, August 24, 2020. Available online: https://www.whitecase.com/publications/insight/regulation-artificial-intelligence-europe-and-japan [Accessed on 22 May 2024].

Mitieka, D., Luke, R., Twinomurinzi, H., & Mageto, J. (2023). Smart Mobility in Urban Areas: A Bibliometric Review and Research Agenda. *Sustainability (Basel)*, 15(8), 6754. DOI: 10.3390/su15086754

Moemen, Y. S., Alshater, H., & El-Sayed, I. E. T. (2023). Clean Energy Management Based on Internet of Things and Sensor Networks for Climate Change Problems. In *The Power of Data: Driving Climate Change with Data Science and Artificial Intelligence Innovations* (pp. 117–136). Springer Nature Switzerland.

Mohabeer, R., & Roberts, J. L. (2020). Promoting the Blue Economy: The Challenge. In *Shaping the Future of Small Islands: Roadmap for Sustainable Development* (pp. 251–268). Springer Nature Singapore.

Mohammed, M. N., Aljibori, H. S. S., Al-Tamimi, A., Alani, Z. N., Hameed, M. S., Ibrahim, F. M., . . . Alfiras, M. (2023, October). Toward Sustainable Smart Cities in Bahrain: A Groundbreaking Approach to Marine Renewable Energy Harnessing Sea Tides and Waves for a Greener Energy Future. In *2023 IEEE 8th International Conference on Engineering Technologies and Applied Sciences (ICETAS)* (pp. 1-5). IEEE. DOI: 10.1109/ICETAS59148.2023.10346346

Mohapatra, H., & Rath, A. K. (2023). Designing of fault-tolerant models for wireless sensor network-assisted smart city applications. In *Intelligent Technologies: Concepts, Applications, and Future Directions* (Vol. 2, pp. 25–43). Springer Nature Singapore. DOI: 10.1007/978-981-99-1482-1_2

Mohring, K. A. (2018). *Smart streetlights: a feasibility study* (Doctoral dissertation, James Cook University).

Mondejar, M. E., Avtar, R., Diaz, H. L. B., Dubey, R. K., Esteban, J., Gómez-Morales, A., Hallam, B., Mbungu, N. T., Okolo, C. C., Prasad, K. A., She, Q., & Garcia-Segura, S. (2021). Digitalization to achieve sustainable development goals: Steps towards a Smart Green Planet. *The Science of the Total Environment*, 794, 148539. DOI: 10.1016/j.scitotenv.2021.148539 PMID: 34323742

Monteleoni, C., Schmidt, G., Saroha, S., & Asplund, E. (2011). Tracking climate models. *Statistical Analysis and Data Mining*, 4(4), 372–392. DOI: 10.1002/sam.10126

Mori, M. (2021). AI-powered virtual assistants in the realms of banking and financial services. In *Virtual Assistant*. IntechOpen. DOI: 10.5772/intechopen.95813

Mukhopadhyay, D., Bhattacharyya, S., Krishnan, B., & Roy, S. (Eds.). (2022). *Blockchain for IoT*. CRC Press. DOI: 10.1201/9781003188247

Mulazzani, , Trevisi, R., Manrique, R., & Malorgio, G. (2016). Blue growth and the relationship between ecosystem services and human activities: The Salento artisanal fisheries case study. *Ocean and Coastal Management*, 134, 120–128. DOI: 10.1016/j.ocecoaman.2016.09.019

Mulvaney, K. (2019). Climate change report card: These countries are reaching targets. *National Geographic,* September 19, 2019. Available online: https://www.nationalgeographic.com/environment/article/climate-change-report-card-co2-emissions [Accessed on 21 May 2024].

Muraleedharan, S. (2021). Role of artificial intelligence in environmental sustainability. *EcoMENA,* January 30, 2021. Available online: https://www.ecomena.org/artificial-intelligence-environmental-sustainability/ [Accessed on 21 May 2024].

Mylrea, M., & Robinson, N. (2023). Artificial Intelligence (AI) Trust Framework and Maturity Model: Applying an Entropy Lens to Improve Security, Privacy, and Ethical AI. *Entropy (Basel, Switzerland)*, 25(10), 1429. DOI: 10.3390/e25101429 PMID: 37895550

Nagy, S., & Hajdú, N. (2021). Consumer acceptance of the use of artificial intelligence in online shopping: Evidence from Hungary. *Amfiteatru Economic*, 23(56), 155–173. DOI: 10.24818/EA/2021/56/155

Nahar, S. (2024). Modeling the effects of artificial intelligence (AI)-based innovation on sustainable development goals (SDGs): Applying a system dynamics perspective in a cross-country setting. *Technological Forecasting and Social Change*, 201, 123203. DOI: 10.1016/j.techfore.2023.123203

Nalini, M., Radhakrishnan, D. P., Yogi, G., Santhiya, S., & Harivardhini, V. (2021). Impact of Machine leaning (ML) on marketing. *International Journal of Aquatic Science*, 12(2), 3159–3167.

Nashold, D. B., Jr. (2020). *Trust in Consumer Adoption of Artificial Intelligence-Driven Virtual Finance Assistants: A Technology Acceptance Model Perspective* (Doctoral dissertation, The University of North Carolina at Charlotte).

Nasserddine, G., Nassereddine, M., & El Arid, A. A. (2023). Internet of things integration in renewable energy Systems. In *Handbook of Research on Applications of AI, Digital Twin, and Internet of Things for Sustainable Development* (pp. 159-185). IGI Global. DOI: 10.4018/978-1-6684-6821-0.ch010

Nassereddine, M., & Khang, A. (2024). Applications of Internet of Things (IoT) in smart cities. In *Advanced IoT Technologies and Applications in the Industry 4.0 Digital Economy* (pp. 109–136). CRC Press. DOI: 10.1201/9781003434269-6

Nazir, S., Khadim, S., Asadullah, M. A., & Syed, N. (2023). Exploring the influence of artificial intelligence technology on consumer repurchase intention: The mediation and moderation approach. *Technology in Society*, 72, 102190. DOI: 10.1016/j.techsoc.2022.102190

Neo, E. X., Hasikin, K., Lai, K. W., Mokhtar, M. I., Azizan, M. M., Hizaddin, H. F., Razak, S. A., & Yanto, . (2023). Artificial intelligence-assisted air quality monitoring for smart city management. *PeerJ. Computer Science*, 9, e1306. DOI: 10.7717/peerj-cs.1306 PMID: 37346549

New AI-Powered Drone Technology Aids Elephant Conservation | Tufts Now. (2023, March 8). https://now.tufts.edu/2023/03/08/new-ai-powered-drone-technology-aids-elephant-conservation

Nezhad, M. M., Neshat, M., Sylaios, G., & Garcia, D. A. (2024). Marine energy digitalization digital twin's approaches. *Renewable & Sustainable Energy Reviews*, 191, 114065. DOI: 10.1016/j.rser.2023.114065

Nickel, M., Murphy, K., Tresp, V., & Gabrilovich, E. (2015). A review of relational machine learning for Machine learning. *Proceedings of the IEEE*, 104(1), 11–33. DOI: 10.1109/JPROC.2015.2483592

Nitschke, M., Tucker, G. R., Hansen, A. L., Williams, S., Zhang, Y., & Bi, P. (2011). Impact of two recent extreme heat episodes on morbidity and mortality in Adelaide, South Australia: A case-series analysis. *Environmental Health*, 10(1), 1–9. DOI: 10.1186/1476-069X-10-42 PMID: 21592410

Nizetic, S., Arıcı, M., & Hoang, A. T. (2023). Smart and Sustainable Technologies in energy transition. *Journal of Cleaner Production*, 389, 135944. DOI: 10.1016/j.jclepro.2023.135944

Nnaji, U. O., Benjamin, L. B., Eyo-Udo, N. L., & Etukudoh, E. A. (2024). A review of strategic decision-making in marketing through big data and analytics. *Magna Scientia Advanced Research and Reviews, 11*(1), 084-091.

Nost, E., & Colven, E. (2022). Earth for AI: A political ecology of data-driven climate initiatives. *Geoforum*, 130, 23–34. DOI: 10.1016/j.geoforum.2022.01.016

Nourildean, S. W., Hassib, M. D., & Mohammed, Y. A. (2022). Internet of things based wireless sensor network: A review. *Indonesian Journal of Electrical Engineering and Computer Science*, 27(1), 246–261. DOI: 10.11591/ijeecs.v27.i1.pp246-261

Nti, E. K., Cobbina, S. J., Attafuah, E. E., Opoku, E., & Gyan, M. A. (2022). Environmental sustainability technologies in biodiversity, energy, transportation and water management using artificial intelligence: A systematic review. *Sustainable Futures : An Applied Journal of Technology, Environment and Society*, 4, 100068. DOI: 10.1016/j.sftr.2022.100068

Nyangon, J. (2024). Climate-proofing critical energy infrastructure: Smart grids, artificial intelligence, and machine learning for power system resilience against extreme weather events. *Journal of Infrastructure Systems*, 30(1), 03124001. DOI: 10.1061/JITSE4.ISENG-2375

Obaideen, K., Albasha, L., Iqbal, U., & Mir, H. (2024). Wireless power transfer: Applications, challenges, barriers, and the role of AI in achieving sustainable development goals-A bibliometric analysis. *Energy Strategy Reviews*, 53, 101376. DOI: 10.1016/j.esr.2024.101376

Obiako, I. V. (2022). *Toward a Bio-Inspired System Architecting Framework: Simulation of the Integration of Autonomous Bus Fleets and Alternative Fueling Infrastructures in Closed Sociotechnical Environments* (Doctoral dissertation, University of South Alabama).

OECD. (2016). *The Ocean Economy in 2030*. OECD Publishing., DOI: 10.1787/9789264251724-

OECD. (2019). Recommendation of the council on artificial intelligence. OECD Legal Instruments. Available online: https://legalinstruments.oecd.org/api/print?ids=648&lang=en [Accessed on 22 May 2024].

Ojo, B. (2024). Strategies for the Optimization of Critical Infrastructure Projects to Enhance Urban Resilience to Climate Change. *The Journal of Scientific and Engineering Research*, 11, 107–123.

Okorie, G. N., Egieya, Z. E., Ikwue, U., Udeh, C. A., & Adaga, E. M. (2024). Leveraging big data for personalized marketing campaigns: A review. *International Journal of Management & Entrepreneurship Research*, 6(1), 216–242. DOI: 10.51594/ijmer.v6i1.778

Oladimeji, D., Gupta, K., Kose, N. A., Gundogan, K., Ge, L., & Liang, F. (2023). Smart transportation: An overview of technologies and applications. *Sensors (Basel)*, 23(8), 3880. DOI: 10.3390/s23083880 PMID: 37112221

Olawade, D. B., Wada, O. Z., David-Olawade, A. C., Fapohunda, O., Ige, A. O., & Ling, J. (2024). Artificial intelligence potential for net zero sustainability: Current evidence and prospects. *Next Sustainability*, 4, 100041. DOI: 10.1016/j.nxsust.2024.100041

Oluwaseyi, J., Odu, A., & Vincent, J. (2024). Explore how this system integrates with broader smart city initiatives, such as traffic management, IoT (Internet of Things) sensors, and data-driven decision-making for urban planning.

Orumwense, E. F., & Abo-Al-Ez, K. (2023). Internet of Things for smart energy systems: A review on its applications, challenges and future trends. *AIMS Electronics & Electrical Engineering, 7*(1).

Osasona, F., Amoo, O., Atadoga, A., Abrahams, T., Farayola, O., & Ayinla, B. (2024). REVIEWING THE ETHICAL IMPLICATIONS OF AI IN DECISION MAKING PROCESSES. *International Journal of Management & Entrepreneurship Research*, 6(2), 322–335. DOI: 10.51594/ijmer.v6i2.773

Oyekunle, D. O. T., & Boohene, D. (2024). Digital Transformation Potential: The Role of Artificial Intelligence in Business. *International Journal of Professional Business Review*.

Pandey, D. K., Hunjra, A. I., Bhaskar, R., & Al-Faryan, M. A. S. (2023). Artificial intelligence, machine learning and big data in natural resources management: A comprehensive bibliometric review of literature spanning 1975–2022. *Resources Policy*, 86, 104250. DOI: 10.1016/j.resourpol.2023.104250

Paris Agreement—An overview | ScienceDirect Topics. (n.d.). Retrieved August 15, 2024, from https://www.sciencedirect.com/topics/earth-and-planetary-sciences/paris-agreement

Parliament of Ukraine. (2018). On approval of the concept of development of the digital economy and society of Ukraine for 2018-2020 and approval of the action plan for its implementation. Available online: https://www.kmu.gov.ua/npas/pro-shvalennya-koncepciyi-rozvitku-cifrovoyi-ekonomiki-ta-suspilstva-ukrayini-na-20182020-roki-ta-zatverdzhennya-planu-zahodiv-shodo-yiyi-realizaciyi [Accessed 22 May 2024].

Parliament of Ukraine. (2020). The concept of artificial intelligence development in Ukraine. Available online: https://zakon.rada.gov.ua/laws/show/1556-2020-%D1%80#Text [Accessed on 22 May 2024].

Parmesan, C., & Yohe, G. (2003). A globally coherent fingerprint of climate change impacts across natural systems. nature, 421(6918), 37-42.

Parmesan, C. (2006). Ecological and evolutionary responses to recent climate change. *Annual Review of Ecology, Evolution, and Systematics*, 37(1), 637–669. DOI: 10.1146/annurev.ecolsys.37.091305.110100

Patil, S. C., & Gidde, M. R. (2024). An IoT-based waste management approach for environmental sustainability. *Journal of Integrated Science and Technology*, 12(3), 769–769. DOI: 10.62110/sciencein.jist.2024.v12.769

Pearce, F. (2018). Energy hogs: Can world's huge data centers be made more efficient? *Yale Environment 360,* April 3, 2018. Available online: https://perma.cc/J2H3-EL75 [Accessed on 21 May 2024].

Pereira, A. M., Moura, J. A. B., Costa, E. D. B., Vieira, T., Landim, A. R., Bazaki, E., & Wanick, V. (2022). Customer models for artificial intelligence-based decision support in fashion online retail supply chains. *Decision Support Systems*, 158, 113795. DOI: 10.1016/j.dss.2022.113795

Ploennigs, J., Semertzidis, K., Lorenzi, F., & Mihindukulasooriya, N. (2022, October). Scaling Machine learning for automating AI of digital twins. In *International Semantic Web Conference* (pp. 810-826). Cham: Springer International Publishing.

Ponnusamy, M., & Alagarsamy, A. (2023). Traffic monitoring in smart cities using internet of things assisted robotics. *Materials Today: Proceedings*, 81, 290–294. DOI: 10.1016/j.matpr.2021.03.192

Popkova, E., Sergi, B., & Bogoviz, A. (2023). Editorial: Evolution of environmental economics and management in the age of artificial intelligence for sustainable development. *Frontiers in Environmental Science*, 11, 1176612. Advance online publication. DOI: 10.3389/fenvs.2023.1176612

Potla, R. T., & Pottla, V. K. (2024). AI-Powered Personalization in Salesforce: Enhancing Customer Engagement through Machine Learning Models. *Valley International Journal Digital Library*, 1388-1420.

Pozova, D. (2017). Prospects of legal regulation of artificial intelligence under EU legislation. *Chasopys tsyvilistyky,* 27, 116-120.

Prakash, J., Murali, L., Manikandan, N., Nagaprasad, N., & Ramaswamy, K. (2024). A vehicular network based intelligent transport system for smart cities using machine learning algorithms. *Scientific Reports*, 14(1), 468. DOI: 10.1038/s41598-023-50906-7 PMID: 38172149

Pramanik, S., Pandey, D., Joardar, S., Niranjanamurthy, M., Pandey, B. K., & Kaur, J. (2023, October). An overview of IoT privacy and security in smart cities. In *AIP Conference Proceedings* (Vol. 2495, No. 1). AIP Publishing. DOI: 10.1063/5.0123511

Qiao, W., Ju, Y., Dong, P., & Tiong, R. L. (2024). How to realize value creation of digital transformation? A system dynamics model. *Expert Systems with Applications*, 244, 122667. DOI: 10.1016/j.eswa.2023.122667

Raab, C. D. (2020). Information privacy, impact assessment, and the place of ethics. *Computer Law & Security Report*, 37, 105404. DOI: 10.1016/j.clsr.2020.105404

Raban, D., Gordon, A., & Geifman, D. (2011). The information society. *Information Communication and Society*, 14(3), 375–399. DOI: 10.1080/1369118X.2010.542824

Rabby, F., Chimhundu, R., & Hassan, R. (2021). Artificial intelligence in digital marketing influences consumer behaviour: A review and theoretical foundation for future research. *Academy of Marketing Studies Journal*, 25(5), 1–7.

Radutnyi, A. (2018). Subjectivity of artificial intelligence in criminal law. *Pravo Ukrayiny*, 1, 123–136.

Raghav, Y. Y., Tipu, R. K., Bhakhar, R., Gupta, T., & Sharma, K. (2024). The Future of Digital Marketing: Leveraging Artificial Intelligence for Competitive Strategies and Tactics. In *The Use of Artificial Intelligence in Digital Marketing: Competitive Strategies and Tactics* (pp. 249-274). IGI Global.

Raihan, A. (2024). A review of recent advances, challenges, and potential future directions of climate-smart agriculture. In *Proceedings of The International Conference on Climate-Smart Agriculture.*

Raj, E. F. I. (2023). A Detailed Review on Wind and Solar Hybrid Green Energy Technologies for Sustainable Smart Cities. *Polytechnica*, 6(1), 1. DOI: 10.1007/s41050-023-00040-0

Rajkumar, Y., & Santhosh Kumar, S. V. N. (2024). A comprehensive survey on communication techniques for the realization of intelligent transportation systems in IoT based smart cities. *Peer-to-Peer Networking and Applications*, 17(3), 1–46. DOI: 10.1007/s12083-024-01627-9

Raman, R., Pattnaik, D., Lathabai, H. H., Kumar, C., Govindan, K., & Nedungadi, P. (2024). Green and sustainable AI research: An integrated thematic and topic modeling analysis. *Journal of Big Data*, 11(1), 55. DOI: 10.1186/s40537-024-00920-x

Ramírez-Moreno, M. A., Keshtkar, S., Padilla-Reyes, D. A., Ramos-López, E., García-Martínez, M., Hernández-Luna, M. C., & Lozoya-Santos, J. D. J. (2021). Sensors for sustainable smart cities: A review. *Applied Sciences (Basel, Switzerland)*, 11(17), 8198. DOI: 10.3390/app11178198

Rane, N. (2023). Roles and challenges of ChatGPT and similar generative artificial intelligence for achieving the sustainable development goals (SDGs). *Available atSSRN* 4603244. DOI: 10.2139/ssrn.4603244

Rani, P., & Sharma, R. (2023). Intelligent transportation system for internet of vehicles based vehicular networks for smart cities. *Computers & Electrical Engineering*, 105, 108543. DOI: 10.1016/j.compeleceng.2022.108543

Rani, V. S., & Sundaram, N. (2022). Collaborative social media marketing in small scale business using artificial intelligence. *ECS Transactions*, 107(1), 5175–5182. DOI: 10.1149/10701.5175ecst

Rao, A. N., & Korres, N. E. (2023). Climate Change and Ecologically Based Weed Management. *Ecologically-Based Weed Management: Concepts, Challenges, and Limitations*, 23-48.

Rao, P. M., & Deebak, B. D. (2023). Security and privacy issues in smart cities/industries: Technologies, applications, and challenges. *Journal of Ambient Intelligence and Humanized Computing*, 14(8), 10517–10553. DOI: 10.1007/s12652-022-03707-1

Rao, P. M., Pedada, S., Jangirala, S., Das, A. K., & Rodrigues, J. J. (2024). Role of IoT in the ages of digital to smart cities: Security challenges and countermeasures. *IEEE Internet of Things Magazine*, 7(1), 56–64. DOI: 10.1109/IOTM.001.2300124

Raska, M., & Bitzinger, R. A. (Eds.). (2023). *The AI Wave in Defence Innovation: Assessing Military Artificial Intelligence Strategies, Capabilities, and Trajectories*. Taylor & Francis. DOI: 10.4324/9781003218326

Rasp, S., Pritchard, M., & Gentine, P. (2018). Deep learning to represent subgrid processes in climate models. *Proceedings of the National Academy of Sciences of the United States of America*, 115(39), 9684–9689. DOI: 10.1073/pnas.1810286115 PMID: 30190437

Rawat, A., Kumar, D., & Khati, B. S. (2024). A review on climate change impacts, models, and its consequences on different sectors: A systematic approach. *Journal of Water and Climate Change*, 15(1), 104–126. DOI: 10.2166/wcc.2023.536

Rêgo, B. S., Lourenço, D., Moreira, F., & Pereira, C. S. (2024). Digital transformation, skills and education: A systematic literature review. *Industry and Higher Education*, 38(4), 336–349. DOI: 10.1177/09504222231208969

Regona, M., Yigitcanlar, T., Hon, C., & Teo, M. (2024). Artificial intelligence and sustainable development goals: Systematic literature review of the construction industry. *Sustainable Cities and Society*, 108, 105499. DOI: 10.1016/j.scs.2024.105499

Reichstein, M., Camps-Valls, G., Stevens, B., Jung, M., Denzler, J., Carvalhais, N., & Prabhat, . (2019). Deep learning and process understanding for data-driven earth system science. *Nature*, 566(7743), 195–204. DOI: 10.1038/s41586-019-0912-1 PMID: 30760912

Research Journal, 5(5), 1178–1209. DOI: 10.51594/csitrj.v5i5.1156

Rezvani, S. M., de Almeida, N. M., & Falcão, M. J. (2023). Climate Adaptation Measures for Enhancing Urban Resilience. *Buildings*, 13(9), 2163. DOI: 10.3390/buildings13092163

Richter, N. F., & Tudoran, A. A. (2024). Elevating theoretical insight and predictive accuracy in business research: Combining PLS-SEM and selected machine learning algorithms. *Journal of Business Research*, 173, 114453. DOI: 10.1016/j.jbusres.2023.114453

Rissland, E., Ashley, K., & Loui, R. (2003). AI and law: A fruitful synergy. *Artificial Intelligence*, 150(1-2), 1–15. DOI: 10.1016/S0004-3702(03)00122-X

Ritchie, H. (2019). Number of people in the world without electricity falls below one billion. *Our World in Data,* January 18, 2019. Available online: https://ourworldindata.org/number-of-people-in-the-world-without-electricity-access-falls-below-one-billion [Accessed on 21 May 2024].

Roach, J. (2020). Microsoft finds underwater datacenters are reliable, practical and use energy sustainably. September 14, 2020. Available online: https://news.microsoft.com/innovation-stories/project-natick-underwater-datacenter/ [Accessed on 21 May 2024].

Robles, P., & Mallinson, D. J. (2023). Artificial intelligence technology, public trust, and effective governance. *Review of Policy Research*, ropr.12555. DOI: 10.1111/ropr.12555

Rodgers, W., & Nguyen, T. (2022). Advertising benefits from ethical artificial intelligence algorithmic purchase decision pathways. *Journal of Business Ethics*, 178(4), 1043–1061. DOI: 10.1007/s10551-022-05048-7

Rolnick, D., Donti, L., Kaack, L., Kochanski, K., Lacoste, A., Sankaran, K., Slavin Ross, A., Milojevic-Dupont, N., Jaques, N., Waldman-Brown, A., Luccioni, A., Maharaj, T., Sherwin, D. E., Mukkavilli, S. K., Kording, K. P., Gomes, C., Ng, A. Y., Hassabis, D., Platt, J. C., . . . Bengio, Y. (2019). Tackling climate change with machine learning. Available online: https://arxiv.org/abs/1906.05433 [Accessed on 21 May 2024].

Rosayyan, P., Paul, J., Subramaniam, S., & Ganesan, S. I. (2023). An optimal control strategy for emergency vehicle priority system in smart cities using edge computing and IOT sensors. *Measurement. Sensors*, 26, 100697. DOI: 10.1016/j.measen.2023.100697

Rustholkarhu, S., Toukola, S., Aarikka-Stenroos, L., & Mahlamäki, T. (2022). Managing B2B customer journeys in digital era: Four management activities with artificial intelligence-empowered tools. *Industrial Marketing Management*, 104, 241–257. DOI: 10.1016/j.indmarman.2022.04.014

Rutenberg, . (2021). *"Use and impact of artificial intelligence on climate change adaptation in Africa," in African Handbook of Climate Change Adaptation* (Oguge, N., Ayal, D., Adeleke, L., & da Silva, I., Eds.). Springer International Publishing.

SaberiKamarposhti, , MWhy, , N. KYadollahi, , MKamyab, , HCheng, , JKhorami, , M. (2024). Cultivating a sustainable future in the artificial intelligence era: A comprehensive assessment of greenhouse gas emissions and removals in agriculture. *Environmental Research*, 118528. PMID: 38403150

Saini, S., Chauhan, A., Thakur, G., & Sapra, L. (2023). Challenges and opportunities in secure smart cities for enhancing the security and privacy. *Enabling Technologies for Effective Planning and Management in Sustainable Smart Cities*, 1-27.

Salama, R., & Al-Turjman, F. (2023). Sustainable energy production in smart cities. *Sustainability (Basel)*, 15(22), 16052. DOI: 10.3390/su152216052

Salman, M. Y., & Hasar, H. (2023). Review on environmental aspects in smart city concept: Water, waste, air pollution and transportation smart applications using Machine Learning techniques. *Sustainable Cities and Society*, 94, 104567. DOI: 10.1016/j.scs.2023.104567

Salunke, M. B., Kulkarni, S. V., Ovale, S., Manjre, B. M., Limkar, S., Shaikh, F., & Sai, V. (2024). A Sustainable IoT-Based Smart Transportation System for Urban Mobility. In *WSN and IoT* (pp. 144–164). CRC Press. DOI: 10.1201/9781003437079-6

Samaei, S. R., & Ghahfarrokhi, M. A. (2023). Optimizing Coastal Hydro Turbines Integrating Artificial Intelligence for Sustainable Energy Conversion. In *2th International Conference on Creative achievements of architecture, urban planning, civil engineering and environment in the sustainable development of the Middle East.*

Sani, M. S., Iranmanesh, S., Salarian, H., Raad, R., & Jamalipour, A. (2024). BIDS: Blockchain-Enabled Intrusion Detection System in Smart Cities. *IEEE Internet of Things Magazine*, 7(2), 107–113. DOI: 10.1109/IOTM.001.2300191

Saranya, M., & Amutha, B. (2024). A Survey of Innovative Machine Learning Approaches in Smart City Applications. In *Innovative Machine Learning Applications for Cryptography* (pp. 231–241). IGI Global. DOI: 10.4018/979-8-3693-1642-9.ch013

Sarker, , Bhuyan, M. A. H., Rahman, M. M., Islam, M. A., Hossain, M. S., Basak, S. C., & Islam, M. M. (2018). From science to action: Exploring the potentials of Blue Economy for enhancing economic sustainability in Bangladesh. *Ocean and Coastal Management*, 157, 180–192. DOI: 10.1016/j.ocecoaman.2018.03.001

Sarker, I. H. (2021). Machine Learning: Algorithms, Real-World Applications and Research Directions. *SN Computer Science*, 2(3), 160. DOI: 10.1007/s42979-021-00592-x PMID: 33778771

Sarker, I. H. (2022). AI-Based Modeling: Techniques, Applications and Research Issues Towards Automation, Intelligent and Smart Systems. *SN Computer Science*, 3(2), 158. DOI: 10.1007/s42979-022-01043-x PMID: 35194580

Sarker, I. H. (2024). *AI-driven cybersecurity and threat intelligence: cyber automation, intelligent decision-making and explainability*. Springer Nature. DOI: 10.1007/978-3-031-54497-2

Sarma, H. H., Paul, A., Kakoti, M., Talukdar, N., & Hazarika, P. (2024). Climate resilient agricultural strategies for enhanced sustainability and food security: A review. *Plant Archives*, 24(1), 787–792. DOI: 10.51470/PLANTARCHIVES.2024.v24.no.1.108

Sartal, A., Carou, D., & Davim, J. P. (Eds.). (2020). *Enabling technologies for the successful deployment of industry 4.0*. CRC Press. DOI: 10.1201/9780429055621

Satpathy, I., Nayak, A., Jain, V., & Padmadas, S. S. (2024). Applying Data Into Action: AI-Powered Solutions for Mitigating Climate Change and Fostering Sustainable Future. In *AI Applications for Clean Energy and Sustainability* (pp. 50-74). IGI Global.

sBalancing the scale: Navigating ethical and practical challenges of artificial intelligence (AI) integration in legal practices | Discover Artificial Intelligence.* (n.d.). Retrieved August 15, 2024, from https://link.springer.com/article/10.1007/s44163-024-00121-8

Schier, M., Reinders, C., & Rosenhahn, B. (2023). Deep reinforcement learning for autonomous driving using high-level heterogeneous ML representations. In *International Conference on Robotics and Automation (ICRA).*

Schlesinger, W. H., Reynolds, J. F., Cunningham, G. L., Huenneke, L. F., Jarrell, W. M., Virginia, R. A., & Whitford, W. G. (1990). Biological feedbacks in global desertification. *Science*, 247(4946), 1043–1048. DOI: 10.1126/science.247.4946.1043 PMID: 17800060

Sehgal, V., Gaur, N., & Mohanty, B. P. Global Flash Drought Monitoring using Surface Soil.

Seidl, R., Schelhaas, M. J., Rammer, W., & Verkerk, P. J. (2014). Increasing forest disturbances in Europe and their impact on carbon storage. *Nature Climate Change*, 4(9), 806–810. DOI: 10.1038/nclimate2318 PMID: 25737744

Selby, J., Cox, E., & Royston, S. (2016). Impact of non-energy policies on energy systems. UK Energy Research Centre, London, November 2016. Available online: https://ukerc.ac.uk/publications/impact-of-non-energy-policies-on-energy-systems/ [Accessed on 21 May 2024].

Shackelford, S. J. (2020). *The Internet of Things: What Everyone Needs to Know*. Oxford University Press. DOI: 10.1093/wentk/9780190943813.001.0001

Shahzad, F., Javed, A. R., Zikria, Y. B., Rehman, S., & Jalil, Z. (2021). Future smart cities: requirements, emerging technologies, applications, challenges, and future aspects. *TechRxiv*.

Shaik, A. S., Alshibani, S. M., Jain, G., Gupta, B., & Mehrotra, A. (2024). Artificial intelligence (AI)-driven strategic business model innovations in small-and medium-sized enterprises. Insights on technological and strategic enablers for carbon neutral businesses. *Business Strategy and the Environment*, 33(4), 2731–2751. DOI: 10.1002/bse.3617

Sharma, A., & Singh, B. (2022). Measuring Impact of E-commerce on Small Scale Business: A Systematic Review. *Journal of Corporate Governance and International Business Law*, 5(1).

Sharma, R., & Arya, R. (2023). Security threats and measures in the Internet of Things for smart city infrastructure: A state of art. *Transactions on Emerging Telecommunications Technologies*, 34(11), e4571. DOI: 10.1002/ett.4571

Sharma, S., Islam, N., Singh, G., & Dhir, A. (2022). Why do retail customers adopt Machine leaning (ML) based autonomous decision-making systems? *IEEE Transactions on Engineering Management.*

Shen, S., Wei, Z. Q., Sun, L. J., Su, Y. Q., Wang, R. C., & Jiang, H. M. (2018). The shared bicycle and its network—internet of shared bicycle (IOSB): A review and survey. *Sensors (Basel)*, 18(8), 2581. DOI: 10.3390/s18082581 PMID: 30087263

Shivanna, K. R. (2022). Climate change and its impact on biodiversity and human welfare. *Proceedings of the Indian National Science Academy. Part A, Physical Sciences*, 88(2), 160–171. DOI: 10.1007/s43538-022-00073-6

Shukla, S. S., Jain, M., Jain, A., Bajpai, S., Mishra, A. K., Gupta, A., & Pramanik, S. (2024). Industry 5.0: Blue Wind Energy for Sustainable Urbanization and Intelligent Control of Energy. In *Convergence of Human Resources Technologies and Industry 5.0* (pp. 170-187). IGI Global.

Siddiqui, S., Hameed, S., Shah, S. A., Khan, A. K., & Aneiba, A. (2023). Smart contract-based security architecture for collaborative services in municipal smart cities. *Journal of Systems Architecture*, 135, 102802. DOI: 10.1016/j.sysarc.2022.102802

Silvestro, D., Goria, S., Sterner, T., & Antonelli, A. (2022). Improving biodiversity protection through artificial intelligence. *Nature Sustainability*, 5(5), 415–424. DOI: 10.1038/s41893-022-00851-6 PMID: 35614933

Singh, B. (2023). Blockchain Technology in Renovating Healthcare: Legal and Future Perspectives. In *Revolutionizing Healthcare Through Artificial Intelligence and Internet of Things Applications* (pp. 177-186). IGI Global.

Singh, B., & Kaunert, C. (2024). Future of Digital Marketing: Hyper-Personalized Customer Dynamic Experience with AI-Based Predictive Models. *Revolutionizing the AI-Digital Landscape: A Guide to Sustainable Emerging Technologies for Marketing Professionals*, 189.

Singh, B. (2019). Profiling Public Healthcare: A Comparative Analysis Based on the Multidimensional Healthcare Management and Legal Approach. *Indian Journal of Health and Medical Law*, 2(2), 1–5.

Singh, B. (2022). COVID-19 Pandemic and Public Healthcare: Endless Downward Spiral or Solution via Rapid Legal and Health Services Implementation with Patient Monitoring Program. *Justice and Law Bulletin*, 1(1), 1–7.

Singh, B. (2022). Relevance of Agriculture-Nutrition Linkage for Human Healthcare: A Conceptual Legal Framework of Implication and Pathways. *Justice and Law Bulletin*, 1(1), 44–49.

Singh, B. (2023). Blockchain Technology in Renovating Healthcare: Legal and Future Perspectives. In Kaushik, K., Dahiya, S., Aggarwal, S., & Dwivedi, A. (Eds.), *Revolutionizing Healthcare Through Artificial Intelligence and Internet of Things Applications* (pp. 177–186). IGI Global., DOI: 10.4018/978-1-6684-5422-0.ch012

Singh, B. (2023). Federated Learning for Envision Future Trajectory Smart Transport System for Climate Preservation and Smart Green Planet: Insights into Global Governance and SDG-9 (Industry, Innovation and Infrastructure). *National Journal of Environmental Law*, 6(2), 6–17.

Singh, B. (2023). Unleashing Alternative Dispute Resolution (ADR) in Resolving Complex Legal-Technical Issues Arising in Cyberspace Lensing E-Commerce and Intellectual Property: Proliferation of E-Commerce Digital Economy. *Revista Brasileira de Alternative Dispute Resolution-Brazilian Journal of Alternative Dispute Resolution-RBADR*, 5(10), 81–105. DOI: 10.52028/rbadr.v5i10.ART04.Ind

Singh, B. (2024). Cherish Data Privacy and Human Rights in the Digital Age: Harmonizing Innovation and Individual Autonomy. In Pucelj, M., & Bohinc, R. (Eds.), *Balancing Human Rights, Social Responsibility, and Digital Ethics* (pp. 199–226). IGI Global., DOI: 10.4018/979-8-3693-3334-1.ch007

Singh, B. (2024). Evolutionary Global Neuroscience for Cognition and Brain Health: Strengthening Innovation in Brain Science. In Prabhakar, P. (Ed.), *Biomedical Research Developments for Improved Healthcare* (pp. 246–272). IGI Global., DOI: 10.4018/979-8-3693-1922-2.ch012

Singh, B. (2024). Legal Dynamics Lensing Metaverse Crafted for Videogame Industry and E-Sports: Phenomenological Exploration Catalyst Complexity and Future. *Journal of Intellectual Property Rights Law*, 7(1), 8–14.

Singh, B. (2024). Lensing Legal Dynamics for Examining Responsibility and Deliberation of Generative AI-Tethered Technological Privacy Concerns: Infringements and Use of Personal Data by Nefarious Actors. In Ara, A., & Ara, A. (Eds.), *Exploring the Ethical Implications of Generative AI* (pp. 146–167). IGI Global., DOI: 10.4018/979-8-3693-1565-1.ch009

Singh, B. (2024). Social Cognition of Incarcerated Women and Children: Addressing Exposure to Infectious Diseases and Legal Outcomes. In Reddy, K. (Ed.), *Principles and Clinical Interventions in Social Cognition* (pp. 236–251). IGI Global., DOI: 10.4018/979-8-3693-1265-0.ch014

Singh, B., Dutta, P. K., Gautam, R., & Kaunert, C. (2024). Uncapping the Potential of Quantum Computing Towards Manufacturing Optimization: Routing Supply Chain Projecting Sustainability. In Hassan, A., Bhattacharya, P., Dutta, P., Verma, J., & Kundu, N. (Eds.), *Quantum Computing and Supply Chain Management: A New Era of Optimization* (pp. 395–419). IGI Global., DOI: 10.4018/979-8-3693-4107-0.ch024

Singh, B., Dutta, P. K., & Kaunert, C. (2024). Replenish Artificial Intelligence in Renewable Energy for Sustainable Development: Lensing SDG 7 Affordable and Clean Energy and SDG 13 Climate Actions With Legal-Financial Advisory. In Derbali, A. (Ed.), *Social and Ethical Implications of AI in Finance for Sustainability* (pp. 198–227). IGI Global., DOI: 10.4018/979-8-3693-2881-1.ch009

Singh, B., Jain, V., Kaunert, C., Dutta, P. K., & Singh, G. (2024). Privacy Matters: Espousing Blockchain and Artificial Intelligence (AI) for Consumer Data Protection on E-Commerce Platforms in Ethical Marketing. In Saluja, S., Nayyar, V., Rojhe, K., & Sharma, S. (Eds.), *Ethical Marketing Through Data Governance Standards and Effective Technology* (pp. 167–184). IGI Global., DOI: 10.4018/979-8-3693-2215-4.ch015

Singh, B., & Kaunert, C. (2024). AI-Driven Strategies for Customer Engagement, Market Segmentation, and Resource Optimization: Projecting End User Satisfaction and Futuristic Growth of Business. In Nadda, V., Tyagi, P., Singh, A., & Singh, V. (Eds.), *AI Innovations in Service and Tourism Marketing* (pp. 104–128). IGI Global., DOI: 10.4018/979-8-3693-7909-7.ch006

Singh, B., & Kaunert, C. (2024). Aroma of Highly Smart Internet of Medical Things (IoMT) and Lightweight EdgeTrust Expansion Medical Care Facilities for Electronic Healthcare Systems: Fortified-Chain Architecture for Remote Patient Monitoring and Privacy Protection Beyond Imagination. In Hassan, A., Bhattacharya, P., Tikadar, S., Dutta, P., & Sagayam, M. (Eds.), *Lightweight Digital Trust Architectures in the Internet of Medical Things (IoMT)* (pp. 196–212). IGI Global., DOI: 10.4018/979-8-3693-2109-6.ch011

Singh, B., & Kaunert, C. (2024). Augmented Reality and Virtual Reality Modules for Mindfulness: Boosting Emotional Intelligence and Mental Wellness. In Hiran, K., Doshi, R., & Patel, M. (Eds.), *Applications of Virtual and Augmented Reality for Health and Wellbeing* (pp. 111–128). IGI Global., DOI: 10.4018/979-8-3693-1123-3.ch007

Singh, B., & Kaunert, C. (2024). Computational Thinking for Innovative Solutions and Problem-Solving Techniques: Transforming Conventional Education to Futuristic Interdisciplinary Higher Education. In Fonkam, M., & Vajjhala, N. (Eds.), *Revolutionizing Curricula Through Computational Thinking, Logic, and Problem Solving* (pp. 60–82). IGI Global., DOI: 10.4018/979-8-3693-1974-1.ch004

Singh, B., & Kaunert, C. (2024). Embryonic Machine-Deep Learning, Smart Healthcare and Privacy Deliberations in Hospital Industry: Lensing Confidentiality of Patient's Information and Personal Data in Legal-Ethical Landscapes Projecting Futuristic Dimensions. In Kumar, P., Singh, P., Diwakar, M., & Garg, D. (Eds.), *Healthcare Industry Assessment: Analyzing Risks, Security, and Reliability. Engineering Cyber-Physical Systems and Critical Infrastructures* (Vol. 11). Springer., DOI: 10.1007/978-3-031-65434-3_7

Singh, B., & Kaunert, C. (2024). Harnessing Sustainable Agriculture Through Climate-Smart Technologies: Artificial Intelligence for Climate Preservation and Futuristic Trends. In Kannan, H., Rodriguez, R., Paprika, Z., & Ade-Ibijola, A. (Eds.), *Exploring Ethical Dimensions of Environmental Sustainability and Use of AI* (pp. 214–239). IGI Global., DOI: 10.4018/979-8-3693-0892-9.ch011

Singh, B., & Kaunert, C. (2024). Hidden Gems Breakthrough Dynamic Landscape of Adaptive Learning Technologies for Higher Education: Bridging the Gap Between Theoretical and Practical Knowledge Projecting Student Learning Outcomes. In Minh Tung, T. (Ed.), *Adaptive Learning Technologies for Higher Education* (pp. 222–247). IGI Global., DOI: 10.4018/979-8-3693-3641-0.ch010

Singh, B., & Kaunert, C. (2024). Human-Machine Nexus for Digital Rebound Fostering Futuristic Energy-Efficiency: Ecological Footprint of Technology for Smart-Sustainable Urban Mobility. In González-Lezcano, R., & Sansaniwal, S. (Eds.), *Design Strategies for Efficient and Sustainable Building Facilities* (pp. 222–244). IGI Global., DOI: 10.4018/979-8-3693-3200-9.ch010

Singh, B., & Kaunert, C. (2024). Integration of Cutting-Edge Technologies such as Internet of Things (IoT) and 5G in Health Monitoring Systems: A Comprehensive Legal Analysis and Futuristic Outcomes. *GLS Law Journal*, 6(1), 13–20. DOI: 10.69974/glslawjournal.v6i1.123

Singh, B., & Kaunert, C. (2024). Revealing Green Finance Mobilization: Harnessing FinTech and Blockchain Innovations to Surmount Barriers and Foster New Investment Avenues. In Jafar, S., Rodriguez, R., Kannan, H., Akhtar, S., & Plugmann, P. (Eds.), *Harnessing Blockchain-Digital Twin Fusion for Sustainable Investments* (pp. 265–286). IGI Global., DOI: 10.4018/979-8-3693-1878-2.ch011

Singh, B., & Kaunert, C. (2024). Salvaging Responsible Consumption and Production of Food in the Hospitality Industry: Harnessing Machine Learning and Deep Learning for Zero Food Waste. In Singh, A., Tyagi, P., & Garg, A. (Eds.), *Sustainable Disposal Methods of Food Wastes in Hospitality Operations* (pp. 176–192). IGI Global., DOI: 10.4018/979-8-3693-2181-2.ch012

Singh, B., & Kaunert, C. (2024). Vertical Assimilation of Artificial Intelligence and Machine Learning in Safeguarding Financial Data. In Naz, F., & Karim, S. (Eds.), *Safeguarding Financial Data in the Digital Age* (pp. 169–197). IGI Global., DOI: 10.4018/979-8-3693-3633-5.ch010

Singh, B., & Kaunert, C. (2024). Wind and Solar Energy as Renewable Energy for Sustainable Global Future: Projecting Future Multi-Sector Sustainable Policies and Innovation. In Ara, A., & Thakore, R. (Eds.), *Promoting Multi-Sector Sustainability With Policy and Innovation* (pp. 210–245). IGI Global., DOI: 10.4018/979-8-3693-2113-3.ch009

Singh, B., & Kaunert, C. Reinventing Artificial Intelligence and Blockchain for Preserving Medical Data. In *Ethical Artificial Intelligence in Power Electronics* (pp. 77–91). CRC Press. DOI: 10.1201/9781032648323-6

Singh, B., Kaunert, C., & Malviya, R. (2024). Coupling Human-Computer Interface Lensing Artificial Emotional Intelligence: Transforming for Futuristic Cognitive Recognition for Decision-Marking in Industry 5.0. In Kumar, N., Pal, S., Agarwal, P., Rosak-Szyrocka, J., & Jain, V. (Eds.), *Human-Machine Collaboration and Emotional Intelligence in Industry 5.0* (pp. 233–250). IGI Global., DOI: 10.4018/979-8-3693-6806-0.ch013

Sinha, S., & Lee, Y. M. (2024). Challenges with developing and deploying AI models and applications in industrial systems. *Discover Artificial Intelligence*, 4(1), 55. DOI: 10.1007/s44163-024-00151-2

Smart Cities | Free Full-Text | The Making of Responsible Innovation and Technology: An Overview and Framework. (n.d.). Retrieved August 15, 2024, from https://www.mdpi.com/2624-6511/6/4/93

Smith, A., Harvey, J., Goulding, J., Smith, G., & Sparks, L. (2021). Exogenous cognition and cognitive state theory: The plexus of consumer analytics and decision-making. *Marketing Theory*, 21(1), 53–74. DOI: 10.1177/1470593120964947

Snow, J. (2019). How artificial intelligence can tackle climate change. *National Geographic,* July 18, 2019. Available online: https://www.nationalgeographic.com/environment/2019/07/artificial-intelligence-climate-change/ [Accessed on 21 May 2024].

Soma, , van den Burg, S. W. K., Hoefnagel, E. W. J., Stuiver, M., & van der Heide, C. M. (2018). Social innovation – a future pathway for blue growth? *Marine Policy*, 87, 363–370. DOI: 10.1016/j.marpol.2017.10.008

Son, X.. (2021). Vietnam's Regulations to Prevent Pollution from Plastic Waste: A Review Based on the Circular Economy Approach. *Journal of Environmental Law, Oxford University Press*, 33(1), 137–166.

Srinivasan, R., & San Miguel González, B. (2022). The role of empathy for artificial intelligence accountability. *Journal of Responsible Technology*, 9, 100021. DOI: 10.1016/j.jrt.2021.100021

St. John, J. (2018). Texas takes a big step in improving access to smart meter data. *Greentechmedia,* February 6, 2018. Available online: https://perma.cc/G4ZJ-L4LT [Accessed on 21 May 2024].

Stahl, B. C., Rodrigues, R., Santiago, N., & Macnish, K. (2022). A European Agency for Artificial Intelligence: Protecting fundamental rights and ethical values. *Computer Law & Security Report*, 45, 105661. DOI: 10.1016/j.clsr.2022.105661

Stein, A. L. (2020). Artificial intelligence and climate. *Yale Journal on Regulation*, 37(3), 890–939.

Stone, B., Hess, J. J., & Frumkin, H. (2010). Urban form and extreme heat events: Are sprawling cities more vulnerable to climate change than compact cities? *Environmental Health Perspectives*, 118(10), 1425–1428. DOI: 10.1289/ehp.0901879 PMID: 21114000

Stone, M., Aravopoulou, E., Ekinci, Y., Evans, G., Hobbs, M., Labib, A., & Machtynger, L. (2020). Machine leaning (ML) in strategic marketing decision-making: A research agenda. *The Bottom Line (New York, N.Y.)*, 33(2), 183–200. DOI: 10.1108/BL-03-2020-0022

Strengthening international cooperation on artificial intelligence. (n.d.). Brookings. Retrieved August 15, 2024, from https://www.brookings.edu/articles/strengthening-international-cooperation-on-artificial-intelligence/

Strubell, E., Ganesh, A., & McCallum, A. (2019). Energy and policy considerations for deep learning in NLP. Available online: https://arxiv.org/abs/1906.02243 [Accessed on 21 May 2024]. DOI: 10.18653/v1/P19-1355

Subramaniam, S., Raju, N., Ganesan, A., Rajavel, N., Chenniappan, M., Prakash, C., Pramanik, A., Basak, A. K., & Dixit, S. (2022). Artificial Intelligence Technologies for Forecasting Air Pollution and Human Health: A Narrative Review. *Sustainability (Basel)*, 14(16), 9951. DOI: 10.3390/su14169951

Su, C.-H., & Yang, Y.-T. C. (2022). Robotics and AI applications in education: Implications for teaching and learning practices in K-12 settings worldwide. *Computers & Education*, 182, 104136.

Sumit, C., Chhillar, R. S., Dalal, S., Dalal, S., Lilhore, U. K., & Samiya, S. (2024). A dynamic and optimized routing approach for VANET communication in smart cities to secure intelligent transportation system via a chaotic multi-verse optimization algorithm. *Cluster Computing*, 27(5), 1–26. DOI: 10.1007/s10586-024-04322-9

Sun, Q., Miao, C., Hanel, M., Borthwick, A. G., Duan, Q., Ji, D., & Li, H. (2019). Global heat stress on health, wildfires, and agricultural crops under different levels of climate warming. *Environment International*, 128, 125–136. DOI: 10.1016/j.envint.2019.04.025 PMID: 31048130

Sustainable AI: AI for sustainability and the sustainability of AI | AI and Ethics. (n.d.). Retrieved August 15, 2024, from https://link.springer.com/article/10.1007/s43681-021-00043-6

Taghikhah, F., Erfani, E., Bakhshayeshi, I., Tayari, S., Karatopouzis, A., & Hanna, B. (2022). Artificial intelligence and sustainability: solutions to social and environmental challenges. In *Artificial intelligence and data science in environmental sensing* (pp. 93–108). Academic Press.

Tang, X., Sun, T., Zhu, R., & Wang, S. (2021, January). CKG: dynamic representation based on context and AI. In *2020 25th International Conference on Pattern Recognition (ICPR)* (pp. 2889-2895). IEEE.

Taylor, J., & Feng, M. (2022). A deep learning model for forecasting global monthly mean sea surface temperature anomalies. *Frontiers in Climate*, 4, 932932. DOI: 10.3389/fclim.2022.932932

Thaler, R., & Sunstein, C. (2008). *Nudge: Improving decisions about health, wealth and happiness*. Yale University Press.

THE 17 GOALS | Sustainable Development. (n.d.). Retrieved August 15, 2024, from robin

Thomas, C. D., Cameron, A., Green, R. E., Bakkenes, M., Beaumont, L. J., Collingham, Y. C., Erasmus, B. F. N., de Siqueira, M. F., Grainger, A., Hannah, L., Hughes, L., Huntley, B., van Jaarsveld, A. S., Midgley, G. F., Miles, L., Ortega-Huerta, M. A., Townsend Peterson, A., Phillips, O. L., & Williams, S. E. (2004). Extinction risk from climate change. *Nature*, 427(6970), 145–148. DOI: 10.1038/nature02121 PMID: 14712274

Trevisan, L. V., Eustachio, J. H. P. P., Dias, B. G., Filho, W. L., & Pedrozo, E. Á. (2024). Digital transformation towards sustainability in higher education: State-of-the-art and future research insights. *Environment, Development and Sustainability*, 26(2), 2789–2810. DOI: 10.1007/s10668-022-02874-7 PMID: 36624732

Tsaples, G., Papathanasiou, J., & Manou, D. (2024). Synergies and Challenges: Exploring Organizational Perspectives on Digital Transformation and Sustainable Development in the Context of Skills and Education. *Buildings*, 14(2), 395. DOI: 10.3390/buildings14020395

Ullah, A., Qi, G., Hussain, S., Ullah, I., & Ali, Z. (2024). The role of llms in sustainable smart cities: Applications, challenges, and future directions. *arXiv preprint arXiv:2402.14596.*

Ullah, A., Qi, G., Hussain, S., Ullah, I., & Ali, Z. (2024). The Role of LLMs in Sustainable Smart Cities: Applications, Challenges, and Future Directions. *arXiv preprint arXiv:2402.14596.*

Ullah, F., Sepasgozar, S. M., & Wang, C. (2018). A systematic review of smart real estate technology: Drivers of, and barriers to, the use of digital disruptive technologies and online platforms. *Sustainability (Basel)*, 10(9), 3142. DOI: 10.3390/su10093142

Umamaheswari, D. D. (2024). Role of Artificial Intelligence in Marketing Strategies and Performance. *Migration Letters : An International Journal of Migration Studies*, 21(S4), 1589–1599.

UNCTAD. (2014). United Nations Conference on Trade and Development, The Ocean Economy: Opportunities and Challenges for Small Isalnd Developing States, available at https://unctad.org/en/publicationslibrary/ditcted2014d5_en.pdf

UNESCO. (2019a). An integrated system for global real-time precipitation observation using PDIR. Available online: https://en.unesco.org/news/irain-newmobile-app-promote-citizen-science-and-support-water-management [Accessed on 21 May 2024].

UNESCO. (2019b). Artificial intelligence for sustainable development: Challenges and opportunities for UNESCO's science and engineering programmes. UNESCO Digital Library. Available online: https://unesdoc.unesco.org/ark:/48223/pf0000368028.locale=en [Accessed on 21 May 2024].

United Nations Conference on Trade and Development (UNCTAD). (2020). The COVID-19 Pandemic and the Blue Economy: New challenges and prospects for recovery and resilience (Geneva: UNCTAD).

United Nations. (1992). Framework convention on climate change. Available online: https://www.un.org/ru/documents/decl_conv/conventions/climate_framework_conv.shtml [Accessed on 21 May 2024].

United Nations. (1997). Kyoto protocol to the United Nations framework convention on climate change. Available online: https://www.un.org/ru/documents/decl_conv/conventions/kyoto.shtml [Accessed on 21 May 2024].

United Nations. (2014). Blue Economy Concept Paper Available at https://sustainabledevelopment.un.org/concent/documents/2978BEconcept.pdf

United Nations. (2015). The Paris agreement. Available online: https://www.un.org/ru/climatechange/paris-agreement [Accessed on 21 May 2024].

United States Environmental Protection Agency. (2020). Sources of greenhouse gas emissions. Available online: https://www.epa.gov/ghgemissions/sources-greenhouse-gas-emissions [Accessed 21 May 2024].

Vadivel, G., Hussain, M. J. M., & Sangeetha, S. T. (2023). SMART TRANSPORTATION SYSTEMS: IOT-CONNECTED WIRELESS SENSOR NETWORKS FOR TRAFFIC CONGESTION MANAGEMENT. *INTERNATIONAL JOURNAL OF ADVANCES IN SIGNAL AND IMAGE SCIENCES*, 9(1), 40–49. DOI: 10.29284/IJASIS.9.1.2023.40-49

van Esch, P., & Cui, Y. (2021). Does consumer promiscuity influence purchase intent? The role of Machine leaning (ML), change seeking, and pride. *Journal of the Association for Consumer Research*, 6(3), 394–401. DOI: 10.1086/714503

Van Hoang, T. (2024). Impact of Integrated Artificial Intelligence and Internet of Things Technologies on Smart City Transformation. *Journal ofTechnical Education Science*, 19(1), 64–73. DOI: 10.54644/jte.2024.1532

Van Wynsberghe, A. (2021). Sustainable AI: AI for sustainability and the sustainability of AI. *AI and Ethics*, 1(3), 213–218. DOI: 10.1007/s43681-021-00043-6

Varadharajan, V. (2021). Next Generation of Intelligent Cities: Case Studies From Europe. In *Developing and Monitoring Smart Environments for Intelligent Cities* (pp. 87-111). IGI Global.

Venčkauskas, A., Taparauskas, M., Grigaliūnas, Š., & Brūzgienė, R. (2024). Enhancing Communication Security an In-Vehicle Wireless Sensor Network. *Electronics (Basel)*, 13(6), 1003. DOI: 10.3390/electronics13061003

Verma, J., Sandys, L., Matthews, A., & Goel, S. (2024). Readiness of artificial intelligence technology for managing energy demands from renewable sources. *Engineering Applications of Artificial Intelligence*, 135, 108831. DOI: 10.1016/j.engappai.2024.108831

Vijayalakshmi, S. Savita, & Durgadevi, P. (2023). AI and Machine Learning in Improving Resilience of Smart Energy Infrastructure. In *AI-Powered Machine Learning in the Energy Industry: Digital Technology and Sustainable Energy Systems* (pp. 189-213). Cham: Springer International Publishing.

Vijayalakshmi, S. Savita, Genish, T., & George, J. P. (2023). The role of artificial intelligence in renewable energy. In *AI-Powered Machine Learning in the Energy Industry: Digital Technology and Sustainable Energy Systems* (pp. 253-269). Cham: Springer International Publishing.

Vinodhini, M., Rajkumar, S., & Subramaniam, S. K. (2023). Real-time Internet of LoRa Things (IoLT)-based accident detection and prevention system in vehicular networks towards smart city. *International Journal of Communication Systems*.

Vinuesa, R., Azizpour, H., Leite, I., Balaam, M., Dignum, V., Domisch, S., & Nerini, F. (2020). The role of artificial intelligence in achieving the sustainable development goals. *Nature Communications*, 11(1), 233. Advance online publication. DOI: 10.1038/s41467-019-14108-y PMID: 31932590

vinx*IUCN*. (n.d.). Retrieved August 15, 2024, from https://iucn.org/

Vohra, S. K., Kumar, V. S., Krishnamoorthy, R., Mahesh, P., Haralayya, B., Soni, N., & Gupta, S. K. (2024). Original Research Article The 5G revolution: Tackling challenges in smart cities and intelligent transportation systems. *Journal of Autonomous Intelligence*, 7(2). Advance online publication. DOI: 10.32629/jai.v7i2.1342

Vos, J. (2009). Actions speak louder than words: Greenwashing in corporate America. *Notre Dame Journal of Law, Ethics & Public Policy*, 23(2), 673–697.

Wang, A., Wang, P., Miao, X., Li, X., Ye, N., & Liu, Y. (2020). A review on non-terrestrial wireless technologies for Smart City Internet of Things. *International Journal of Distributed Sensor Networks*, 16(6), 1550147720936824. DOI: 10.1177/1550147720936824

Wang, H., Wen, C., Duan, L., Li, X., Liu, D., & Guo, W. (2024). Sustainable energy transition in cities: A deep statistical prediction model for renewable energy sources management for low-carbon urban development. *Sustainable Cities and Society*, 107, 105434. DOI: 10.1016/j.scs.2024.105434

Wang, X., Ji, K., & Xie, T. (2023). AI carbon footprint management with multi-agent participation: A tripartite evolutionary game analysis based on a case in China. *Sustainability (Basel)*, 15(11), 9013. DOI: 10.3390/su15119013

Wang, Z., & Zhu, X. (2024). Enhancing nanocrystal synthesis: A comparative study of online artificial intelligence optimization and offline high-throughput experimentation in chemical material discovery. *ACS Applied Nano Materials*, 7(6), 6499–6505. DOI: 10.1021/acsanm.4c00255

Watts, N., Adger, W. N., Agnolucci, P., Blackstock, J., Byass, P., Cai, W., Chaytor, S., Colbourn, T., Collins, M., Cooper, A., Cox, P. M., Depledge, J., Drummond, P., Ekins, P., Galaz, V., Grace, D., Graham, H., Grubb, M., Haines, A., & Costello, A. (2015). Health and climate change: Policy responses to protect public health. *Lancet*, 386(10006), 1861–1914. DOI: 10.1016/S0140-6736(15)60854-6 PMID: 26111439

Westerling, A. L., Hidalgo, H. G., Cayan, D. R., & Swetnam, T. W. (2006). Warming and earlier spring increase western US forest wildfire activity. science, 313(5789), 940-943.

Whitman, S., Good, G., Donoghue, E. R., Benbow, N., Shou, W., & Mou, S. (1997). Mortality in Chicago attributed to the July 1995 heat wave. *American Journal of Public Health*, 87(9), 1515–1518. DOI: 10.2105/AJPH.87.9.1515 PMID: 9314806

William J. Snape III. (2010). *Joining The Convention On Biological Diversity: A Legal and Scientific Overview of Why the United States Must Wake Up.*

Williams-Wynn, M. D., & Naidoo, P. (2020). A review of the treatment options for marine plastic waste in South Africa. *Marine Pollution Bulletin*, 161, 111785. Advance online publication. DOI: 10.1016/j.marpolbul.2020.111785 PMID: 33128983

Wirtz, J., & Lovelock, C. (2021). *Services marketing: People, technology, strategy*. World Scientific. DOI: 10.1142/y0024

WMO (World Meteorological Organization). (2019). The global climate in 2015–2019. Available online: https://library.wmo.int/index.php?lvl=notice_display&id=21522#.YNb_cmgzbIU [Accessed on 22 May 2024].

Wolff-Anthony, L., Kanding, B., & Selvan, R. (2020). Carbon Tracker: Tracking and predicting the carbon footprint of training deep learning models. Available online: https://arxiv.org/abs/2007.03051 [Accessed on 21 May 2024].

Wong, L. W., Tan, G. W. H., Ooi, K. B., Lin, B., & Dwivedi, Y. K. (2022). Artificial intelligence-driven risk management for enhancing supply chain agility: A deep-learning-based dual-stage PLS-SEM-ANN analysis. *International Journal of Production Research*, •••, 1–21.

World Bank. (2017). *The potential of the Blue Economy: Increasing long-term benefits of the sustainable use of marine resources for small island developing states and coastal least developed countries*. World Bank.

World Economic Forum. 2016. The New Plastics Economy: rethinking the future of plastics. https://www.ellenmacarthurfoundation.org/assets/downloads/EllenMacArthurFoundation_TheNewPlasticsEconomy_Pages.pdf

Wu, C., Zhang, R., Kotagiri, R., & Bouvry, P. (2023). Strategic decisions: Survey, taxonomy, and future directions from artificial intelligence perspective. *ACM Computing Surveys*, 55(12), 1–30. DOI: 10.1145/3571807

Xiao, D., Dianati, M., Geiger, W. G., & Woodman, R. (2023). Review of ML-based hazardous event detection methods for autonomous driving systems. *IEEE Transactions on Intelligent Transportation Systems*, 24(5), 4697–4715. DOI: 10.1109/TITS.2023.3240104

Xie, H., Huang, R., Sun, H., Han, Z., Jiang, M., Zhang, D., Goh, H. H., Kurniawan, T. A., Han, F., Liu, H., & Wu, T. (2023). Wireless energy: Paving the way for smart cities and a greener future. *Energy and Building*, 297, 113469. DOI: 10.1016/j.enbuild.2023.113469

Xue, L., & Pang, Z. (2022). Ethical governance of artificial intelligence: An integrated analytical framework. *Journal of Digital Economy*, 1(1), 44–52. DOI: 10.1016/j.jdec.2022.08.003

Xu, J., Li, M., He, Z., & Anwlnkom, T. (2023). Security and privacy protection communication protocol for Internet of vehicles in smart cities. *Computers & Electrical Engineering*, 109, 108778. DOI: 10.1016/j.compeleceng.2023.108778

Xu, W., Feng, L., & Ma, J. (2022). Understanding the domain of driving distraction with Machine learning. *PLoS One*, 17(12), e0278822. DOI: 10.1371/journal.pone.0278822 PMID: 36490240

Xu, Y., Liu, X., Cao, X., Huang, C., Liu, E., Qian, S., Liu, X., Wu, Y., Dong, F., Qiu, C.-W., Qiu, J., Hua, K., Su, W., Wu, J., Xu, H., Han, Y., Fu, C., Yin, Z., Liu, M., & Zhang, J. (2021). Artificial intelligence: A powerful paradigm for scientific research. *Innovation (Cambridge (Mass.))*, 2(4), 100179. DOI: 10.1016/j.xinn.2021.100179 PMID: 34877560

Yang, G., Ji, G., & Tan, K. H. (2022). Impact of artificial intelligence adoption on online returns policies. *Annals of Operations Research*, 308(1-2), 1–24. DOI: 10.1007/s10479-020-03602-y

Yashi, Y., Nijhawan, R., & Bhargava, D. (2024). Intelligent Communication Systems for Urban Things. In *Computational Intelligence in Urban Infrastructure* (pp. 155–168). CRC Press.

Yau, K. L. A., Saad, N. M., & Chong, Y. W. (2021). Artificial intelligence marketing (AIM) for enhancing customer relationships. *Applied Sciences (Basel, Switzerland)*, 11(18), 8562. DOI: 10.3390/app11188562

Yigitcanlar, T., Mehmood, R., & Corchado, J. M. (2021). *Green Artificial Intelligence: Towards an Efficient*. Sustainable and Equitable Technology for Smart Cities and Futures.

Yousef, L. A., Yousef, H., & Rocha-Meneses, L. (2023). Artificial intelligence for management of variable renewable energy systems: A review of current status and future directions. *Energies*, 16(24), 8057. DOI: 10.3390/en16248057

Yu, C., Wang, X., Xu, X., Zhang, M., Ge, H., Ren, J., Sun, L., Chen, B., & Tan, G. (2019). Distributed multiagent coordinated learning for autonomous driving in highways based on dynamic coordination MLs. *IEEE Transactions on Intelligent Transportation Systems*, 21(2), 735–748. DOI: 10.1109/TITS.2019.2893683

Zakeri, , Naji, A., Akbarzadeh, A., & Uddin, S. (2020). Microplastic ingestion in important commercial fish in the southern Caspian Sea. *Marine Pollution Bulletin*, 160, 111598. Advance online publication. DOI: 10.1016/j.marpolbul.2020.111598 PMID: 32871433

Zakutynskyi, I., Sibruk, L., & Kokarieva, A. (2023). IoT System for Monitoring and Managing Public Transport Data. *WSEAS Transactions on Systems*, 22, 242–248. DOI: 10.37394/23202.2023.22.25

Zearah, S. A., Al-Tahee, M., Mohammed, A., Shnawa, A. H., Mahdi, H. H. J., & Majeed, Z. (2023, December). Smart City Application based on Internet-of-Things and Artificial Intelligence for Energy Management Dynamic Memory Modelling. In *2023 International Conference on Emerging Research in Computational Science (ICERCS)* (pp. 1-6). IEEE. DOI: 10.1109/ICERCS57948.2023.10434196

Zhang, P., Xiao, Y., Li, Y., Ge, X., Shi, G., & Yang, Y. (2024). Toward net-zero carbon emissions in network AI for 6G and beyond. *IEEE Communications Magazine*, 62(4), 58–64. DOI: 10.1109/MCOM.003.2300175

Zhang, , Ruiz-Menjivar, J., Luo, B., Liang, Z., & Swisher, M. E. (2020). Predicting climate change mitigation and adaptation behaviors in agricultural production: A comparison of the theory of planned behavior and the Value-Belief-Norm Theory. *Journal of Environmental Psychology*, 68, 101408. DOI: 10.1016/j.jenvp.2020.101408

Zhao, J., Liu, D., & Huang, R. (2023). A review of climate-smart agriculture: Recent advancements, challenges, and future directions. *Sustainability (Basel)*, 15(4), 3404. DOI: 10.3390/su15043404

Zhao, W., Liang, Z., & Li, B. (2022). Realizing a Rural Sustainable Development through a Digital Village Construction: Experiences from China. *Sustainability (Basel)*, 14(21), 14199. DOI: 10.3390/su142114199

Zheng, Z., Zhou, B., Zhou, D., Soylu, A., & Kharlamov, E. (2022, October). Executable AI for transparent machine learning in welding monitoring at bosch. In *Proceedings of the 31st ACM International Conference on Information & Knowledge Management* (pp. 5102-5103).

Zhou, Y., & Liu, J. (2024). Advances in emerging digital technologies for energy efficiency and energy integration in smart cities. *Energy and Building*, 315, 114289. DOI: 10.1016/j.enbuild.2024.114289

Zuiderwijk, A., Chen, Y.-C., & Salem, F. (2021). Implications of the use of artificial intelligence in public governance: A systematic literature review and a research agenda. *Government Information Quarterly*, 38(3), 101577. DOI: 10.1016/j.giq.2021.101577

Zuo, J., Pullen, S., Palmer, J., Bennetts, H., Chileshe, N., & Ma, T. (2015). Impacts of heat waves and corresponding measures: A review. *Journal of Cleaner Production*, 92, 1–12. DOI: 10.1016/j.jclepro.2014.12.078

About the Contributors

Bhupinder Singh working as Professor at Sharda University, India. Also, Honorary Professor in University of South Wales UK and Santo Tomas University Tunja, Colombia. His areas of publications as Smart Healthcare, Medicines, fuzzy logics, artificial intelligence, robotics, machine learning, deep learning, federated learning, IoT, PV Glasses, metaverse and many more. He has 3 books, 139 paper publications, 163 paper presentations in international/national conferences and seminars, participated in more than 40 workshops/FDP's/QIP's, 25 courses from international universities of repute, organized more than 59 events with international and national academicians and industry people's, editor-in-chief and co-editor in journals, developed new courses. He has given talks at international universities, resource person in international conferences such as in Nanyang Technological University Singapore, Tashkent State University of Law Uzbekistan; KIMEP University Kazakhstan, All'ah meh Tabatabi University Iran, the Iranian Association of International Criminal law, Iran and Hague Center for International Law and Investment, The Netherlands, Northumbria University Newcastle UK, Taylor's University Malaysia, AFM Krakow University Poland, European Institute for Research and Development Georgia, Business and Technology University Georgia, Texas A & M University US name a few. His leadership, teaching, research and industry experience is of 16 years and 3 Months. His research interests are health law, criminal law, research methodology and emerging multidisciplinary areas as Blockchain Technology, IoT, Machine Learning, Artificial Intelligence, Genome-editing, Photovoltaic PV Glass, SDG's and many more.

Christian Kaunert is Professor of International Security at Dublin City University, Ireland. He is also Professor of Policing and Security, as well as Director of the International Centre for Policing and Security at the University of South Wales. In addition, he is Jean Monnet Chair, Director of the Jean Monnet Centre of Excellence and Director of the Jean Monnet Network on EU Counter-Terrorism.

Previously, he served as Academic Director and Professor at the Institute for European Studies, Vrije Universiteit Brussel, a Professor of International Politics, Head of Discipline in Politics, and the Director of the European Institute for Security and Justice, a Jean Monnet Centre for Excellence, at the University of Dundee. Prof Kaunert has researched and taught in many international universities, such as University of Shandong, Jinan, China, IBEI Barcelona, Spain, Diplomatic Academy Yerevan, Armenia – United Nations Development Mission, University of Cairo, Egypt, Collegio Carlo Alberto, Turin, Italy, Ecole nationale d'administration Paris, European Institute of Public Administration Maastricht, Netherlands, European Studies Institute (ESI) at Moscow, University of Porto, Portugal, etc. He was previously Senior Lecturer at the University of Dundee, Marie Curie Senior Research Fellow at the European University Institute Florence, and Senior Lecturer in EU Politics & International Relations, University of Salford. Christian has been invited expert for the European Institute for Public Administration (EIPA), the Counter-Terrorism Committee of the European Parliament, the European Union Institute for Security Studies (an agency of the EU), the United Nations Development Programme (UNDP) and the North Atlantic Treaty Organisation (NATO). Over his career, he has been awarded seven prestigious Marie Curie Fellowships, four Jean Monnet Chairs, and, finally, three Jean Monnet Centre of Excellence, as well as one Jean Monnet Network of Excellence 'EUCTER'. Additionally, he was awarded a large Horizon 2020 research grant on Terrorist Radicalisation processes – Mindb4Act. He is currently Editor of the Journal of Contemporary European Studies, was also previously an elected member of the national (UK) Executive Committee of the University Association for Contemporary European Studies (UACES).

Sahil Lal is a Research Scholar at School of Law, Sharda University. He pursued his graduation from National Law University, Lucknow and completed his Master's in Law from Gujarat National Law University, Gandhinagar. Sahil has completed a certificate course on 'AI & Law' from Lund University, Sweden and on 'Cyber Law' from Indian Law Institute, New Delhi. Sahil Lal is a qualified UGC-NET & JRF candidate. He has a 10+ year Industry Experience and is an active member of Family Dispute Resolution Clinic, Mediation Center, Greater Noida. Sahil has attended 20+ International & National Conferences along with Paper Presentation and has also participated in more than 5+ Faculty Development Programme. Mr. Sahil has more than 50+ publications in various esteemed Journals & Books. Mr. Sahil has a keen interest in area of Artificial Intelligence, Machine Learning, Robotics, Block-Chain Technology and Intellectual Property Rights.

Manmeet Kaur Arora is a Research Scholar at the School of Law, Sharda University, Greater Noida. She is a graduate of Law and has completed her Master's in Law from Amity University, Noida. Manmeet Kaur is a qualified Company Secretary Executive Programme. She has over 8+ years of Industry Experience and 1 year of experience as an Associate at a reputed IPR Law Firm. Manmeet Kaur has attended 20+ National & International Conferences, Webinars, Seminars, Symposiums, and Paper Presentations and participated in 5+ Faculty Development programs. She has over 50+ publications in various esteemed Journals and Books. She is active at the Family Dispute Resolution Clinic, Meditation Centre, Greater Noida. Manmeet Kaur has a keen interest in the areas of Corporate Law, Intellectual Property Rights, Insolvency, Artificial Intelligence, Machine Learning, and Blockchain Technology.

Anjali Raghav is a legal professional and accomplished academic, currently a Research Scholar at the Sharda School of Law, Sharda University, Greater Noida, India. With a solid educational foundation, she holds a Master in Law from Sharda School of Law, where she cultivated her expertise in Criminal Law, Intellectual Property Rights, Artificial Intelligence, Machine Learning, Blockchain Technology, Sustainable Energy, etc. Her three years of experience as a criminal practicing lawyer have equipped her with practical legal insights that she seamlessly integrates into her academic and editorial endeavors. She has a robust portfolio of publications in esteemed journals and books, reflecting her dedication to contributing to the body of knowledge in her fields of interest. Her research has been recognized on various platforms, as evidenced by her participation in national and international conferences, seminars, webinars, and symposiums. Also, in her academic and research pursuits, Anjali is a skilled editor, having successfully overseen the editorial process of several publications.

* * *

Abhishek Basak is pursuing an M.Sc. in Data Science at VIT-AP University, Amaravati, Andhra Pradesh. He completed a Bachelor of Science in Mathematics (Hons) from Calcutta University. With a solid foundation in mathematical principles and a keen interest in data-driven decision-making, Abhishek's academic journey is marked by a dedication to advancing knowledge in data science. Mr. Abhishek Basak has developed a robust skill set in statistical analysis, machine learning, and data visualization, and he is actively planning to engage in research that leverages advanced data science techniques. He is passionate about uncovering insights from complex datasets and leveraging advanced analytical techniques to solve real-world problems. As part of his ongoing studies, Mr. Abhishek Basak is engaged in research areas, focusing on data science and artificial intelligence. He is committed to contributing to the data science community through scholarly research and practical applications. His goal is to bridge the gap between theoretical knowledge and practical implementation, ultimately driving innovation and excellence in data science.

Navonita Mallick, a dedicated and accomplished legal professional, is a seasoned Lawyer and Legal Scholar with a rich background in criminal law, commercial law, and legal research. Currently pursuing a PhD in Child Sexual Abuse and the implementation of legal frameworks in India at KIIT Law School, Bhubaneswar, Navonita brings a unique blend of academic rigour and practical experience to her role. Her passion for legal education and research is evident through her various publications and contributions to the field. She has authored several well-received books, including "Comparative Public Law", "MCQs on Law of Criminal Procedure," "Rajasthan Judicial Service Examination: Previous Years Solved MCQs." and "MCQ's on Indian Penal Code 1860," all published by Legis Orbis. Her expertise extends beyond books, as she has also penned insightful articles published in reputable legal platforms.

Pratiti Nayak, Ph.D (Law), serves as Associate Professor- I & Associate Dean (Research) at School of Law, KIIT-DU. She has an extensive experience in teaching, academia, and research for over a period of 12 years. Her areas of expertise, interest & research are primarily focused on Environmental Social Governance (ESG), Sustainability, Circular Economy, Corporate Governance, Securities Law, etc. besides other facets of Business Laws. She has successfully completed her doctoral research in "Corporate Governance & ESG Norms", further enhancing her expertise in the field.

Soumyajit Pal is currently working as an Assistant Professor in the Post Graduate and Research Department of Computer Science at St. Xavier's College (Autonomous), Kolkata. He is also pursuing his PhD as an external scholar in the Department of Computer Science and Engineering at the Indian Institute of Information Technology, Kalyani (IIITK). His research interest lies in the Applications of Deep Learning techniques in Atmospheric Sciences.

Sanghamitra Patnaik is a Professor, HoD, Department of Social Sciences and Humanities, School of Law, KIIT University, Bhubaneswar, obtained her M.Phil and Ph.D from American Studies, SIS, Jawaharlal Nehru University, New Delhi,India. She has Masters Degree in Political Science and Public Administration from Utkal University. She has wide experience in the field of academics and research. She was awarded with various grants and scholarships instituted by Indian Council of Social Science and Research (ICSSR), New Delhi, Government of Odisha and American Studies Research Centre (ASRC), Hyderabad. Her published works cover Indo-US Relations, Environmental Politics, Climate Change, Indigenous people and Culture and Human Resource Management. She was invited to share her expertise by the Advisory Committee, Advanced Centre for American Studies (UGC Funded), Osmania University, Hyderabad, Andhra Pradesh. She is a member of the Advisory Committee of IPSA RC-35. She represented India in SUSI-2013 sponsored /nominated by Department of State, United States of America. She has recently attended/participated in the Study of the U.S. Institutes for Scholars on American Politics and Political Thought 2019 Reunion Conference-"Institutions, Ideas, and the Challenge of Democracy in the 21st Century" co -organized by SUSI and University of Economics, Prague in September 2019.

Shayut Pavapanunkul (Original Name: Chavana Pavaganun). A regular expert innovative politics commentator Thai Nation TV 22. He hold 3 Adjunct Research Professor from three famous universities in Indonesia: Namely: Universitas Muhammadiyah Makassar, Indonesia; Universitas Muhammadiyah Sinjai, Indonesia and Universitas Muhammadiyah Sidenreng Rappang, Indonesia. He is talented and showed his outstanding achievements in many field knowledge, his real affairs, includes his crea-innovation concept ideas such as: Former weapons of mass destruction member, security service antiterrorism assistance program, special agent, Bureau of Diplomatic Security, U.S. Department of State. Additionally, he worked as a co-creator and co-producer "Thailand political reform movie", including the only one copywriter for Thailand political reform, reader team, sticker and outdoor billboards nationwide during the General Chavalit yongchaiyut era, joint the reader team 109 books of former Prime Minister Thaksin Shinawatra, and served as an National Security Board, more than 20 years, and advisory at Prime Minister Office such as Former Prime Minister General Surayut Chulanont, Thaksin Shinawatra etc. In addition, he recently retired from his position as an advisor to the Deputy Minister of Defense. Support work for Minister of Defense Prawit Wongsuwan and Prime Minister General Prayut Chan-o-cha. He is so proudly his cultural reproduction concept be honored to Chapter 9 contents of social psychology of culture textbook Chi-Yue-Chin and Ying-Yi Hong (Eds.) University of Illinios, Urbana, Champaign Published (2006, p.226), 7 pocket books best sellers and has been reprinted three times and three publishers. In relative to entertainment success both as an actor and director for his work such as he received golden trophy awards "Best Thai Movie Supporter" (1995) from It's like this world has soy milk and a teacher named "Rabieb" (In Thai Baebwa Loak Nee Mee Num Tao Hoo and Khru Rabeib). Then in 1997 his excellence second unit director contributors were recognized from "The King of Kick Boxer" Hollywood and Hong Kong, a prestigious Thai martial art movie (after theatre movie Hollywood established in television series for 3 seasons 1997-1999). Furthermore, he is recognized for his talented skill in naming over

3,000 movies and video from every language in Thai. For the well-known movies title namely "Resident Evil" (In Thai "Phi Chiwa" or "Ghost Biology") Jackie Chan movie "The Meldalion" (in Thai "Fad Ammata" or "Immortal Fight") Chow Yun Fat "Treasure Hunt" (In Thai "Taea Theu…Lok Taek Nae" or "Touch her…The World will End") and Andy Lan (Liu De Hua) "Casino Tycoon" (In Thai "Fa Nee Yai Dai Kon Deiw" or "This sky can be big alone") etc. Trumps up and appreciate the courage that he and Nasaran Mahittichatkul co-created a new avatar concept to presents as a research methodology for market-ing. Both of them jointly produce an innovation beyond textbook as the first in Thailand and dare to say may the first in the world as well because explored and no one used it yet. Concerning to this book called "Avatar Market-ing Research Methodology". Moreover, this text book is the first text book in Thailand to be promoted on TV show and it is the only text book that been promoted on a popular TV show "Guess My Age" Channel 7 on-air September 10, 2020, 6 p.m. Even more surprisingly, 500 copies of this book were sold out in two days. Finally, In fact, that he has various outstanding achievements of work is probably because he is fluent in world tradition languages such as magick, the picture language, Latin, Spanish, Italian etc. He has a total of 15 degrees. Both bachelor's degree with honors in Journalism and Mass Communication and Political Science, master degrees in various fields, such as Political Science (Government), Public Administration, Business Administration, Public and Private Administration. In relation to doctoral (Thailand) degrees, he holds 7 Ph.D. and 1 D.M.S., refer to important about this Public Administration, Psychology (G.P.A.4) Philosophy (G.P.A.4) (Europe), Comparative Philosophy and Religion (Thailand) Information Technology and Strategic Management (Europe), Political Science (Asia) Doctor of Management Science (The Philippines). In addition, he also gained knowledge from famous universities in the United States and England to Australia, including Asia as well, such as University of California Berkeley, University of California, University of London, University of Westminster, University of Royal Halloway University of London, University of Woolongong, University of Pune, etc.

Shashwata Sahu, a distinguished research scholar currently pursuing her Ph.D. in Law at KIIT University, Bhubaneswar, and serving as a dedicated member of the Law Faculty at Utkal University, embodies an unwavering passion for the field of law. With a profound interest in Constitutional Law, she stands as a self-driven scholar, firmly rooted in the principles of altruism. With an impressive literary portfolio, Shashwata has authored nine impactful books, each delving into crucial aspects of law and society. Her works include "The Growing Significance of Artificial Intelligence," "The role of Indian media in covering crimes: a critical analysis," "The use of Amicus Curiae Briefs at the International Criminal Tribunals," "Assessing the legal landscape of bio-medical waste management in India," "Journey of Law:

From Deities to Decrees," "Evolution of Local-Self Government in India", "The Invisible Manual Scavengers of India", "Law and Justice in a Globalizing World: An LL.M. Companion" and "Comparative Public Law: An LLM. Companion." Beyond her authored books, Shashwata's contributions extend to numerous publications in both international and national journals.

Anant Vijay Maria is a distinguished Advocate on Record with over a decade of experience in civil, commercial, arbitration, insolvency, and company law. He holds a BA LLB(H) from Guru Gobind Singh Indraprastha University, an LLM in Business and Commercial Laws from National Law University, Delhi, and a PhD in Corporate Rehabilitation and Insolvency and Bankruptcy Law. Dr. Maria has represented clients in the Supreme Court and Delhi High Court, authored international articles, delivered guest lectures, and served as a legal advisor and consultant. His judicial clerkship with the Delhi High Court enriched his practice, and he continues to advocate and consult across India.

Kittisak Wongmahesak is a lecturer at the Faculty of Political Science, North Bangkok University, and the Director of the Association of Legal & Political Studies in Thailand. He also holds the position of Research Fellow at Shinawatra University, Thailand, and Universiti Sultan Zainal Abidin, Malaysia. His influence extends to Indonesia, where he is an Adjunct Professor at Sekolah Tinggi Ilmu Administrasi Abdul Haris and an Adjunct Research Professor at several Indonesian universities, including Universitas Muhammadiyah Sinjai, Universitas Muhammadiyah Makassar, and Universitas Muhammadiyah Sidenreng Rappang. With a wealth of academic experience and a robust set of qualifications, Kittisak Wongmahesak has held the position of Full Professor of Public Administration at Huanghe Business School, Henan University of Economics and Law in China, and the University of City Island in Cyprus. He has also contributed as an Adjunct Researcher at Ton Duc Thang University and Duy Tan University in Vietnam. His academic qualifications include a Ph.D. in Social Sciences, an M.A., and a B.A. in Political Science, all from Kasetsart University. Additionally, he earned an M.B.A. and an M.P.A. from Ramkhamhaeng University and a B.Pol.Sc. in International Relations and Comparative Government and Politics from Sukhothai Thammathirat Open University. Kittisak Wongmahesak's expertise spans Political Science, Public and Private Management, International Political Economy, and Social Research. He is the editor of numerous prestigious journals worldwide, including Sustainability, Frontiers in Psychology, PLOS One, World Food Policy, Asian Academy of Management Journal, International Journal of Public Policy and Administration Research, International Journal of Innovative Research and Scientific Studies, and Recent Advances in Computer Science and Communications. His global recognition as one of the world's top 2% of scientists by Stanford University since 2020 is a testament to his influence. His extensive publication record, with over 450 indexed Scopus/Web of Science documents and an h-index of 56 with more than 10,000 citations, is a wealth of knowledge that informs his significant scholarly contributions.

Index

U

W

www.ingramcontent.com/pod-product-compliance
Lightning Source LLC
LaVergne TN
LVHW061550290926
849162LV00004B/969